Lecture Notes in Computer Science 13345

More information about this series at https://link.springer.com/bookseries/558

Ying Tan · Yuhui Shi · Ben Niu (Eds.)

Advances in Swarm Intelligence

13th International Conference, ICSI 2022
Xi'an, China, July 15–19, 2022
Proceedings, Part II

 Springer

Editors
Ying Tan 🄳
Peking University
Beijing, China

Yuhui Shi
Southern University of Science
and Technology
Shenzhen, China

Ben Niu
Shenzhen University
Shenzhen, China

ISSN 0302-9743   ISSN 1611-3349 (electronic)
Lecture Notes in Computer Science
ISBN 978-3-031-09725-6   ISBN 978-3-031-09726-3 (eBook)
https://doi.org/10.1007/978-3-031-09726-3

This Springer imprint is published by the registered company Springer Nature Switzerland AG
The registered company address is: Gewerbestrasse 11, 6330 Cham, Switzerland

Preface

This book and its companion volumes, LNCS vols. 13344 and 13345, constitute the proceedings of The Thirteenth International International Conference on Swarm Intelligence (ICSI 2022) held during July 15–19, 2022 in Xi'an, China, both onsite and online.

The theme of ICSI 2022 was "Serving Life with Swarm Intelligence." ICSI 2022 provided an excellent opportunity for academics and practitioners to present and discuss the latest scientific results and methods, innovative ideas, and advantages in theories, technologies, and applications in swarm intelligence. The technical program covered a number of aspects of swarm intelligence and its related areas. ICSI 2022 was the thirteenth international gathering for academics and researchers working on aspects of swarm intelligence, following successful events in Qingdao (ICSI 2021), Serbia (ICSI 2020) virtually, ChiangMai (ICSI 2019), Shanghai (ICSI 2018), Fukuoka (ICSI 2017), Bali (ICSI 2016), Beijing (ICSI-CCI 2015), Hefei (ICSI 2014), Harbin (ICSI 2013), Shenzhen (ICSI 2012), Chongqing (ICSI 2011), and Beijing (ICSI 2010), which provided a high-level academic forum for participants to disseminate their new research findings and discuss emerging areas of research. The conference also created a stimulating environment for participants to interact and exchange information on future challenges and opportunities in the field of swarm intelligence research.

Due to the continuous global COVID-19 pandemic, ICSI 2022 provided both online and offline presentations. On one hand, ICSI 2022 was normally held in Xi'an, China. On the other hand, the ICSI 2022 technical team enabled the authors of accepted papers who were restricted from traveling overseas to present their work through an interactive online platform or video replay. The presentations by accepted authors were available to all registered attendees onsite and online.

The host city of ICSI 2022, Xi'an in China, is the capital of Shaanxi Province. A sub-provincial city on the Guanzhong Plain in Northwest China, it is one of the oldest cities in China, the oldest prefecture capital, and one of the Chinese Four Great Ancient Capitals, having held the position under several of the most important dynasties in Chinese history, including Western Zhou, Qin, Western Han, Sui, Northern Zhou, and Tang. The city is the starting point of the Silk Road and home to the UNESCO World Heritage site of the Terracotta Army of Emperor Qin Shi Huang.

The ICSI 2022 received a total of 171 submissions and invited submissions from about 368 authors in 15 countries and regions (Brazil, China, the Czech Republic, Germany, India, Italy, Japan, Mexico, Portugal, Russia, South Africa, Taiwan (China), Thailand, the UK, and the USA) across five continents (Asia, Europe, North America, South America, and Africa). Each submission was reviewed by at least 2 reviewers, and on average 2.6 reviewers. Based on rigorous reviews by the Program Committee members and reviewers, 85 high-quality papers were selected for publication in this proceedings volume with an acceptance rate of 49.7%. The papers are organized into 13 cohesive sections covering major topics of swarm intelligence research and its development and applications along with a competition session entitled "Competition on Single Objective Bounded Optimization Problems (ICSI-OC 2022)."

On behalf of the Organizing Committee of ICSI 2022, we would like to express our sincere thanks to the International Association of Swarm and Evolutionary Intelligence (IASEI), which is the premier international scholarly society devoted to advancing the theories, algorithms, real-world applications, and developments of swarm intelligence and evolutionary intelligence (iasei.org). We would also like to thank Peking University, Xi'an Jiaotong University, Shaanxi Normal University, Xi'dan University, Xi'an University of Posts & Telecommunications, and the Southern University of Science and Technology for their co-sponsorships, the Computational Intelligence Laboratory of Peking University and IEEE Beijing Chapter for their technical co-sponsorships, and Nanjing Kanbo iHealth Academy for its technical and financial co-sponsorship, as well as our supporters: the International Neural Network Society, the World Federation on SoftComputing, MDPI's journal 'Entropy', the Beijing Xinghui Hi-Tech Co., and Springer.

We would also like to thank the members of the Advisory Committee for their guidance, the members of the international Program Committee and additional reviewers for reviewing the papers, and the members of the Publication Committee for checking the accepted papers in a short period of time. We are particularly grateful to Springer for publishing the proceedings in the prestigious series of Lecture Notes in Computer Science. Moreover, we wish to express our heartfelt appreciation to the plenary speakers, session chairs, and student helpers. In addition, there are many more colleagues, associates, friends, and supporters who helped us in immeasurable ways; we express our sincere gratitude to them all. Last but not the least, we would like to thank all the speakers, authors, and participants for their great contributions that made ICSI 2022 successful and all the hard work worthwhile.

May 2022

Ying Tan
Yuhui Shi
Ben Niu

Organization

Honorary Co-chairs

Zongben Xu   Xi'an Jiaotong University, China
Russell C. Eberhart   IUPUI, USA

General Chair

Ying Tan   Peking University, China

Program Committee Chair

Yuhui Shi   Southern University of Science and Technology, China

Advisory Committee Chairs

Xingui He   Peking University, China
Gary G. Yen   Oklahoma State University, USA

Technical Committee Co-chairs

Haibo He   University of Rhode Island, USA
Kay Chen Tan   City University of Hong Kong, China
Nikola Kasabov   Auckland University of Technology, New Zealand
Ponnuthurai Nagaratnam   Nanyang Technological University, Singapore
  Suganthan
Xiaodong Li   RMIT University, Australia
Hideyuki Takagi   Kyushu University, Japan
M. Middendorf   University of Leipzig, Germany
Yaochu Jin   University of Surrey, UK
Qirong Tang   Tongji University, China
Milan Tuba   Singidunum University, Serbia

Plenary Session Co-chairs

Andreas Engelbrecht   University of Pretoria, South Africa
Chaoming Luo   University of Mississippi, USA

Invited Session Co-chairs

Andres Iglesias	University of Cantabria, Spain
Haibin Duan	Beihang University, China

Special Sessions Co-chairs

Ben Niu	Shenzhen University, China
Yan Pei	University of Aizu, Japan
Shaoqiu Zheng	China Electronics Technology Group Corporation, China

Tutorial Co-chairs

Junqi Zhang	Tongji University, China
Gaige Wang	Ocean University of China, China

Publications Co-chairs

Swagatam Das	Indian Statistical Institute, India
Radu-Emil Precup	Politehnica University of Timisoara, Romania
Pengfei Guo	Xi'an Jiaotong University Press, China

Publicity Co-chairs

Yew-Soon Ong	Nanyang Technological University, Singapore
Carlos Coello	CINVESTAV-IPN, Mexico
Mengjie Zhang	Victoria University of Wellington, New Zealand
Dongbin Zhao	Institute of Automation, CAS, China
Rossi Kamal	GERIOT, Bangladesh

Finance and Registration Chairs

Andreas Janecek	University of Vienna, Austria
Suicheng Gu	Google Corporation, USA

Local Arrangement Chairs

Liangjun Ke	Xi'an Jiaotong University, China
Shi Cheng	Shanxi Normal University, China
Yongzhi Zhe	Xi'an University of Posts and Telecommunications, China

Conference Secretariat

Yifan Liu   Peking University, China

Program Committee

Abdelmalek Amine   Tahar Moulay University of Saida, Algeria
Sabri Arik   Istanbul University, Turkey
Helio Barbosa   Laboratório Nacional de Computação Científica,
                                     Spain
Carmelo J. A. Bastos Filho   University of Pernambuco, Brazil
Heder Bernardino   Universidade Federal de Juiz de Fora, Brazil
Sandeep Bhongade   G.S. Institute of Technology, India
Sujin Bureerat   Khon Kaen University, Thailand
Angelo Cangelosi   University of Manchester, UK
Mu-Song Chen   Da-Yeh University, Taiwan, China
Walter Chen   National Taipei University of Technology,
                                     Taiwan, China
Long Cheng   Institute of Automation, CAS, China
Shi Cheng   Shaanxi Normal University, China
Prithviraj Dasgupta   U. S. Naval Research Laboratory, USA
Khaldoon Dhou   Texas A&M University–Central Texas, USA
Haibin Duan   Beijing University of Aeronautics and
                                     Astronautics, China
Wei Fang   Jiangnan University, China
Liang Feng   Chongqing University, China
Philippe Fournier-Viger   Shenzhen University, China
Hongyuan Gao   Harbin Engineering University, China
Shangce Gao   University of Toyama, Japan
Zhigao Guo   Queen Mary University of London, UK
Guosheng Hao   Jiangsu Normal University, China
Mo Hongwei   Harbin Engineering University, China
Changan Jiang   Osaka Institute of Technology, Japan
Mingyan Jiang   Shandong University, China
Qiaoyong Jiang   Xi'an University of Technology, China
Colin Johnson   University of Nottingham, UK
Yasushi Kambayashi   Nippon Institute of Technology, Japan
Liangjun Ke   Xi'an Jiaotong University, China
Waqas Haider Khan   University of Gujrat, Pakistan
Vivek Kumar   Università degli Studi di Cagliari, Italy
Germano Lambert-Torres   PS Solutions, USA
Xiujuan Lei   Shaanxi Normal University, China
Bin Li   University of Science and Technology of China,
                                     China

Jing Liang	Zhengzhou University, China
Fernando B. De Lima Neto	University of Pernambuco, Brazil
Peng Lin	Capital University of Economics and Business, China
Jia Liu	University of Surrey, UK
Ju Liu	Shandong University, China
Qunfeng Liu	Dongguan University of Technology, China
Wenlian Lu	Fudan University, China
Chaomin Luo	Mississippi State University, USA
Dingsheng Luo	Peking University, China
Wenjian Luo	Harbin Institute of Technology, China
Lianbo Ma	Northeastern University, China
Chengying Mao	Jiangxi University of Finance and Economics, China
Yi Mei	Victoria University of Wellington, New Zealand
Bernd Meyer	Monash University, Australia
Carsten Mueller	Baden-Wuerttemberg Cooperative State University, Mosbach, Germany
Sreeja N. K.	PSG College of Technology, India
Qingjian Ni	Southeast University, China
Ben Niu	Shenzhen University, China
Lie Meng Pang	Southern University of Science and Technology, China
Bijaya Ketan Panigrahi	IIT Delhi, India
Endre Pap	Singidunum University, Serbia
Om Prakash Patel	Mahindra University, India
Mario Pavone	University of Catania, Spain
Yan Pei	University of Aizu, Japan
Danilo Pelusi	University of Teramo, Italy
Radu-Emil Precup	Politehnica University of Timisoara, Romania
Quande Qin	Guangzhou University, China
Robert Reynolds	Wayne State University, USA
Yuji Sato	Hosei University, Japan
Carlos Segura	Centro de Investigación en Matemáticas, A.C. (CIMAT), Mexico
Ke Shang	Southern University of Science and Technology, China
Zhongzhi Shi	Institute of Computing Technology, CAS, China
Joao Soares	Polytechnic Institute of Porto, Portugal
Wei Song	North China University of Technology, China
Yifei Sun	Shaanxi Normal University, China
Ying Tan	Peking University, China

Qirong Tang	Tongji University, China
Eva Tuba	University of Belgrade, Serbia
Mladen Veinović	Singidunum University, Serbia
Kaifang Wan	Northwestern Polytechnical University, China
Gai-Ge Wang	Ocean University of China, China
Guoyin Wang	Chongqing University of Posts and Telecommunications, China
Lei Wang	Tongji University, China
Liang Wang	Northwestern Polytechnical University, China
Yuping Wang	Xidian University, China
Ka-Chun Wong	City University of Hong Kong, Hong Kong SAR, China
Man Leung Wong	Lingnan University, Hong Kong SAR, China
Ning Xiong	Mälardalen University, Sweden
Benlian Xu	Changshu Institute of Technology, China
Rui Xu	Hohai University, China
Yu Xue	Nanjing University of Information Science & Technology, China
Xuesong Yan	China University of Geosciences, China
Yingjie Yang	De Montfort University, UK
Guo Yi-Nan	China University of Mining and Technology, China
Peng-Yeng Yin	National Chi Nan University, Taiwan, China
Jun Yu	Niigata University, Japan
Ling Yu	Jinan University, China
Zhi-Hui Zhan	South China University of Technology, China
Fangfang Zhang	Victoria University of Wellington, New Zealand
Jie Zhang	Newcastle University, UK
Junqi Zhang	Tongji University, China
Xiangyin Zhang	Beijing University of Technology, China
Xingyi Zhang	Anhui University, China
Zili Zhang	Deakin University, Australia
Xinchao Zhao	Beijing University of Posts and Telecommunications, China
Shaoqiu Zheng	Peking University, China
Yujun Zheng	Zhejiang University of Technology, China
Miodrag Zivkovic	Singidunum University, Serbia

Additional Reviewers

Cai, Gaocheng
Cao, Zijian
Chen, Guoyu
Fang, Junchao
Guo, Chen
Guo, Weian
Huang, Yao

Jiang, Yi
Li, Junqing
Liang, Gemin
Liu, Yuxin
Luo, Wei
Nawaz, M. Saqib
Nguyen, Kieu Anh

Qiu, Haiyun
Song, Xi
Wang, Yixin
Xue, Bowen
Yang, Qi-Te
Zhou, Zheng
Zhu, Zhenhao

Contents – Part II

Machine Learning

Data Mining

Other Optimization Applications

ICSI-OC'2022: Competition on Single Objective Bounded Optimization Problems

Contents – Part I

Brain Storm Optimization Algorithm

Swarm Intelligence Approach-Based Applications

Multi-objective Optimization

Swarm Robotics and Multi-agent System

A Bio-Inspired Neural Network Approach to Robot Navigation and Mapping with Nature-Inspired Algorithms

Tingjun Lei[1], Timothy Sellers[1], Chaomin Luo[1(✉)], and Li Zhang[2]

[1] Department of Electrical and Computer Engineering, Mississippi State University, Mississippi State, MS 39762, USA
Chaomin.Luo@ece.msstate.edu
[2] Department of Poultry Science,
Mississippi State University, Mississippi State, MS 39762, USA

Abstract. Nature-inspired algorithms have been successfully applied to autonomous robot path planning, vision and mapping. However, concurrent path planning and mapping with replanning feature is still a challenge for autonomous robot navigation. In this paper, a new framework in light of the replanning-based methodology of concurrent mapping and path planning is proposed. It initially performs global path planning through a developed Gravitational Search Algorithm (GSA) to generate a global trajectory. The surrounding environment can then be described through a monocular framework and transformed into occupancy grid maps (OGM) for autonomous robot path planning. With updated moving obstacles and road conditions, the robot can replan the trajectory with the GSA based on the updated map. Local trajectory in the vicinity of the obstacles is generated by a developed bio-inspired neural network (BNN) method integrated with speed profile mechanism, and safe border patrolling waypoints. Simulation and comparative studies demonstrate the effectiveness and robustness of the proposed model.

Keywords: Autonomous robot path planning · Gravitational Search Algorithm (GSA) · Bio-inspired Neural Networks (BNN) · Occupancy Grid Maps (OGM) · Replanning-based path planning

1 Introduction

Modeling, perception, and understanding of the environment are critical for autonomous robots with the assistance of nature-inspired algorithms. It still remains a daunting task to achieve safe motion planning for autonomous robots while robustly and efficiently sensing, mapping, and modeling the surrounding environment. In recent decades, there have been numerous studies to resolve navigation and motion planning issues, such as sampling-based algorithm [1], graph-based method [3,5], and neural networks models [11,12,19]. Adiyatov *et al.* [1] proposed a modified sampling-based algorithm for rapidly

© Springer Nature Switzerland AG 2022
Y. Tan et al. (Eds.): ICSI 2022, LNCS 13345, pp. 3–16, 2022.
https://doi.org/10.1007/978-3-031-09726-3_1

exploring random tree (RRT) to generate a collision-free robot path. A graph theory-based model has been proposed in Lei *et al.* [5] to conduct point-to-point robot navigation with smooth modulation through a workspace, while simultaneously avoiding obstacles. Luo and Yang [11] developed a coverage path planning algorithm for real-time map building and robot navigation in unknown environments.

Most recently, nature-inspired algorithms advance autonomous navigation, such as fireworks algorithm (FWA) [6], ant colony optimization (ACO) [7,16], bat-pigeon algorithm (BPA) [8], *etc.* A hybrid FWA-based navigation model is proposed in [6], which features concurrent mapping and navigation. It can effectively navigate the robot to the final target. Lei *et al.* [7] proposed an ACO model in conjunction with a variable speed module, which can decrease the robot motion speed in vicinity of the obstacles. To efficiently interact with road conditions, a Bat-Pigeon algorithm with a speed modulation is proposed. It adaptively adjusts speed and refines the generated path according to updated road conditions and decelerate on cracked roads [8]. However, few studies have considered concurrent mapping and path planning in light of monocular techniques.

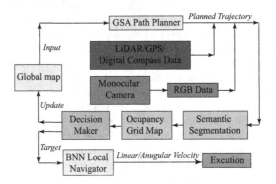

Fig. 1. The overall framework of our proposed method. The components enclosed in dashed box represent our proposed method for an autonomous robot.

In this paper, an autonomous robot trajectory planning method based on gravitational search algorithm (GSA) with replanning feature is proposed. In trajectory planning, local maps in the form of occupancy grid maps (OGM) originated from a Bird-Eye-View (BEV) intermediate representation through a monocular camera are used to generate replanned trajectories. A replanning-based method of concurrent mapping and path planning is proposed as shown in Fig. 1. Collision-free trajectories are generated by a bio-inspired neural network (BNN) local navigator via the local occupancy grid method converted from monocular frames in which the surrounding environment is described.

2   Proposed Algorithm for Autonomous Path Planning

The gravitational search algorithm (GSA) is a nature-inspired optimization algorithm proposed by Rashedi *et al.* [13]. Thanks to the optimization capability of

GSA, the trajectory with minimum time is obtained through the workspace information. In GSA, agents are treated as objects and their performance is measured by their mass. Each agent is affected by the gravitational pull of the other agents in the workspace and produces accelerations toward heavier massive agents. The agents with lighter mass gradually approach the optimal solution of the optimization problem in the process of approaching the heavier mass. The GSA has convincing exploitation capability to search the entire workspace.

Algorithm 1: GSA-based Path Planning Algorithm

Initialization:

Load the $N_x \times N_y$ map; Set (P_{sx}, P_{sy}) as the start point, (P_{gx}, P_{gy}) as the target point; Set $f(i, j) = 0$, if (i, j) is free space; Set $f(i, j) = 1$, if (i, j) is obstacle; Set $Pathbest = \emptyset$ and the maximum number of iteration as $\mathcal{T}_{max}$.

Random generate N_p agents in the workspace $\forall\ 1 \leq i \leq N_x, 1 \leq j \leq N_y$ with $f(i, j) = 0$; Initialize the global best as $Fbest$, local best as $\mathcal{X}(best_X, :)$.

for $\tau = 1 : \mathcal{T}_{max}$ do

    $\mathcal{X} = spacebound(\mathcal{X}, up, low)$; // Check the location of the agent in the workspace

    $fitness = evaluateF(X, F_{index})$; // Evaluate the fitness for each agent

    $[best, best_X] = min(fitness)$;

    if $best < Fbest$ then

        $Fbest = best; Lbest = X(best_X, :)$ // Minimization

    End

    $G(\tau) = G_0 \times e^{-\lambda \tau / T_{\max}}$; // Update the gravitational constant

    for $i \leq N_p$ do

        $\mathcal{M}_i(\tau) = \dfrac{m_i(\tau)}{\sum_{j=1}^{N} m_j(\tau)}$; // Update mass for each agent

        $F_i^k(\tau) = \sum_{j=1, j\neq i}^{N} rand_j\ F_{ij}^k(\tau)$; // Update the force for each agent

        $a_i^k(\tau) = \dfrac{F_i^k(\tau)}{\mathcal{M}_i(\tau)}$; // Update acceleration for each agent

        $v_i^k(\tau + 1) = rand_i \times v_i^k(\tau) + a_i^k(\tau); x_i^k(\tau + 1) = x_i^k(\tau) + v_i^k(\tau + 1)$; // Update velocity and position for each agent

    End

    $Pathbest = best_X$

End

Since the real-time map is dynamically updated in light of the visual images in the subsequent Sect. 3, the grid-based map with OGM presented in Sect. 3 is utilized for environment modeling. The proposed method integrates the local search heuristic to search and find the short and reasonable trajectory with the GSA in a graph. The first stage of seeking the optimal path is to find a feasible solution since the unfeasible ones are not realistic. In order to improve the performance of finding the optimal path through the GSA algorithm, the path is gradually constructed from the generated random points. On the basis of Dijkstra's algorithm to find the shortest path in a graph, the path is built from the start point, whereas the next path point is chosen from the N randomly generated points. Each point is recursively connected with the other points, while the distances of connection lines passing through obstacles are assigned with infinite number. As a result, the point-to-point navigations with obstacles are excluded out, and the feasible solutions are retained. The shortest paths between each pair of points are selected from those feasible solutions. The model of the GSA-based autonomous robot path planning is presented in Algorithm 1.

3 Monocular Camera and Semantic Occupancy Grid Mapping

Autonomous robots estimate the surrounding free space to determine whether they encounter any obstacles (such as pedestrians, curbs or other robots) [18]. There are various sensors to estimate free space, such as Radar, LiDAR, or cameras. Monocular vision, inspired by only one eye used in animals and humans, enhances field of view of surrounding environments of autonomous robots while depth perception is limited. The 2D image provided by the RGB monocular camera is integrated with an occupancy grid to estimate the available space through semantic segmentation, and the output is the semantic grid of the environment from a bird's eye view perspective.

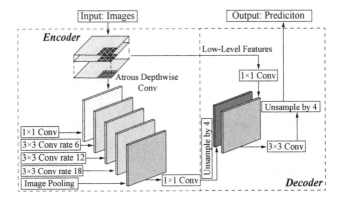

Fig. 2. The proposed network model by employing an encoder-decoder structure.

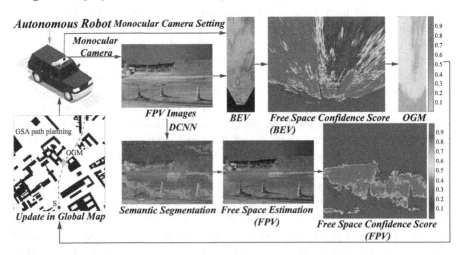

Fig. 3. The framework of the proposed model and the diagram of information flow connected with GSA model (BEV is bird's-eye view; OGM is occupancy gird map).

Segmentation is essential for tasks of image analysis in our applications [17]. Semantic segmentation describes the relevance between each pixel of an image and category labels, such as some geometric structures (sky, ground, buildings, *etc.*). With convolutional neural networks (CNNs) in image processing demonstrating the power of accurately estimating dense depth maps from a single image, it needs to estimate free space (road) in this paper by performing semantic image segmentation on images obtained during the operation of our autonomous robot. The deep convolutional neural network (DCNN) has a significant improvement over other systems on the benchmark task. We utilize a simple and effective *decoder* module as shown in Fig. 2. The CamVid dataset of Cambridge University is utilized for training [2]. This data set is a collection of images containing street views obtained while driving, and 32 semantic classes including cars, pedestrians, and roads provide pixel-level labels. The obtained OGM is then updated in the global map for further replanning by the GSA model and local navigation by the BNN model in Sect. 4, depicted in Fig. 3.

4 Biologically Inspired Dynamic Path Planning

The biologically inspired neural network (BNN) model is derived for trajectory planning of an autonomous robot [4]. The topologically organized neural networks with nonlinear analog neurons are effective for collision-free navigation. Each neuron is locally connected with adjacent neurons to transmit neural activities to each other, and its receptive field is regarded as a circular area with a radius $\mathcal{R}$. The symmetric connection weights w_{ij} between the ith neuron and the jth neuron and the dynamics of neural network is described as

$$w_{ij} = \begin{cases} e^{-\rho|\boldsymbol{x}_i - \boldsymbol{x}_j|}, & |\boldsymbol{x}_i - \boldsymbol{x}_j| \leqslant \mathcal{R} \\ 0, & |\boldsymbol{x}_i - \boldsymbol{x}_j| > \mathcal{R} \end{cases} ; \quad a_i^{\tau+1} = g\left(I_i^\tau + \sum_{j \in N(i)} w_{ij} a_j^\tau\right), \quad (1)$$

where $|\boldsymbol{x}_j - \boldsymbol{x}_i|$ is the Euclidian distance between the ith neuron to the jth neuron, $\rho > 0$ is a constant coefficient. g is the transfer function and $N(m)$ is the neighborhood set in the radius $\mathcal{R}$. I_i^τ is the external stimulus at time τ. If it is a target, $I_i = E$, which is excitatory input for the corresponding grid; If it is an obstacle, whereas $I_i = -E$, which is inhibitory input for the purpose of obstacle avoidance. Otherwise, $I_i = 0$. The transfer function g is designed to normalize the activity landscape of the whole neural network. Therefore, the function is redefined as

$$a_i^{\tau+1} = \begin{cases} \dfrac{I_i^\tau + \sum_{j \in N(i)} w_{ij} a_j^\tau}{\sum_{i=1, I_i^\tau \geqslant 0}^{M} \left(I_i^\tau + \sum_{j \in N(i)} w_{ij} a_j^\tau\right)}, & I_i^\tau \geqslant 0 \\ 0, & I_i^\tau < 0 \end{cases}. \quad (2)$$

Real-time collision-free robot motion is planned based on the dynamic activity landscape of the neural network and the location of the robot. The activity landscape of a neural network changes dynamically due to different external inputs from targets and obstacles, as well as internal activity propagation between neurons. While the location of the robot is simultaneously updated via varying environments. For a given current robot location in $\mathcal{C}$ denoted by $\mathcal{L}_\mathcal{C}$, the next robot location $\mathcal{L}_\mathcal{N}$ is obtained by

$$\mathcal{L}_\mathcal{N} = \underset{i,j}{\mathrm{argmax}}(\boldsymbol{x}(i,j)) \in \{\mathcal{N}_k \mid (i,j)\}. \tag{3}$$

Since the BNN model is based on the grid-based map obtained in Sect. 3. The computational complexity of the BNN depends on the workspace size. The number of neurons required is equal to $N = N_x \times N_y$. Thus, the size of the local navigation workspace should be restricted in a small size [10].

To ensure that the resulting path is safe and consistent with the kinetic of autonomous robots, several safe reference waypoints based on an obstacle safe border patrolling paradigm is proposed (Fig. 4). We refer to the classic model of this robot $\dot{x} = v\cos\theta$, $\dot{y} = v\sin\theta$, and $\dot{\theta} = u \in [-\Omega, \Omega]$, where x, y is the pair of the robot's Cartesian coordinates, θ gives its orientation, v and u are the linear and angular velocity, respectively. The maximal angular velocity Ω is given. The minimal turning radius of the robot at speed v equals $R_{\min} = \frac{v}{\Omega}$. The obstacle $\mathcal{O}$ is assumed to have a C^3-smooth boundary $\partial\mathcal{O}$. Let $\psi(\mathfrak{o})$ represent the signed curvature of the obstacle boundary at point $\mathfrak{o} \in \partial\mathcal{O}$ and the unsigned curvature radius $R_\psi(\mathfrak{o}) := |\psi(\mathfrak{o})|^{-1}$. The point $\mathfrak{p}$ on the robot path at time τ as $\mathfrak{p} = \mathfrak{p}(\tau)$ and related minimum-distance point $\tilde{\mathfrak{o}} \in \partial\mathcal{O}$. Let $\mho(s) \in \mathbb{R}^2$ represent a regular parametric of the boundary $\partial\mathcal{O}$ in a vicinity of $\tilde{\mathfrak{o}} = \mho(\tilde{s})$ where s is the arc length. Let $\vec{T} = \mho'$ be the unit tangent vector and $\vec{N}$ be the unit normal vector directed inwards $\mathcal{O}$. The vector $\mathfrak{p} - \mathfrak{o}$ is perpendicular to a common nonzero vector $\dot{\mathfrak{p}}(\tau)$. There exists a minimum-distance point $\tilde{\mathfrak{o}} \in \partial\mathcal{O}$ such that $d_s = \|\mathfrak{p} - \tilde{\mathfrak{o}}\| = R_\psi(\tilde{\mathfrak{o}})$, $\psi(\tilde{\mathfrak{o}}) < 0$. We have $\tilde{\mathfrak{o}} - \mathfrak{p} = d_s\vec{N}(\tilde{s})$. In view of Frenet-Serret formulas [14], thereby we have $\vec{T}' = \psi\vec{N}$ and $\vec{N}' = -\psi\vec{T}$.

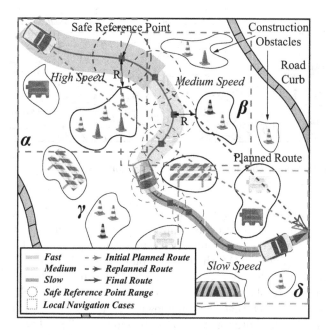

Fig. 4. Illustration of the proposed navigation framework for real-time obstacle avoidance on a construction site. When encountering obstacles, the trajectory will be replanned by the GSA and dynamically connected by the BNN model in local navigation. It is further refined through safe reference points according to the robot speed.

Since the trajectory of the robot is parametrically represented by $\mathfrak{p}(s) := \mathfrak{V}(s) - d_s \vec{N}(s)$, it yields that $\mathfrak{p}' = (1 + \psi d_s)\,\vec{T}$, $\mathfrak{p}'' = \psi' d_s \vec{T} + \psi\,(1 + \psi d_s)\,\vec{N}$. Let $\mathfrak{J} := \begin{pmatrix} 0 & 1 \\ -1 & 0 \end{pmatrix}$ and let ^{T} stand for the transposition. Then the curvature radius $\mathfrak{R}$ of the trajectory is given by

$$\mathfrak{R} = \frac{\|\mathfrak{p}'\|^3}{\left|(\mathfrak{p}')^{\vec{T}} \mathfrak{J}\mathfrak{p}''\right|} = \frac{|1 + \psi d_s|^3}{|\psi|\,|1 + \psi d_s|^2} = |d_s + R_\psi \operatorname{sgn}\psi|. \tag{4}$$

Note that $\mathfrak{R}$ should not be less than the minimum turning radius of the robot. Therefore, we utilized the safe reference points are utlized to ensure the robot travels at the safe distance d_s from the boundary of obstacles at the speed $v > 0$. $|d_s + R_\psi \mathbf{sgn}\psi| \geq \frac{v}{\Omega}$ is required for any point on its trajectory, where $\frac{v}{\Omega}$ is the minimal turning radius of the robot for given speed v, and the curvature ψ and the curvature radius R_ψ are computed at the minimum distance point. **sgn** is the signum function.

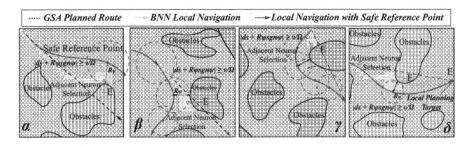

Fig. 5. Illustration of the details of local navigation via the BNN model in Fig. 4.

GSA plans an initial global route while the autonomous robot traverses along the planned path, it simultaneously updates the OGM through the on-board monocular camera. When it is found that the previously planned GSA path fails to be executed through OGM, the GSA re-plans the global route in light of the newly observed obstacles (barrier, cones). The global path obtained is intermittent with multiple re-planning shown in Fig. 4. Each time the GSA re-planning occurs, it indicates that there are unknown obstacles, and local navigation is required to assist the autonomous robot to navigate. The detecting point of each re-planning phase is considered as the starting point of local navigation. The local navigation is dynamically guided by the BNN model. BNN dynamically plans the trajectory *locally* in the space of 30×30. It makes the decision of the next step by comparing the neural activities of nearby neurons of the current neuron. The neural activities of neurons in the field is dynamically updated on the basis of the obtained local environment information to guide the robot to avoid unknown obstacles. Afterwards, the path is refined based on the current speed of the robot and the shape of the obstacle to obtain the safe reference point, which achieves a safe patrolling distance from the obstacle at different speeds. Figure 4 shows a case of an autonomous robot passing through a construction area with many obstacles in the workspace. The pink dotted line is the initial planned GSA path. After encountering the obstacles blocked the road, the GSA re-plans the trajectory. For instance, a blue dotted GSA path is a re-planned path. The detailed local navigation is shown by four windows α, β, γ and δ in Figs. 4 and 5. The local navigation function through neurons in the δ graph in light of the target point is updated within a range of 30×30, which ensures its operation efficiently. On the way of planning, the robot selects the neuron with the highest neural activities in the adjacent neurons. After planning the local path by the BNN model, the safe point is obtained from the shape of the obstacle and the current estimated speed of the autonomous robot, as shown by the red circles in Fig. 5. A safe and collision-free trajectory is finally achieved by the proposed framework.

5   Multi-speed Profile Method

Algorithm 2: Overall Procedure of Proposed Framework

Load the $N_x \times N_y$ global map;
while *The robot is not at the target* **do**
    Obtain the *Pathbest* from Algorithm 1;
    Update the global map with OGM from Monocular Camera in Section 3;
    if *Road is blocked* **then**
        Re-plan the trajectory *Pathbest* with updated global map;
        Update neural activity field $a_i^{\tau+1} = g\left(I_i^{\tau} + \sum_{j \in N(i)} w_{ij} a_j^{\tau}\right)$;
        Obtain the safe reference waypoint $\Re = |d_s + R_\psi \operatorname{sgn} \psi|$;
        Generate the BNN-based local trajectory;
    End
    Update the $a_i^{\ell}, j_i^{\ell}, a_i^{\eta}, j_i^{\eta}$ and current velocity.
End

Let the trajectory generated by the proposed GSA-BNN method on the Cartesian plane be represented by $\mathbf{p}$, which is a one-dimensional manifold parameterized by points q_i evenly sampled in view of the distance of the path, $\mathbf{p}(q) \in \mathbb{R}^2$ shown in Fig. 6. The speed profile is represented by curve $\mathcal{C} = \{(s_i, t_i)\}_i$, which is a monotonic curve. Denote the path length as $s_i := \mathbf{p}(q_i)$ and time interval between s_i and s_{i+1} as $\tau_i = t_{i+1} - t_i$. When the time interval t_i and the distance between q_i and q_{i+1} are relatively small, the speed v_i, acceleration a_i and acceleration rate (jerk) j_i at point q_i can be approximated as

$$v_i = \frac{s_{i+1} - s_i}{\tau_i}, a_i = \frac{2(v_i - v_{i-1})}{\tau_i + \tau_{i-1}}, j_i = \frac{3(a_i - a_{i-1})}{\tau_i + \tau_{i-1} + \tau_{i-2}}. \quad (5)$$

Denote the heading of the robot at point q_i as $\theta_i := \arctan(\dot{\mathbf{p}}(s_i))$. The longitudinal and lateral directions at q_i are denoted as $\ell(\theta_i) := [\cos\theta_i, \sin\theta_i]$ and $\eta(\theta_i) := [\sin\theta_i, -\cos\theta_i]$. The longitudinal velocity and lateral velocity are defined as $v_i^{\ell} = v_i \cdot \ell(\theta_i)$ and $v_i^{\eta} = v_i \cdot \eta(\theta_i)$. The longitudinal acceleration and lateral acceleration are defined as $a_i^{\ell} = a_i \cdot \ell(\theta_i)$ and $a_i^{\eta} = a_i \cdot \eta(\theta_i)$. The longitudinal jerk and lateral jerk are defined as $j_i^{\ell} = j_i \cdot \ell(\theta_i)$ and $j_i^{\eta} = j_i \cdot \eta(\theta_i)$.

To improve the operating quality of the autonomous robots, a_i^{ℓ} and j_i^{ℓ} should be minimized for longitudinal stability, a_i^{η} and j_i^{η} should be minimized for lateral stability. The reference longitudinal velocity v_r is the autonomous robot safe operating speed. Therefore, the time optimization model can be described as

$$\min_{t_1, \cdots, t_n} \omega_1 \sum |a_i^{\ell}|^2 + \omega_2 \sum |a_i^{\eta}|^2 + \omega_3 \sum |j_i^{\ell}|^2$$

$$+ \omega_4 \sum_i |j_i^{\eta}|^2 + \omega_5 \sum (v^r - v_i^{\ell})^2 \quad (6)$$

$$\text{s.t. } \tau_i \in [t_i^{\min}, t_i^{\max}], |a_i| \leq \Omega,$$

where w_1, w_2, w_3, w_4 and w_5 are positive weights. Ω represents acceleration limits. $[t_i^{\min}, t_i^{\max}]$ represents the feasible time stamps for robot to pass through without collision with other moving obstacles. The cost function in Eq. (6) could be treated as $\mathcal{F}_1(\tau)$ and $\mathcal{F}_2(\mathbf{y})$, where $\mathcal{F}_1(\tau)$ is strictly convex. $\mathcal{F}_2(\mathbf{y})$ reaches minimum at $\mathbf{y} = 0$ and it is symmetric also strictly convex. Since the optimization problem satisfies the preliminaries, it could be converted into a quadratic program sequence within the framework of the slack convex feasible set (SCFS) algorithm [9]. With the assistance of the speed profile method, the optimal speed change based on the time is obtained during the navigation.

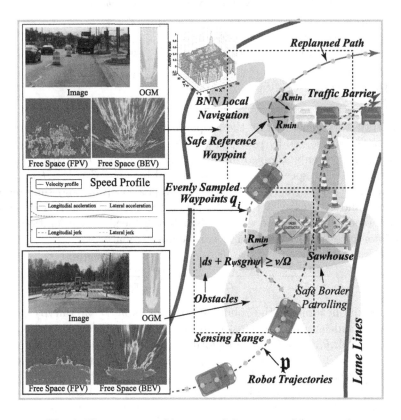

Fig. 6. The system architecture of the proposed framework.

The architecture of the proposed framework is shown in Fig. 6. The blue dotted line is the originally planned path. When the obstacle sawhorse is found, the purple path is replanned by the GSA based on the updated map. Inside the rectangular black dashed border is the local navigation of the BNN. A waypoint is generated according to the current speed of the robot in the pink circle outside the safe distance from the obstacle to assist the robot to avoid the obstacles. The speed profile is utilized to control the speed of the robot. The overall procedure of the proposed framework is summarized in Algorithm 2.

6   Simulation and Comparison Studies

In this section, two simulated experiments and comparison studies are conducted. In the first experiment, the proposed GSA is applied using the benchmarks in comparison with other state-of-the-art path planning algorithms. In the second experiment, the algorithm is applied to navigation in real environments.

6.1   Comparison Studies with Other Path Planning Models

To validate the adaptability and efficiency of our algorithms in various environments, the London and Sydney city maps from the benchmark [15] are selected with resolution 256 × 256 for simulation and comparative studies. Among these maps, the initial and target positions of the robot are randomly set. The white and black nodes represent obstacle-free nodes and obstacle nodes, respectively. The path smoothness rate sm in the Table 1 is described in [8].

Table 1. Comparison of path length, path smoothness rate, autonomous robot travel time, and success rate of PRM, RRT*, A*, and the proposed GSA. The values are reported as mean ± standard deviation for 20 executions.

Map name	Map size	Model name	Minimum path length	Path length (m)	Smoothness rate (rad)	Success rate $(\%)$
London	256 × 256	PRM	338.45	383.29 ± 35.65	11.62 ± 2.68	55
		RRT*	384.53	496.86 ± 68.41	86.75 ± 15.50	95
		A*	336.931	336.931 ± 0	53.40 ± 0	100
		GSA	328.205	331.67 ± 10.66	7.51 ± 0.37	100
Sydney	256 × 256	PRM	378.80	394.14 ± 10.97	9.87 ± 2.82	55
		RRT*	448.75	496.85 ± 44.40	85.56 ± 8.75	100
		A*	368.51	368.51 ± 0	40.05 ± 0	100
		GSA	347.43	348.79 ± 2.44	15.37 ± 2.22	100

Three state-of-the-art path planning algorithms are utilized for comparison: Probabilistic roadmap (PRM), Rapidly-exploring random tree star (RRT*), and A* search algorithm. The number of PRM sampling points is set to 800. The maximum iteration time of RRT* is set to 500000 and the maximum connection distance is set to 4 in all environments. We iteratively perform 20 executions to compute the mean and standard deviation of each metric. The results show that the proposed GSA method outperforms PRM, RRT* and A* in Fig. 7. These results may be explained by the fact that our proposed GSA model can consistently generate better and more robust routes in terms of length, smoothness, and safety. The performance of each model is summarized in Table 1.

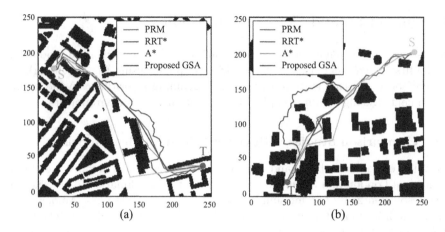

Fig. 7. Illustration of two city maps from [15], composed of 256 × 256 nodes. (a) London, (b) Sydney.

6.2   Autonomous Navigation Under Real-World Scenarios

The simulation in this section aims to validate the GSA path planning model with fusion of semantic information generating OGMs. The real map at the Mississippi State University scene is adopted. The workspace has a size of 900 m × 500 m, which is topologically organized as a grid-based map. In order to achieve BNN-based real-time local navigation, the workspace is restricted to 30 × 30 to update surrounding varying environment and obtain safe trajectory with time-optimized speed control. The proposed GSA path planner initially obtains the dashed dark blue trajectory with the real world map as shown in Figs. 8 (a) and (b). Along the obtained path, while the autonomous robot approaches the location as shown in yellow squares as road blocks in the Fig. 8(b), the on-board monocular camera captures the image of road blocks.

For multiple roadblocks encountered on the planned road, the GSA planer along with the BNN local navigator obtains the replanned paths by updating the map as shown in dashed green and solid orange trajectories in Fig. 8. The simulated experiments show that our proposed model is effective and robust.

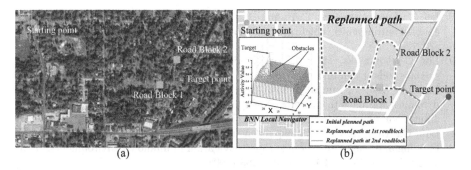

Fig. 8. Illustration of the proposed framework for replanning in real-world scenarios. (a) Real world map from Google Maps. (b) Initial planned trajectory and replanned trajectory with updated map by GSA path planner and BNN local navigator.

7   Conclusion

A new framework for concurrent mapping and path planning was presented in this paper. Collision-free trajectory is generated in light of occupancy grid map method converted from monocular frames in which surrounding environment is described. Once obstacles are detected, the robot can replan its trajectory and navigate locally in light of built OGM. A global path planning has been achieved by GSA method to create a global trajectory. The BNN local navigator integrated with speed profile and safe border patrolling waypoint method ensures the robot traverses at the safe speed locally with obstacle avoidance.

References

1. Adiyatov, O., Varol, H.A.: Rapidly-exploring random tree based memory efficient motion planning. In: 2013 IEEE International Conference on Mechatronics and Automation, pp. 354–359 (2013)
2. Brostow, G.J., Fauqueur, J., Cipolla, R.: Semantic object classes in video: a high-definition ground truth database. Pattern Recogn. Lett. **30**(2), 88–97 (2009)
3. Chen, J., Luo, C., Krishnan, M., Paulik, M., Tang, Y.: An enhanced dynamic Delaunay triangulation-based path planning algorithm for autonomous mobile robot navigation. In: Intelligent Robots and Computer Vision XXVII: Algorithms and Techniques, vol. 7539, pp. 253–264. SPIE (2010)
4. Glasius, R., Komoda, A., Gielen, S.C.: Neural network dynamics for path planning and obstacle avoidance. Neural Netw. **8**(1), 125–133 (1995)
5. Lei, T., Luo, C., Ball, J.E., Rahimi, S.: A graph-based ant-like approach to optimal path planning. In: 2020 IEEE Congress on Evolutionary Computation (CEC), pp. 1–6. IEEE (2020)
6. Lei, T., Luo, C., Ball, J.E., Bi, Z.: A hybrid fireworks algorithm to navigation and mapping. In: Handbook of Research on Fireworks Algorithms and Swarm Intelligence, pp. 213–232. IGI Global (2020)
7. Lei, T., Luo, C., Jan, G.E., Fung, K.: Variable speed robot navigation by an ACO approach. In: Tan, Y., Shi, Y., Niu, B. (eds.) ICSI 2019. LNCS, vol. 11655, pp. 232–242. Springer, Cham (2019). https://doi.org/10.1007/978-3-030-26369-0_22

8. Lei, T., Luo, C., Sellers, T., Rahimi, S.: A bat-pigeon algorithm to crack detection-enabled autonomous vehicle navigation and mapping. Intell. Syst. Appl. **12**, 200053 (2021)

9. Liu, C., Tomizuka, M.: Real time trajectory optimization for nonlinear robotic systems: relaxation and convexification. Syst. Control Lett. **108**, 56–63 (2017)

10. Luo, C., Gao, J., Murphey, Y.L., Jan, G.E.: A computationally efficient neural dynamics approach to trajectory planning of an intelligent vehicle. In: International Joint Conference on Neural Networks (IJCNN), pp. 934–939 (2014)

11. Luo, C., Yang, S.X.: A bioinspired neural network for real-time concurrent map building and complete coverage robot navigation in unknown environments. IEEE Trans. Neural Netw. **19**(7), 1279–1298 (2008)

12. Luo, C., Yang, S.X., Krishnan, M., Paulik, M.: An effective vector-driven biologically-motivated neural network algorithm to real-time autonomous robot navigation. In: 2014 IEEE International Conference on Robotics and Automation (ICRA), pp. 4094–4099 (2014)

13. Rashedi, E., Nezamabadi-Pour, H., Saryazdi, S.: GSA: a gravitational search algorithm. Inf. Sci. **179**(13), 2232–2248 (2009)

14. Sternberg, S.: Lectures on Differential Geometry, vol. 316. American Mathematical Society (1999)

15. Sturtevant, N.R.: Benchmarks for grid-based pathfinding. IEEE Trans. Comput. Intell. AI Games **4**(2), 144–148 (2012)

16. Wang, L., Luo, C., Li, M., Cai, J.: Trajectory planning of an autonomous mobile robot by evolving ant colony system. Int. J. Robot. Autom. **32**(4), 406–413 (2017)

17. Wang, P., Xue, B., Liang, J., Zhang, M.: Multiobjective differential evolution for feature selection in classification. IEEE Trans. Cybern. (2021)

18. Zhao, W., et al.: A privacy-aware Kinect-based system for healthcare professionals. In: IEEE International Conference on Electro Information Technology (EIT), pp. 0205–0210 (2016)

19. Zhu, D., Tian, C., Jiang, X., Luo, C.: Multi-AUVs cooperative complete coverage path planning based on GBNN algorithm. In: 29th Chinese Control and Decision Conference (CCDC), pp. 6761–6766 (2017)

Hybrid Topology-Based Particle Swarm Optimizer for Multi-source Location Problem in Swarm Robots

JunQi Zhang[1,2], Yehao Lu[1,2], and Mengchu Zhou[3(✉)]

[1] Department of Computer Science and Technology, Tongji University, Shanghai, China
zhangjunqi@tongji.edu.cn
[2] Key Laboratory of Embedded System and Service Computing, Ministry of Education, Shanghai, China
[3] Department of Electrical and Computer Engineering, New Jersey Institute of Technology, Newark, NJ 07102, USA
zhou@njit.edu

Abstract. A multi-source location problem aims to locate sources in an unknown environment based on the measurements of the signal strength from them. Vast majority of existing multi-source location methods require such prior environmental information as the signal range of sources and maximum signal strength to set some parameters. However, prior information is difficult to obtain in many practical tasks. To handle this issue, this work proposes a variant of Particle Swarm Optimizers (PSO), named as Hybrid Topology-based PSO (HT-PSO). It combines the advantages of multimodal search capability of a ring topology and rapid convergence of a star topology. HT-PSO does not require any prior knowledge of the environment, thus it has stronger robustness and adaptability. Experimental results show its superior performance over the state-of-the-art multi-source location method.

Keywords: Swarm robots · Multi-source location problem · Particle Swarm Optimizer

1 Introduction

A source location problem is an area of growing interest [6,8,9,13,15]. It arises in many application scenarios, such as chemical spill investigation [8], fire spot discovery [12], disaster zone rescue [3] and target search for military use [7]. Since these tasks may be in a dangerous environment or pose a risk to humans, there has been great interest in deploying autonomous systems to solve it. In order to cope with such problem, the similarity between a particle swarm and a robot swarm is considered. They both try to find the optimal solutions in a certain area through the cooperation of individuals in a swarm. The former is to find multiple optima in a multi-dimensional search space. The latter is to find

© Springer Nature Switzerland AG 2022
Y. Tan et al. (Eds.): ICSI 2022, LNCS 13345, pp. 17–24, 2022.
https://doi.org/10.1007/978-3-031-09726-3_2

multiple source locations with the highest signal strength in a physical space. Thus, by performing PSO as a self-organized strategy, a robot swarm can find sources in reality.

In order to achieve better performance, most existing multi-source location methods require some environmental prior information to set parameters for the algorithm. For example, in [9], the signal threshold parameter is used to construct subgroups. The robots whose obtained signal value exceeds this threshold are selected as candidate cluster centers. However, such prior information is difficult to obtain in practical tasks. When an environment changes, these methods may no longer work well. In this work, a novel PSO variant named as Hybrid Topology-based PSO (HT-PSO) for a robot swarm is proposed. HT-PSO utilizes a ring topology to select elite particles as cluster centers for adaptive grouping in each generation. The remaining robots choose their closest cluster center to form groups. Each group uses a star topology to accelerate the convergence. HT-PSO does not require any prior knowledge of the environment, thus it has stronger robustness and adaptability. In addition, this hybrid topology achieves finer search performance and faster convergence speed than its competitors.

The rest of this paper is arranged as follows. Section 2 reviews the related work of PSO and source location methods. Section 3 develops HT-PSO in detail. Section 4 experimentally validates HT-PSO in different type of environments by comparing it with its peers' performance and gives the visual simulation results. The conclusion is drawn in Sect. 5.

2   Related Work

In this section, we introduce the related work about the description of a multi-source location problem and the framework of standard PSO in detail.

2.1   Multi-source Location Problem Description

A multi-source location problem considers several sources which are located on a plane and continuously transmit signals. A robot swarm explores the environment to locate them. The robots have the capability to measure the signal strength at their current locations. However, they have no information about the exact locations of sources, the number of sources and their decay profiles. The objective of robots is to find the locations of these sources, i.e., the locations where the signal strengths are the maximum. Meanwhile, robots are expected to converge around the sources in the end of their entire search mission.

2.2   Standard PSO

Standard PSO [1] contains two classic topological structures, i.e., star topology and ring one. The fomer allows information sharing between any pair of particles.

Thus, it tends to lead to the fastest convergence. Each particle maintains two attributes: velocity v_i and position x_i, which are updated as:

$$v_i^d = \chi(v_i^d + c_1 r_1^d(\hat{p}_i^d - x_i^d) + c_2 r_2^d(\hat{g}^d - x_i^d)), \tag{1}$$

$$\chi = \frac{2}{|2 - \varphi - \sqrt{\varphi^2 - 4\varphi}|}, \quad \varphi = c_1 + c_2, \tag{2}$$

$$x_i^d = x_i^d + v_i^d \tag{3}$$

where $d \in \{1, 2, ..., D\}$ and D means the dimension of the solution space; $\hat{p}_i$ is the personal historical best position discovered by the i-th particle; $\hat{g}$ is the global historical best position found so far by the whole particle swarm; c_1 and c_2 are two acceleration coefficients to weigh the relative importance of $\hat{p}_i$ and $\hat{g}$, respectively; r_1^d and r_2^d two random numbers sampled from the uniform distribution over $[0, 1]$; and χ represents the constriction factor which is derived from the existing constants in the velocity update equation. When $\varphi > 4$, convergence would be fast and guaranteed according to [1].

A ring topology aims to increase diversity during PSO's search process. It allows each particle to interact only with its immediate left and right neighbors. Each particle's velocity is updated as

$$v_i^d = \chi(v_i^d + c_1 r_1^d(\hat{p}_i^d - x_i^d) + c_2 r_2^d(\hat{q}_i^d - x_i^d)) \tag{4}$$

$$\hat{q}_i = \mathbf{argmax}(f(x_i), f(\overleftarrow{x}), f(\overrightarrow{x})) \tag{5}$$

where $\overleftarrow{x}$ and $\overrightarrow{x}$ are the positions of particle i's immediate left and right neighbors, $\hat{q}_i$ is the neighborhood historical best position of particle i.

3 Proposed Algorithm

In this section, an hybrid topology-based PSO, called HT-PSO, is proposed to solve a multi-source location problem with swarm robots.

3.1 Cluster Centers Selection

In HT-PSO, there is a one-to-one correspondence between robots and particles. At first, a robot swarm is randomly initialized in a search space. In each generation, robots measure the signal strength $f(x)$ at their current positions. Each robot updates its personal historical best position $\hat{p}_i$ as follows:

$$\hat{p}_i = \mathbf{argmax}(f(x_i), f(\hat{p}_i)) \tag{6}$$

Grouping is a crucial step in multi-source location. The number of subswarms and their search directions determine the number of sources that can be located. In HT-PSO, a ring topology is utilized to select elite robots as cluster centers for adaptive grouping. The robots whose personal historical best positions are same with their local historical best positions, i.e.,

$$\hat{p}_i = \hat{q}_i \tag{7}$$

are considered as cluster center candidates. If the signal strength value detected by these candidates in two adjacent generations does not change, they are excluded. Because these robots are likely to be in a no-signal area.

In the meanwhile, a cluster center aging mechanism is proposed to enhance the convergence performance of the swarm. Each robot has an aging attribute θ_i, which represents the remaining time for robot i as a cluster center and is initialized to 0. Once a robot satisfies the above condition, its age θ_i is increased by 1. Otherwise, θ_i is decreased by 1. Note that θ_i is not less than 0.

$$\theta_i = \begin{cases} \theta_i + 1, & \hat{p}_i = \hat{q}_i \\ \mathbf{max}(0, \theta_i - 1), & \hat{p}_i \neq \hat{q}_i \end{cases} \tag{8}$$

All robots with $\theta_i > 0$ are selected as cluster centers. This process is adaptive and does not require any environmental prior information.

3.2 Fast Convergence

The remaining robots choose their closest cluster center to form groups. Each group utilizes a star topology structure to achieve fast convergence, which allows information sharing among all robots in the group. The global historical best position $\hat{g}_i$ is calculated for each group Γ_i as:

$$\hat{g}_i = \mathbf{argmax}_{\forall r_k \in \Gamma_i}(f(\hat{p}_k)) \tag{9}$$

where r_k represents robot k and $\hat{p}_k$ is r_k's personal historical best position. Finally, robots update their velocities and positions by (1) and (3) respectively.

3.3 Overall Framework

HT-PSO utilizes a ring topology to select cluster centers for adaptive grouping in each generation. Then, the remaining robots choose their closest cluster center to form groups. Each group uses a star topology to accelerate the convergence. This hybrid topology aims to help HT-PSO achieve fine search performance and fast convergence. HT-PSO is realized via Algorithm 1.

Algorithm 1. The pseudo code of overall framework
1: Initialize the swarm robots randomly
2: Initialize $t = 1$, $\theta = [0, ..., 0]$
3: **while** $t \leq G$ **do**
4:    Swarm robots measure the signal strength by sensors
5:    Calculate $\hat{p}_i$ and $\hat{q}_i$ for each robot by (6) and (5), respectively
6:    Calculate cluster centers
7:    Form groups and achieve fast convergence
8:    $t = t + 1$
9: **end while**

4   Experimental Settings

4.1   Environment Description

In this work, the environment is set as a 16×16 square area. There are an unknown number of signal sources randomly distributed in the environment. Each source has a signal range and its signal strength gradually weakens with the distance to the source. In order to verify the effectiveness and robustness of the proposed strategy, four different environments are designed, as shown in Fig. 1. The environmental parameters are presented in Table 1, where N means the number of sources, R denotes the signal range of each source and W represents the maximum signal strength. In practice, the signal strength is measured by the robots' sensors. In experiments, for each position x_i, it can be calculated as:

$$\check{d} = \min_{S_i \in S} d_{x_i, S_i}, \tag{10}$$

$$f(x_i) = \begin{cases} W - \check{d} * (W/R), & \check{d} \leq R \\ 0, & \check{d} > R \end{cases} \tag{11}$$

where S represents the set of source locations, $\check{d}$ represents the distance between x_i and the nearest source. To reduce statistical error, each test is repeated for 1000 times independently and the locations of signal sources in each experiment are random.

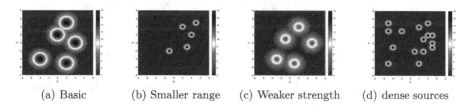

(a) Basic   (b) Smaller range   (c) Weaker strength   (d) dense sources

Fig. 1. Four different environments.

Table 1. Different environments

Environment	W	R	N
Basic environment	18	3	5
Smaller signal range environment	18	1	5
Weaker signal strength environment	13	3	5
Dense sources environment	18	1	16

4.2 Experimental Results

In this section, the proposed algorithm is compared with the state-of-the-art multi-source location method to show its effectiveness and robustness. Searching Auxiliary Points-based Constriction Factors PSO (SAP-CFPSO) [9] is the latest multi-source location method. It is a two-stage search algorithm. The first stage makes robots explore the whole searched area randomly and groups them by estimating the number and location of targets roughly. The second stage makes robots exploit their nearby area in groups guided by CFPSO. The parameter settings are given in Table 2, where NA means not applicable, n is the number of robots, G represents the total number of generations and V_{max} means the maximum velocity of robots. Besides, η is the proportion of the random search stage that is related to the size of the search environment, f_{th} is the threshold parameter for dividing elite robot swarm that is related to the maximum source signal strength, l_d is the distance parameter, which is designed for grouping and related to the influence range of sources. All the parameters are set by following [7]. These parameters affect the performance significantly. However, they are typically difficult to set because the environmental information is unknown. Especially, when the environment changes, these parameters may no longer work well. In comparison, our proposed strategy eliminates such parameters successfully.

Table 2. Parameter Settings

Method	c_1	c_2	n	G	V_{max}	η	f_{th}	l_d
SAP-CFPSO	2.1	2.1	80	100	0.2	0.2	10	6
HT-PSO	2.1	2.1	80	100	0.2	NA	NA	NA

In order to evaluate the performance, a multi-source location task has two basic metrics, i.e., locating accuracy "A" and convergence distance "Δ". If the distance between a robot and a signal source is less than 0.01 when the search ends, it means that this source has been successfully located.

"A" means the success rate of finding sources:

$$A = \frac{N'}{N} \tag{12}$$

where N' donates the number of found sources and N donates the number of total sources.

"Δ" is the average of the distance between each robot and the source closest to it when the search ends:

$$\Delta = \sum_{i=1}^{n} d_{x_i, S_{r_i}} / n \tag{13}$$

where n is the number of robots and $d_{x_i, S_{r_i}}$ represents the distance between x_i and the source closest to it.

The total comparison results are presented in Table 3, which indicate that HT-PSO outperforms the state-of-the-art multi-source location method. HT-PSO improves the search accuracy by 46.75% and reduces the convergence distance by 34.21%. It significantly improves the locating accuracy and reduces the convergence distance. Meanwhile, it eliminates the sensitive environmental parameters in a search process. Thus, HT-PSO exhibits stronger robustness and applicability.

Table 3. Comparison results

Method	Basic		Smaller range		Weaker strength		Dense sources	
	A	Δ	A	Δ	A	Δ	A	Δ
SAP-CFPSO	88.1%	0.0304	54.5%	0.1741	43.8%	0.5330	48.7%	**0.1074**
HT-PSO	**93.9%**	**0.0194**	**87.2%**	**0.1558**	**94.2%**	**0.0182**	**51.3%**	0.1143

5   Conclusion

In this work, a novel multi-source location approach called Hybrid Topology-based PSO (HT-PSO) is proposed. It has two distinct features. First, HT-PSO significantly improves the locating accuracy and greatly accelerates the convergence speed. Second, HT-PSO does not require any prior knowledge of the environment to set some parameters, thus it has stronger robustness and adaptability. Compared with the state-of-the-art multi-source location method, the search accuracy is improved by 46.75% and the convergence distance is reduced by 34.21%. Future work should focus on more benchmark studies with other recently proposed intelligent optimization methods [2,4,5,10,11,14].

Acknowledgements. This work was supported by Innovation Program of Shanghai Municipal Education Commission (202101070007E00098) and Shanghai Industrial Collaborative Science and Technology Innovation Project (2021-cyxt2-kj10). This work was also supported in part by the National Natural Science Foundation of China (51775385, 61703279, 62073244, 61876218) and the Shanghai Innovation Action Plan under grant no. 20511100500.

References

1. Bratton, D., Kennedy, J.: Defining a standard for particle swarm optimization. In: IEEE Swarm Intelligence Symposium (2007)
2. Cao, Z., Lin, C., Zhou, M., Huang, R.: Scheduling semiconductor testing facility by using cuckoo search algorithm with reinforcement learning and surrogate modeling. IEEE Trans. Autom. Sci. Eng. **16**(2), 825–837 (2019)
3. Doroodgar, B., Liu, Y., Nejat, G.: A learning-based semi-autonomous controller for robotic exploration of unknown disaster scenes while searching for victims. IEEE Trans. Cybern. **44**(12), 2719–2732 (2014)
4. Fu, Y., Zhou, M., Guo, X., Qi, L.: Scheduling dual-objective stochastic hybrid flow shop with deteriorating jobs via bi-population evolutionary algorithm. IEEE Trans. Syst. Man Cybern.: Syst. **50**(12), 5037–5048 (2020)
5. Hua, Y., Liu, Q., Hao, K., Jin, Y.: A survey of evolutionary algorithms for multi-objective optimization problems with irregular pareto fronts. IEEE/CAA J. Autom. Sinica **8**(2), 303–318 (2021)
6. Li, R., Wu, H.: Multi-robot source location of scalar fields by a novel swarm search mechanism with collision/obstacle avoidance. IEEE Trans. Intell. Transp. Syst. (2020). https://doi.org/10.1109/TITS.2020.3010056
7. Schumacher, C., Chandler, P.: Task allocation for wide area search munitions. In: Proceedings of American Control Conference, pp. 1917–1922 (2002)
8. Sinha, A., Kumar, R., Kaur, R., Bhondekar, A.P.: Consensus-based odor source localization by multiagent systems. IEEE Trans. Cybern. **49**(12), 4450–4459 (2018)
9. Tang, Q., Ding, L., Yu, F., Zhang, Y., Li, Y., Tu, H.: Swarm robots search for multiple targets based on an improved grouping strategy. IEEE/ACM Trans. Comput. Biol. Bioinf. **15**(6), 1943–1950 (2018)
10. Wang, Y., Gao, S., Zhou, M., Yu, Y.: A multi-layered gravitational search algorithm for function optimization and real-world problems. IEEE/CAA J. Autom. Sinica **8**(1), 94–109 (2021)
11. Wei, G., Wu, Q., Zhou, M.: A hybrid probabilistic multiobjective evolutionary algorithm for commercial recommendation systems. IEEE Trans. Comput. Soc. Syst. **8**(3), 589–598 (2021)
12. Wu, P., Chu, F., Che, A., Zhou, M.: Bi-objective scheduling of fire engines for fighting forest fires: new optimization approaches. IEEE Trans. Intell. Transp. Syst. **19**(4), 1140–1151 (2018)
13. Zhang, J., Lu, Y., Che, L., Zhou, M.: Moving-distance-minimized PSO for mobile robot swarm. IEEE Trans. Cybern. (2022). https://doi.org/10.1109/TCYB.2021.3079346
14. Zhang, J., Zhu, X., Wang, Y., Zhou, M.: Dual-environmental particle swarm optimizer in noisy and noise-free environments. IEEE Trans. Comput. Soc. Syst. **49**(6), 2011–2021 (2019)
15. Zou, R., Kalivarapu, V., Winer, E., Oliver, J., Bhattacharya, S.: Particle swarm optimization-based source seeking. IEEE Trans. Autom. Sci. Eng. **12**(3), 865–875 (2015)

Advances in Cooperative Target Searching by Multi-UAVs

Changjian Wang[1], Xiaoming Zhang[1(✉)], Yingbo Lei[2], Hang Wu[1], Hang Liu[1], and Lele Xie[1]

[1] Institutes of Physical Science and Information Technology, Anhui University, Hefei, China
xmzhang@ustc.edu

[2] Beijing Droneyee Intelligent Technology Co., Ltd., Beijing, China

Abstract. With the development of multi-UAVs technology, detecting and searching unknown areas by using autonomous multi-UAVs have become a frontier research direction with difficult in this field. To illustrate the progress of cooperative target searching by multi-UAVs, firstly, the significance of cooperative searching and its application in military and civil fields are systematically described. Then the current research status of the multi-UAVs cooperative searching is described, and three aspects are analyzed, including environment modeling, cooperation architecture and searching methods. Finally, the conclusions are made in terms of the autonomous and cooperation ability of multi-UAVs. The future research trend of improving the efficiency of multi-UAVs cooperative search is discussed and prospected from the perspectives of UAVs' perception, cognition, autonomy, as well as human-machine cooperative technology.

Keywords: Multi-UAVs · Swarm intelligence · Target searching · Environment modeling · Cooperative architecture · Search method

1 Introduction

Swarm UAVs technology involves the design, construction and deployment of large-scale UAV group to solve problems or perform tasks in a cooperative manner [1]. Due to the limited mission capabilities of a single UAV, the performance of multi-UAVs system is superior to that of an independent individual UAV system through effective cooperation between UAVs [2]. Multi-UAVs system has unique system attributes and functions such as high robustness, scalability and flexibility, which makes the execution of complex tasks more effective and reliable [3]. With the development of multi-UAVs technology, it plays an increasingly important role in military and civil field. In military, with the development of science and technology and the deepening of information warfare, robots have become the best choice to perform dull, hash and dangerous tasks. For example, in recent years, the frequent occurrence of Predator, Global Eagle and other combat UAVs in the Afghan and Syrian wars shows the advantages of UAVs in reconnaissance and search. In civil, the rapid development of multi-UAVs technology has gradually changed people's lifestyle and working style, such as in forest areas where fires occur,

Y. Tan et al. (Eds.): ICSI 2022, LNCS 13345, pp. 25–34, 2022.
https://doi.org/10.1007/978-3-031-09726-3_3

multi-UAVs plays an important role in the troubleshooting of thermal power plant [5]. Therefore, multi-UAVs cooperative search technology is a crucial technology for modern war, civil rescue and other activities.

The autonomous control system of UAV means that the system can automatically generate optimal control strategies, complete various strategic tasks, and have fast and effective task adaptive ability without human intervention through online environment perception and information processing [6]. Challenges faced by UAV systems include complex, unstructured dynamic environments, and various real-time external threats and accidents. The U.S. Department of Defense believes that the first technology for future UAVs is to increase autonomy and collaboration, all tasks of UAV systems depend on autonomous environmental awareness. Therefore, how to achieve high efficiency cooperation of swarm autonomous UAVs in complex environment is the key to the research of multi-UAVs cooperative search. This paper will review the research progress of multi-UAVs cooperative target searching technology around the environment modeling, cooperative architecture and searching method in multi-UAVs cooperative target searching.

2   Advances of Key Technologies for Multi-UAVs Cooperative Target Searching

The US military proposed the concept of swarm UAVs operation in the late 1990s. At present, the requirement for intelligent control of UAV is increasing, as shown in Fig. 1, the U.S. Department of Defense classifies the Autonomous Control Level (ACL) of an UAVs system into 10 levels [7]. With the improvement of the autonomy of the UAVs system, the UAVs system has different requirements from remote control (level 1) to full autonomy (level 10), the corresponding perception, coordination and decision-making of the UAVs system have different requirements. In the area of UAVs system, US Predators (RQ-1/MQ-1) and Global Eagles (RQ-4) can achieve ACL 2 to 3, while Joint UAV(J-UCAS) and X47-B can achieve ACL 5 to 6. Unmanned combat armed rotator (UCAR) achieves ACL 7 to 9. However, most UAVs systems are still at a low level of autonomy, so it is necessary to combine swarm UAVs with specific task requirements to improve the autonomy of the UAV and the efficiency of performing operational tasks.

At present, the multi-UAVs technology has become the frontier hot spot in the research of swarm intelligence [8]. Most of the literatures on the theory of swarm UAVs cooperation focus on the research of the algorithm of swarm UAVs cooperation. For research on algorithm and simulation of swarm UAVs cooperation, the research directions include swarm UAVs autonomous formation [9], swarm UAVs cooperative target searching [10], swarm UAV cooperative area coverage [11], swarm UAVs cooperative target allocation [12]. For the problem of multi-UAVs cooperative search, due to a variety of uncertain factors, the search environment is dynamic. Many fruitful works have been carried out on the research of multi-UAVs cooperative target searching technology in complex environment, most of which are focused on three aspects: environment modeling, cooperation architecture and search method.

For the research of swarm UAVs cooperation technology, we must start from the characteristics and task requirements of UAVs. Firstly, we need to model the environment

Fig. 1. The autonomous control level of UAV defined by AFRL and the development of UAV autonomy system.

according to the task requirements, and then improve the efficiency of UAVs completing the task through appropriate swarm UAVs cooperation architecture and search method. This paper analyzes and discusses the environment modeling, cooperation architecture and search method in the cooperative search of swarm UAVs.

2.1 Environment Modeling

The main methods of environment modeling of swarm UAVs cooperation include grid method, artificial potential field method and graph theory method. Grid method divides the task environment of UAVs into several grids at fixed interval, as shown in Fig. 2, in the task area of $L_X \times L_y$, the task area is divided into $M \times N$ grids according to the Δd interval distance, the size of a single grid directly affects the memory required by the computer to store environmental information. Environment grid search maps mainly include target probability map, digital pheromone map, return map, and search maps generated by different methods. Artificial potential field is a method to model the task environment based on the attractiveness of the target and the repulsion of the threat, swarm UAVs can achieve obstacle avoidance search of the target under the combined action of target gravity and threat repulsion. Graph theory method divides the area into several sub-areas based on the terrain and environmental characteristics of the area, and combines the task characteristics of each sub-area to assign search tasks and plan flight routes for UAVs. The more common method is based on Voronoi, as shown in Fig. 3.

Song et al., proposed a swarm robots search algorithm based on neural network, corresponding the neural network to the grid environment, and constructed a digital pheromone map [13]. Pehlivanoglu et al., and Zhang et al., constructed multiple robots cooperative area models based on Voronoi to achieve search coverage for unknown environment [14, 15]. Zheng et al., divided the area to be searched into several sub-areas, calculated the probability of the existence of the target in each area, constructed the target probability map, and achieved the multi-UAVs cooperative search dynamic target

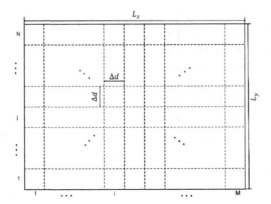

Fig. 2. Area division based on grid method.

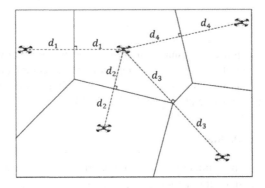

Fig. 3. Area division based on Voronoi.

in unknown environment [16]. Zhen et al., rasterized the UAV task area and represented the threat area in a grid map, built the environmental cognitive model of target attraction field and threat repulsion field, and built the target probability map based on the grid map [17].

2.2 Cooperation Architecture

In order for swarm UAVs to accomplish the complex tasks in a cooperative manner, an architecture for controlling its movements must be established. The main task of the autonomous control system for UAVs is to connect each sub-system into a whole, manage and dispatch each sub-system in a unified way, so that each sub-system can complete the overall task in unison, and an excellent autonomous control system can improve the efficiency of multi-UAVs task completion. Multi-UAVs control systems are generally divided into centralized architecture, distributed architecture and hybrid architecture.

As shown in Fig. 4, it is the centralized architecture, which has a control center, all the information of the UAVs should be sent to the control center centrally, and all the information in the system will be processed centrally. The centralized architecture is relatively simple and the system management is convenient, but when the number of UAVs is large, the computational complexity of the centralized control method increases significantly, resulting in a significant reduction in the work efficiency of UAVs. When the control center fails, the whole system will be paralyzed [18]. Wei et al., adopted an architecture of centralized control of multi-UAVs in the case of a small number of UAVs, and realized the coverage search of the task area by operating the actions of multi-UAVs through the control center [19].

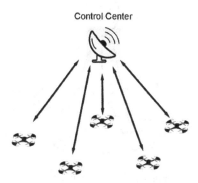

Fig. 4. Centralized architecture.

As shown in Fig. 5, it is a distributed architecture, which adopts the way of autonomy and cooperation, the complex solution problem is divided into sub problems that can be solved by each UAV. There is an equal cooperative relationship between UAVs, and they can communicate directly. Distributed architecture can increase the number of robots and has high flexibility. It is suitable for workspace in dynamic environment. At present, most multi-UAVs cooperation methods use distributed architecture. Li et al., and Bakhshipour et al., used the distributed architecture on the cooperative search of multi-UAVs [20, 21].

Fig. 5. Distributed architecture.

The hybrid architecture is the integration of the first two architectures, with a distributed architecture between the swarm UAVs, and a centralized architecture between the swarm UAVs and the control center, as shown in Fig. 6. The hybrid architecture combines the advantages of the two structures to make the swarm UAVs more intelligent. Hou et al., adopted a hybrid architecture to divide the swarm UAVs into multiple groups, the UAVs in each group can communicate with each other, and the leading UAV in each group can communicate with each other, so as to realize the communication between groups, and each group is centrally controlled by the control center [22].

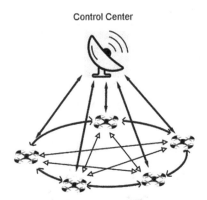

Fig. 6. Hybrid architecture.

2.3   Cooperative Searching Method

For the multi-UAVs cooperative target searching task in unknown environment, from the perspective of scalability and applicability, this paper focuses on the scanning search method, dynamic search method and intelligent optimization method.

Scanning search method mainly searches the task area for the purpose of area coverage. This method can ensure that UAVs can traverse the whole area. Common scanning search methods include parallel search and spiral reconnaissance [23]. As a simple and practical coverage search method, the scanning search method has the advantages in searching for static targets, but it has poor adaptability to dynamic environment. Because it mainly focuses on area coverage, the time efficiency under the target search task is not as efficient as other methods.

Dynamic search methods mainly include discrete Markov decision-making process and distributed Model Predictive Control (DMPC). Dynamic search algorithm is widely used in complex multi-UAVs cooperative target searching task because of its simple calculation and fast response to dynamic events. Elamvazhuthi et al., and Zhou et al., divided and modeled the environment, added information to the environment based on discrete Markov chain, and combined the constructed model with other advanced algorithms to complete the cooperative search task of multi-UAVs [24, 25].

Model Predictive Control (MPC) uses the control system model and optimization technology to design the optimal control input of the system for a predictive cycle. The core idea is to solve the problem by rolling optimization. DMPC uses a distributed structure to improve the decision-making speed of the whole system. Zhao et al., Lun et al., and Yu et al., controlled the motion of UAVs based on DMPC, and solved the problem of cooperative search of multi-UAVs in unknown environment [26–28].

In the multi-UAVs cooperative search problem, the current related research is mostly based on intelligent optimization methods. Traditional optimization methods have less application because of the rapid increase of time and space complexity and poor solution results when solving large-scale problems. The concept of swarm intelligence first appeared in cellular robotic systems proposed by Beni and Wang in 1989 [29]. Intelligent optimization methods derive from intelligent behaviors of cooperation between biological groups, for example, inspired by the behavior of social groups of organisms such as bird flocks, ants, bats, wolves, and plant populations [30]. Numerous studies have shown that intelligent optimization algorithms can be applied cooperative control of multi-UAVs, and can improve the efficiency of UAVs in solving problems in complex environments. Common intelligent optimization algorithms include particle swarm algorithm, ant colony algorithm, bat algorithm, gray wolf optimizer, bean optimization algorithm and bacterial colony algorithm.

MD et al., proposed a new particle swarm algorithm based on motion coding for multi-UAVs searching for dynamic targets [31]. Hta et al., proposed a new adaptive robot grey wolf optimizer algorithm to solve the cooperative search target problem of swarm robots in unknown environment. An adaptive speed adjustment strategy was used to track the dynamic target, and compared with other methods, this method has obvious advantages in the efficiency of target searching [32]. Wang et al., constructed an intrusion model of dynamic targets, and proposed an improved bat algorithm for his model to solve the trajectory optimization problem of UAVs tracking the intrusive targets [33]. Not limited to the research on social animal groups, the adaptive strategy of plant population also provides novel ideas for the research of swarm intelligence and swarm robots, and a swarm UAVs search algorithm based on the evolution of plant population distribution is constructed [34, 35], and effective simulation experiments are carried out for the problem of pollution source search. Kyriakakis et al., used moving peak to simulate the movement of multiple UAVs searching for multiple moving targets [36]. An online multi-population framework was constructed, which is suitable for most intelligent optimization methods and meets the experimental requirements of multi-UAVs searching for multiple targets in unknown environments.

3   Research Summary and Prospect

The main problem of multi-UAVs cooperative target searching technology is how to make full use of the autonomy, cooperation and reduce the time and space complexity of the algorithm. In recent years, swarm intelligence and machine learning technology have been gradually introduced into the multi-UAVs cooperative problem. The application of these technologies makes the UAVs have stronger self-learning and self-organization ability in complex environment. The complexity of the algorithm is lower than other traditional algorithms. The combination of distributed architecture and centralized architecture makes UAVs fleet more cooperative, enabling swarm UAVs to remain cooperative to complete tasks even when a single UAV fails or is threatened by the environment.

With the increasing complexity of task environment, there is no effective means to achieve the autonomous control of UAVs in dynamic unknown complex environment and time-sensitive situation, and to achieve fast and effective acquisition and processing of information and the control ability of the platform. The contents that need further research on the autonomous control technology of swarm UAVs systems include:

(1) The comprehensive environmental awareness and intelligent battlefield situation awareness capabilities of the UAVs. Due to the harshness and complexity of the battlefield environment, the impact of emergencies such as single UAV failures and threat in the environment cannot be ignored when performing target searching tasks. In the future, UAVs need to have a more comprehensive environmental awareness, be able to timely perceive and respond to changes in the battlefield environment, improve the awareness of the battlefield situation, and achieve better search results.

(2) Human-machine intelligence integration and learning adaptability. Operators should give proper guidance to multi-UAVs systems in order to achieve more accurate control, how to use the characteristics of each human-machine effectively to achieve the integration of human-machine intelligence, and how to design a more efficient control structure of human-machine cooperation system to improve the usability and overall operational efficiency of multi-UAVs systems is the further research content to improve the cooperative efficiency of multi-UAVs systems.

(3) The autonomous navigation, planning and control capabilities of the UAVs under complex conditions. In order to realize the in-depth cognition of complex environment of swarm UAVs, future UAVs systems should establish intelligent development mechanism, which can make UAVs systems have similar stable learning and intelligent development mechanisms to human beings, improve the self-learning, self-reasoning and self-organization capabilities of UAVs systems, and give full play to the real-time decision-making advantages of swarm UAVs. Enhance the ability to cope with complex environment, such as combining traditional methods with swarm intelligence and machine learning, to make UAVs more autonomous.

4 Conclusion

This paper systematically combs the problem of multi-UAVs cooperative target searching. The key technologies of target searching, including environment modeling, cooperative architecture and search method, are discussed in depth. The development trend of multi-UAVs cooperative target searching is analyzed. The key to improve the efficiency of multi-UAVs cooperative search is to improve the autonomy and cooperative capability of multi-UAVs system. Multi-UAVs system has made great progress in self-organization and cooperative control of UAVs. The integration of swarm intelligence and other methods enables multi-UAVs to have certain autonomous decision-making ability. In the future, for complex unknown environment, improving environment modeling method, improving UAVs perception and cognitive ability, designing a more reasonable cooperative architecture, realizing more efficient human-machine cooperative control technology, building swarm self-learning and intelligent development mechanism will greatly improve the effectiveness and efficiency of multi-UAVs cooperative target searching.

Acknowledgements. This research was funded by Qinghai Natural Science Foundation project under grant number 2020-zj-913, Anhui Natural Science Foundation project under grant number s202202a04021815, Anhui Graduate Scientific Research project under grant number yjs20210087.

References

1. Dorigo, M., Theraulaz, G., Trianni, V.: Swarm robotics: past, present, and future. Proc. IEEE **109**(7), 1152–1165 (2021)
2. Raja, G., Anbalagan, S., Subramaniyan, A.G.: Efficient and secured swarm pattern multi-UAV communication. IEEE Trans. Veh. Technol. **70**(7), 7050–7058 (2021)
3. Senanayake, M., Senthooran, I., Barca, J.C.: Search and tracking algorithms for swarms of robots: a survey. Rob. Auton. Syst. **75**, 422–434 (2016)
4. Zhen, Z., Zhu, P., Xue, Y.: Distributed intelligent self-organized mission planning of multi-UAV for dynamic targets cooperative search-attack. Chin. J. Aeronaut. **32**(12), 2706–2716 (2019)
5. Pham, H.X., La, H.M., Feil-Seifer, D.: A distributed control framework of multiple unmanned aerial vehicles for dynamic wildfire tracking. IEEE Trans. Syst. Man Cybern. Syst. **50**(4), 1537–1548 (2018)
6. Zhou, Z., Feng, J., Bo, G.: When mobile crowd sensing meets UAV: energy-efficient task assignment and route planning. IEEE Trans. Commun. **66**(11), 5526–5538 (2018)
7. Haddon, D.R., Whittaker, C.J.: Office of the secretary of defense, unmanned aircraft systems roadmap. Implementation Netw. **8**(14), 263–287 (2005)
8. Oh, H., Shirazi, A.R., Sun, C.: Bio-inspired self-organising multi-robot pattern formation: a review. Rob. Auton. Syst. **91**, 83–100 (2017)
9. Meng, W., He, Z., Su, R.: Decentralized multi-UAV flight autonomy for moving convoys search and track. IEEE Trans. Control Syst. **25**(4), 1480–1487 (2016)
10. Li, L., Zhang, X., Yue, W.: Cooperative search for dynamic targets by multiple UAVs with communication data losses. ISA Trans. **114**, 230–241 (2021)
11. Shang, Z., Bradley, J., Shen, Z.: A co-optimal coverage path planning method for aerial scanning of complex structures. Expert Syst. Appl. **158**(4), 113535 (2020)
12. Kurdi, H.A., Aloboud, E., Alalwan, M.: Autonomous task allocation for multi-UAV systems based on the locust elastic behavior. Appl. Soft Comput. **71**, 110–126 (2018)
13. Song, Y., Fang, X., Liu, B.: A novel foraging algorithm for swarm robotics based on virtual pheromones and neural network. Appl. Soft Comput. **90**, 106156 (2020)
14. Pehlivanoglu, Y.V.: A new vibrational genetic algorithm enhanced with a Voronoi diagram for path planning of autonomous UAV. Aerosp. Sci. Technol. **6**(1), 47–55 (2012)
15. Zhang, X., Ali, M.: A bean optimization-based cooperation method for target searching by swarm UAVs in unknown environments. IEEE Access **8**, 43850–43862 (2020)
16. Zheng, Y., Du, Y., Ling, H.: Evolutionary collaborative human-UAV search for escaped criminals. IEEE Trans. Evol. Comput. **24**(2), 217–231 (2019)
17. Zhen, Z., Chen, Y., Wen, L.: An intelligent cooperative mission planning scheme of UAV swarm in uncertain dynamic environment. Aerosp. Sci. Technol. **100**, 105826 (2020)
18. Yong, Z., Rui, Z., Teng, J.L.: Wireless communications with unmanned aerial vehicles: opportunities and challenges. IEEE Commun. Mag. **54**(5), 36–42 (2016)
19. Wei, M., He, Z., Rong, S.: Decentralized multi-UAV flight autonomy for moving convoys search and track. IEEE Trans. Control Syst. Technol. **25**(4), 1480–1487 (2017)
20. Li, P., Duan, H.: A potential game approach to multiple UAV cooperative search and surveillance. Aerosp. Sci. Technol. **68**, 403–415 (2017)

21. Bakhshipour, M., Ghadi, M.J., Namdari, F.: Swarm robotics search & rescue; a novel artificial intelligence-inspired optimization approach. Appl. Soft Comput. **57**, 708–726 (2017)

22. Hou, Y., Liang, X., He, L.: Time-coordinated control for unmanned aerial vehicle swarm cooperative attack on ground-moving target. IEEE Access **25**(4), 1480–1487 (2019)

23. Chakravorty, S., Ramirez, J.: Fuel optimal maneuvers for multispacecraft interferometric imaging systems. J. Guid. Control Dyn. **30**(1), 227–236 (2007)

24. Elamvazhuthi, K., Kakish, Z., Shirsat, A.: Controllability and stabilization for herding a robotic swarm using a leader: a mean-field approach. IEEE Trans. Robot. **37**(2), 418–432 (2020)

25. Zhou, X., Wang, W., Wang, T.: Bayesian reinforcement learning for multi-robot decentralized patrolling in uncertain environments. IEEE Trans. Veh. Technol. **68**(99), 11691–11703 (2019)

26. Zhao, J., Sun, J., Cai, Z.: Distributed coordinated control scheme of UAV swarm based on heterogeneous roles. Chin. J. Aeronaut. **12**, 1–17 (2022)

27. Lun, Y., Yao, P., Wang, Y.: Trajectory optimization of SUAV for marine vessels communication relay mission. IEEE Syst. J. **14**(4), 5014–5024 (2020)

28. Yu, Y., Wang, H., Liu, S.: Distributed multi-agent target tracking: a nash-combined adaptive differential evolution method for UAV systems. IEEE Trans. Veh. Technol. **70**(8), 8122–8133 (2021)

29. Beni, G., Wang, J.: Swarm intelligence in cellular robotic systems. In: Dario, P., Sandini, G., Aebischer, P. (eds) Robots and Biological Systems: Towards a New Bionics? NATO ASI Series, vol. 102. Springer, Berlin, Heidelberg (1993). https://doi.org/10.1007/978-3-642-58069-7_38

30. Osaba, E., Del Ser, J., Iglesias, A.: Soft computing for swarm robotics: new trends and applications. J. Comput. Sci. **39**, 101049 (2020)

31. Phung, M.D., Ha, Q.P.: Motion-encoded particle swarm optimization for moving target search using UAVs. Appl. Soft Comput. **97**, 106705 (2020)

32. Hta, B., Wei, S.B., Al, C.: A GWO-based multi-robot cooperation method for target searching in unknown environments. Expert Syst. Appl. **186**, 115795 (2021)

33. Li, K., Han, Y., Ge, F., Xu, W., Liu, L.: Tracking a dynamic invading target by UAV in oilfield inspection via an improved bat algorithm. Appl. Soft Comput. **90**, 106150 (2020)

34. Zhang, X., Hu, Y., Li, T.: A novel target searching algorithm for swarm UAVs inspired from spatial distribution patterns of plant population. Int. J. Comput. Intell. Syst. **14**(1), 159–167 (2021)

35. Wang, C., Zhang, X., Liu, H., Wu, H.: RBOA algorithm based on region segmentation and point update. In: 2021 China Automation Congress (CAC), pp. 6983–6988 (2021)

36. Kyriakakis, N.A., Marinaki, M., Matsatsinis, N.: Moving peak drone search problem: an online multi-swarm intelligence approach for UAV search operations. Swarm Evol. Comput. **66**, 100956 (2021)

Artificial Potential Field Method with Predicted State and Input Threshold for Multi-agent System

Dengxiu Yu[1]([✉])[ID], Zhize Du[1], and Zhen Wang[2]

[1] Unmanned System Research Institute, Northwestern Polytechnical University,
Xi'an 710072, China
yudengxiu@126.com
[2] Center for Optical Imagery Analysis and Learning, Northwestern Polytechnical
University, Xi'an 710072, China

Abstract. In this article, we consider the multi-agent control based on the artificial potential field (APF) method with predicted state and input threshold. APF is a very practical and efficient method for multi-agent control. However, the accuracy of APF is susceptible to communication delay. Hence, we introduce the predictive state model to reduce the impact of this delay when the agent is avoiding collisions and maintaining formation. Meanwhile, the input threshold is applied to ensure the safety of the system. The introduction of the predicted state and the input threshold leads to the failure of traditional APF. Therefore, we propose a new controller based on the improved APF. Then, the Lyapunov stability of the designed controller is analyzed. Simulation results show the effectiveness of the proposed controller and its superiority over the original method.

Keywords: Multi-agent system · Artificial potential field · Input threshold · Predicted state model

1 Introduction

Considering the poor performance and the limited load of a single agent, the formation of multiple agents must be dispatched. In a rapidly changing environment, the multi-agent system needs to adjust the current state, which not only requires fast and stability but also needs to ensure that communications of the whole system remain connected [1,2]. At the same time, the multi-agent system needs to avoid obstacles in real-time and avoid collisions between individuals in formation [3]. Therefore, further research is urgently needed by facing the cooperative control problem under many constraints.

Formation control of multi-agent is a vital issue that has been maturely researched [4,5]. Through effective information interaction, the swarm formation can quickly converge to the desired formation and remain unchanged or change

© Springer Nature Switzerland AG 2022
Y. Tan et al. (Eds.): ICSI 2022, LNCS 13345, pp. 35–44, 2022.
https://doi.org/10.1007/978-3-031-09726-3_4

within the allowable error range. The main formation control algorithms include the leader-follower method, artificial potential field (APF) function method, consensus theory-based method, and so on. In [6], authors transform the tracking errors into new virtual error variables and limit the state tracking errors within the prescribed bounds based on a new adaptive fuzzy output-feedback method. In [7], authors respectively define formation errors of position and velocity and design the controller based on the consensus theory. In [8], authors apply consensus-based approaches to deal with the time-varying formation control problems for swarm systems and put forward the conditions needed to form a time-varying center. However, the application of formation flying in the actual environment still faces many problems. We aim to study further and find a more effective method.

Formation control and obstacle avoidance control are often inseparable. For multi-agent formation flying, a minor collision may cause a chain reaction, so how to achieve collision avoidance is a crucial technology to ensure the safe flight of formations [9,10]. Authors in [11] propose a barrier function-based formation navigation controller based on only local information to achieve collision avoidance. It has the characteristics of fast and small consumption but cannot effectively prevent the deadlock problem. Authors in [12] design a novel Barrier Lyapunov functions to limit the position of the agent to achieve the location constraints of the multi-agent system, which is similar to APF based methods. Researchers have proven that the APF method is a mature and efficient real-time path collision avoidance method [13].

Taking into account the communication delay, physical delay, or some state uncertainty in the real situation, the distance between the two agents may be less than the safe distance [14]. Authors in [15] investigate the latest developments in personalized driving assistance systems, and they find that online learning to adjust driving thresholds is the future development direction. Authors in [16] introduce the estimate position obtained by additional information to design the controller and make the agents converge to consensus in a cooperative manner. However, too much information makes it more challenging to achieve. Therefore, we aim to design an APF function based on the predicted state to make the control simpler and more effective.

The contributions of this paper are discussed as follows.

(1) We propose predicted state modeling to improve APF and give the agent more buffer to better complete the tasks of collision avoidance and target tracking, which reduces the impact of delay to a certain extent.

(2) To avoid the uncertainty caused by severe velocity changes, we design the controller so that the speed of each agent is limited to the maximum safe speed.

(3) A new controller is designed to realize collision avoidance and formation maintenance.

2   Problem Statement

2.1   Multi-agent Dynamics Modeling

Consider a multi-agent system with N agents. We define the dynamics model of the agent i in the system as

$$\dot{p}_i = v_i. \\ \dot{v}_i = u_i. \tag{1}$$

where $p_i \in \mathbb{R}^d, v_i \in \mathbb{R}^d$ are the position vector and the velocity vector of the agent i, and $u_i \in \mathbb{R}^d$ is the acceleration vector applied on the ith agent as control input in d-dimensional space, $i = 1, 2, \cdots, N$.

2.2   Predicted State Modeling

In this section, we take into account the actual situation of the movement with certain inertia and define a predicted state as

$$\hat{p}_i(t) \triangleq p_i(t) + \mu_1 v_i(t) + \mu_1 \mu_2 u_i(t-1). \tag{2}$$

where $\mu_1 > 0$ and $\mu_2 > 0$ are the predicted position impact factor and the predicted velocity impact factor, respectively, and their size are related to the delay. $i = 1, 2, \cdots, N$.

2.3   Problem Formulation

The APF method based on the predicted state is designed to realize agents moving with collision avoidance and formation maintenance. The expected distance between the agent is d_{ij} and the collision occurs if the distance is less than r_{in}. The constraints in formation control problems are as follows.

(1) For collision avoidance, $|p_i - p_j| \geq r_{in}$.
(2) For formation maintenance, $|p_i - p_j| = d_{ij}$.

To realize the purpose, we define system potential Lyapunov function as $V(t) = \sum_{i=1}^{N} V_i$, where V_i is the potential function of agent i. Therefor, the maximum upper limit of designed collision avoidance part G_1^{max} and formation maintenance part G_2^{max} should Satisfy

$$V_i(t) < \min \{G_1^{max}, G_2^{max}\}. \tag{3}$$

Assumption 1: The system can provide as much control as we need to ensure the steady work of the multi-agent system.

3   Controller Design

We define the predicted distance between agents i and j is

$$\hat{x} = |\hat{p}_i - \hat{p}_j|. \tag{4}$$

and the measurement distance between agents i and j is

$$x = |p_i - p_j|. \tag{5}$$

for $i, j = 1, 2, \cdots, N$.

3.1   Collision Avoidance

This part is designed to keep each agent at a boundary distance to avoid the collision. The obstacle avoidance term function works when $r_{in} \leq \hat{x} \leq r_{out}$, where r_{out} is collision avoidance range.

The set of all agents entering collision avoidance distances of agent i is designed as

$$N_i^l(t) = \{j \in \{1, \ldots, N\} : \hat{x} \leq r_{out}\}. \tag{6}$$

To make the potential function change smoothly, the potential function is chosen as

$$G_1(\hat{x}) = \begin{cases} k_1 \int_{\hat{x}}^{r_{out}} g(s)ds, & \hat{x} \in [r_{in}, r_{out}] \\ 0, & \text{otherwise} \end{cases}. \tag{7}$$

where k_1 is the coefficients of the collision avoidance potential function, and

$$g_1(\hat{x}) = \frac{1}{2}\left[1 + \cos\left(\pi\frac{\hat{x} - r_{in}}{r_{out} - r_{in}}\right)\right] + 1. \tag{8}$$

Then the collision avoidance control input of agent i can be design as

$$u_i^c = -\sum_{j \in N_i^l(t)} \nabla_{\hat{p}_i} G_1(\hat{x}_i). \tag{9}$$

where $\nabla_{\hat{p}_i}$ represents the gradient along $\hat{p}_i$.

3.2   Formation Maintenance

The formation maintenance part function works when $r_{out} < \hat{x} \leq R$, where R is the maximum communication distance.

The set of communication connections with agent i is design as

$$N_i^n(t) = \{j \in \{1, \ldots, N\} : (i, j)\}. \tag{10}$$

To generate the desired formation, the potential function is chosen as:

$$G_2(\hat{x}) = \begin{cases} k_2 \int_{\hat{x}}^{d} g_2(s)ds, & \hat{x} \in (r_{out}, R] \\ 0, & \text{otherwise} \end{cases}. \tag{11}$$

where k_2 is the coefficients of the formation maintenance potential function, and

$$g_2(\hat{x}) = \begin{cases} \frac{1}{2}\left[1+\cos\left(\pi\frac{\hat{x}-r_{out}}{d_{ij}-r_{out}}\right)\right], & \hat{x} \in (r_{out}, d_{ij}] \\ 1+\cos\left(\pi\frac{\hat{x}-R}{R-d_{ij}}\right), & \hat{x} \in (d_{ij}, R] \end{cases}. \tag{12}$$

Then the formation maintenance control input of agent i can be design as

$$u_i^t = - \sum_{j \in N_i^n(t)} \nabla_{\hat{p}_i} G_2(\hat{x}_i). \tag{13}$$

3.3   Control Law Design

Through the previous explanation, We can design control law of intelligent agent i in p_i as

$$u_i(t) = - \sum_{j \in N_i^l(t)} \nabla_{\hat{p}_i} G_1(\hat{x}) - \sum_{j \in N_i^n(t)} \nabla_{\hat{p}_i} G_2(\hat{x}) - S(e_i^p, v_m) - S(e_i^v, u_m) + u_d. \tag{14}$$

where $-\sum_{j \in N_i^l(t)} \nabla_{\hat{p}_i} G_1(\hat{x})$ is the collision avoidance part, $-\sum_{j \in N_i^n(t)} \nabla_{\hat{p}_i} G_2(\hat{x})$ is the formation maintenance part, u_d is the control input of global leader and

$$S(e_i^p, v_m) \triangleq \begin{cases} \frac{v_m}{\|e_i^p\|} e_i^p, & \|e_i^p\| > v_m \\ e_i^p, & \|e_i^p\| \leq v_m \end{cases}. \tag{15}$$

$$S(e_i^v, u_m) \triangleq \begin{cases} \frac{u_m}{\|e_i^v\|} e_i^v, & \|e_i^v\| > u_m \\ e_i^v, & \|e_i^v\| \leq u_m \end{cases}. \tag{16}$$

where v_m is maximum safe speed, and $u_m = \dot{v}_m$ is the maximum threshold input. $e_i^p = p_i - p_d$, $e_i^v = v_i - v_d$, the target postilion is given as p_d, $v_d = \dot{p}_d$ is the target velocity. Velocity tends to be consistent when agents reach target points.

3.4   Stability Analysis

To prove the convergence of the proposed controller, we design a integral Lyapunov function

$$V = \sum_{i=1}^{N}\left[\sum_{j \in N_i^l} G_1(\hat{x}) + \sum_{j \in N_i^n} G_2(\hat{x}) \right.$$
$$\left. + \int_0^{e_i^p} S(\tau, v_m)^T \, d\tau + \int_0^{e_i^v} S(\tau, u_m)^T \, d\tau \right]. \tag{17}$$

By derivation of V,

$$\dot{V} = \sum_{i=1}^{N}\left[\dot{p}_i^T \sum_{j \in N_i^l} \nabla_{\hat{p}_i} G_1(\hat{x}) + \dot{p}_i^T \sum_{j \in N_i^n} \nabla_{\hat{p}_i} G_2(\hat{x}) \right.$$
$$\left. + S(e_i^p, v_m)^T (v_i - v_d) + S(e_i^v, u_m)^T (u_i - u_d)\right]. \tag{18}$$

According to the definition in (16), $\frac{u_m}{\|e_i^v\|}$ will be a positive number that is always less than 1 so that $e_i^v \geq S(e_i^v, u_m)$, and then substitute $\dot{\hat{p}}_i = \dot{p}_i = v_i$ and u_i in (14), we can get

$$\dot{V} \leq \sum_{i=1}^{N} v_d^T \left[\sum_{j \in N_i^l} \nabla_{\hat{p}_i} G_1(\hat{x}) + \sum_{j \in N_i^n} \nabla_{\hat{p}_i} G_2(\hat{x}) \right] - (e_i^v)^T (e_i^v). \tag{19}$$

Due to the reciprocity of forces, the repulsion between two agents is equal in size and opposite in direction. So there is

$$V_1 = \sum_{j \in N_i^l} \nabla_{\hat{p}_i} G_1(\hat{x}) + \sum_{j \in N_i^n} \nabla_{\hat{p}_i} G_2(\hat{x}) = 0. \tag{20}$$

Substitute the above formula into $\dot{V}$,

$$\dot{V} \leq - (e_i^v)^T (e_i^v) \leq 0. \tag{21}$$

Since $V(t) \geq 0$ and $\dot{V} \leq 0$, it proves that $V(t)$ is bounded which implies $v_i \rightarrow v_d$ and $p_i \rightarrow p_d$ as $t \rightarrow \infty$ for each agent, which means the difference between predicted position and measurement position $\hat{p}_i - p_i$ will also tend to a constant value.

In practice, it will happen that although the actual position has not entered r_{in}, the predicted position has entered r_{in} which will cause the system to freeze. Because of such the uncertainty of the predicted state, we need to limit the predicted position. According to the definition of predicted position in (2), we have

$$|\hat{x}_i| = |x_i| + |\mu_1 [v_i(t) - v_j(t)]| + |\mu_1 \mu_2 [u_i(t-1) - u_j(t-1)]|. \tag{22}$$

Since $u_m = \dot{v}_m$, v_i and v_j will not exceed v_m if $|v_i(0)| \leq v_m$ and $|v_j(0)| \leq v_m$. We set $p_m = 2\mu_1 v_m + 2\mu_1 \mu_2 u_m$, and we can get

$$|x_i| - p_m \leq |\hat{x}_i| \leq |x_i| + p_m. \tag{23}$$

If $|\hat{x}_i| \geq r_{in} + p_m$ we further have

$$r_{in} + p_m \leq |\hat{x}_i| \leq |x_i| + p_m. \tag{24}$$

which means $|x_i| \geq r_{in}$. The same can be obtained that if $|\hat{x}_i| \leq R - p_m$, we have

$$|x_i| - p_m \leq |\hat{x}_i| \leq R - p_m. \tag{25}$$

which means $|x_i| \leq R$.

The collision avoidance part reach the maximum value at $r_{in} + p_m$ and the formation maintenance part at $R - p_m$. We define

$$\begin{cases} \bar{G}_1 \doteq \int_{r_{in}+p_m}^{r_{out}} g_1(s)ds, \\ \bar{G}_2 \doteq \int_{r_{out}}^{R-p_m} g_2(s)ds. \end{cases} \tag{26}$$

We assume the system can provide the largest collision avoidance potential function as $G_1^{max} = k_1 \bar{G}_1$ and the largest formation maintenance potential function as $G_2^{max} = k_2 \bar{G}_2$. To ensure the effectiveness of collision avoidance and formation maintenance, we need to ensure

$$V_i(0) < \min \left\{ k_1 \bar{G}_1, k_2 \bar{G}_2 \right\}. \tag{27}$$

In other words, if we choose a large enough parameter k_1 and k_2 to make the above formula true, we can ensure the designed controller have the functions of collision avoidance and forming a predetermined formation.

4   Simulation

In order to verify the feasibility of the proposed method, simulations have been performed by a multi-agent system of 12 agents. When changing from follower to leader, the agent 8 needs to bypass the agent on the path, while still keeping the overall configuration unchanged. The path of the swarm is shown in Fig. 1.

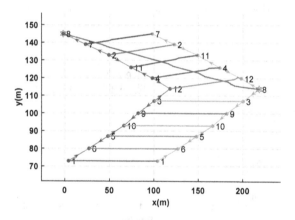

Fig. 1. Initial positions and target positions

We set $r_{in} = 1, r_{out} = 5, R = 20, k_1 = k_2 = 2, v_m = 35, \mu_1 = \mu_2 = 0.35$. Figure 2 reflects the minimum distance between agents with and without the predicted state. The comparison result shows that our proposed method increase the minimum distance between agents and thus achieve better collision avoidance.

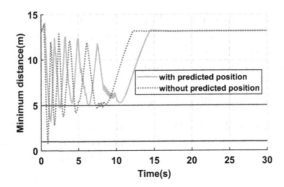

Fig. 2. Minimum distance between agents

Because a stable formation can eventually be formed, both sets of experiments can obtain the same minimum distance. Figure 3 shows the sum of errors between initial positions and target positions, which can prove the iterative convergence of the proposed control laws in this article. It can be see that the method with predicted position does not increase the time to converge to the target location.

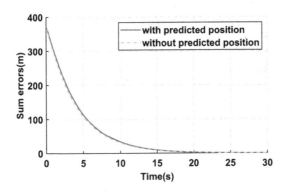

Fig. 3. Sum of errors between current positions and target positions

5   Conclusion

In this paper, we design an improved APF method for multi-agent system control, and the predicted state has been introduced into the new controller, which gives each agent a larger buffer to avoid collisions and prevent the uncertainty of the actual operation of the system. The control input threshold has been set to make agents' speed not exceed the maximum safe speed. The distributed controller has been designed based on the improved APF method to ensure collision avoidance and preserve communication topology during the formation change. The simulation results demonstrate the effectiveness of the designed algorithm.

References

1. Yu, D., Chen, C.L.P.: Automatic leader-follower persistent formation generation with minimum agent-movement in various switching topologies. IEEE Trans. Cybern. **50**(4), 1569–1581 (2020). https://doi.org/10.1109/tcyb.2018.2865803
2. Yu, D., Chen, C.L.P.: Smooth transition in communication for swarm control with formation change. IEEE Trans. Industr. Inform. **16**(11), 6962–6971 (2020). https://doi.org/10.1109/tii.2020.2971356
3. Yu, D., Chen, C.L.P., Ren, C.E., Sui, S.: Swarm control for self-organized system with fixed and switching topology. IEEE Trans. Cybern. **50**(10), 4481–4494 (2020). https://doi.org/10.1109/tii.2020.2971356
4. Dong, X., Hu, G.: Time-varying formation tracking for linear multiagent systems with multiple leaders. IEEE Trans. Autom. Control **62**(7), 3658–3664 (2017). https://doi.org/10.1109/tac.2017.2673411
5. Yu, D., Chen, C.P., Ren, C.E., Sui, S.: Swarm control for self-organized system with fixed and switching topology. IEEE Trans. Cybern. **50**(10), 4481–4494 (2019). https://doi.org/10.1109/tcyb.2019.2952913
6. Tong, S., Sui, S., Li, Y.: Fuzzy adaptive output feedback control of MIMO nonlinear systems with partial tracking errors constrained. IEEE Trans. Fuzzy Syst. **23**(4), 729–742 (2015). https://doi.org/10.1109/TFUZZ.2014.2327987
7. Wen, G., Chen, C.L.P., Liu, Y.J.: Formation control with obstacle avoidance for a class of stochastic multiagent systems. IEEE Trans. Industr. Electron. **65**(7), 5847–5855 (2018). https://doi.org/10.1109/TIE.2017.2782229
8. Dong, X., Yu, B., Shi, Z., Zhong, Y.: Time-varying formation control for unmanned aerial vehicles: theories and applications. IEEE Trans. Control Syst. Technol. **23**(1), 340–348 (2014). https://doi.org/10.1109/tcst.2014.2314460
9. Yang, C., Chen, C., Wang, N., Ju, Z., Fu, J., Wang, M.: Biologically inspired motion modeling and neural control for robot learning from demonstrations. IEEE Trans. Cogn. Develop. Syst. **11**(2), 281–291 (2018)
10. Yu, D., Long, J., Philip Chen, C., Wang, Z.: Bionic tracking-containment control based on smooth transition in communication. Inf. Sci. **587**, 393–407 (2022). https://doi.org/10.1016/j.ins.2021.12.060
11. Fu, J., Wen, G., Yu, X., Wu, Z.G.: Distributed formation navigation of constrained second-order multiagent systems with collision avoidance and connectivity maintenance. IEEE Trans. Cybern. pp. 1–14 (2020). https://doi.org/10.1109/TCYB.2020.3000264
12. Li, D.P., Li, D.J.: Adaptive neural tracking control for an uncertain state constrained robotic manipulator with unknown time-varying delays. IEEE Trans. Syst. Man Cybern. Syst. **48**(12), 2219–2228 (2018). https://doi.org/10.1109/tsmc.2017.2703921
13. Yu, D., Chen, C.P., Xu, H.: Intelligent decision making and bionic movement control of self-organized swarm. IEEE Trans. Industr. Electron. **68**(7), 6369–6378 (2020). https://doi.org/10.1109/tie.2020.2998748
14. Yu, D., Chen, C.L.P., Xu, H.: Fuzzy swarm control based on sliding-mode strategy with self-organized omnidirectional mobile robots system. IEEE Trans. Syst. Man Cybern. Syst., pp. 1–13 (2021). https://doi.org/10.1109/tsmc.2020.3048733

15. Yi, D., et al.: Implicit personalization in driving assistance: state-of-the-art and open issues. IEEE Trans. Intell. Veh. **5**(3), 397–413 (2020). https://doi.org/10.1109/tiv.2019.2960935
16. Xia, Y., Na, X., Sun, Z., Chen, J.: Formation control and collision avoidance for multi-agent systems based on position estimation. ISA Trans. **61**, 287–296 (2016)

An Efficient Scheduling and Navigation Approach for Warehouse Multi-Mobile Robots

Kang Zhao[1,2], Benlian Xu[2(✉)], Mingli Lu[3], Jian Shi[3], and Zhen Li[1,2]

[1] School of Mechanical and Electrical Engineering, Soochow University,
Suzhou, People's Republic of China
zk2671768005@163.com
[2] School of Mechanical Engineering, Changshu Institute of Technology,
Changshu, People's Republic of China
xu_benlian@cslg.edu.cn
[3] School of Electrical and Automatic Engineering, Changshu Institute of Technology,
Changshu, People's Republic of China
luml@cslg.edu.cn

Abstract. Multi-robot scheduling and navigation methods are critical for efficient warehouse handling. In this paper, we propose a Robot Operating System (ROS) based scheduling and navigation method for multi-mobile robots. In order to solve the problem of multi-robot multi-task point assignment in the warehouse environment, we establish a target model that minimizes the total transportation time and propose a hierarchical Genetic Algorithm-Ant Colony Optimization algorithm. By repeating the upper and lower operations, the shortest total transport time allocation scheme for multi-robot multi-tasking can be obtained. In order to realize the multi-robot path planning after task assignment, a multi-robot communication system is designed on the basis of ROS, and the autonomous navigation of mobile robots is employed with the help of SLAM map. The experimental results show that the proposed multi-robot scheduling method can effectively reduce the overall transportation time, realize the reasonable allocation of multi-robots and multi-tasks, and successfully complete the cargo transportation task.

Keywords: Multi-robot scheduling · Genetic Algorithm · Ant Colony Optimization algorithm · Navigation · SLAM

1 Introduction

With the rapid development of computer technology and automation technology, mobile robots have become an irreplaceable part in the industrial production process. In some applications with high task complexity or large number of tasks, many enterprises have replaced the simple and repetitive manual labor links in the traditional production process with the use of various multi-mobile robots

© Springer Nature Switzerland AG 2022
Y. Tan et al. (Eds.): ICSI 2022, LNCS 13345, pp. 45–55, 2022.
https://doi.org/10.1007/978-3-031-09726-3_5

with different functions, which not only reduces the consumption of human resources and improves the efficiency and quality of production.

Multi-mobile robot scheduling is to assign more tasks to relatively few robots and arrange the processing order of the tasks. The solutions to the robot task assignment problem are mainly divided into online methods and offline methods. The offline algorithms mainly include heuristic swarm intelligence algorithms, such as Ant Colony Optimization algorithm [1,2], Genetic Algorithm [3], Particle Swarm Optimization algorithm [4], etc. Nowadays, many scholars combine a variety of intelligent algorithms to solve some problems, which can effectively solve some problems and improve the efficiency and accuracy of the solution. Kong et al. [5] proposed a multi-robot task assignment strategy combining improved particle swarm optimization and greedy algorithm. Tao et al. [6] proposed a Genetic Particle Swarm Optimization (GPSO) to solve the scheduling optimization problem of multiple automated guided vehicles (multi-AGV) in the production workshop. Online algorithms are mainly auction algorithms. Wu et al. [7] considered the dynamic problems related to multi-robot detection and task execution, and proposed an improved auction algorithm. The auction algorithm can realize online task assignment in dynamic scenarios, but cannot obtain the global optimal solution. So far, most of the distribution methods aim to minimize the transportation path, which is not suitable for scenarios with uneven task distribution, which may cause large differences in the running time of each robot, resulting in a longer overall task execution time. This paper considers a warehouse cargo handling scenario, where robots with a load limit can handle multiple cargoes. With the goal of minimizing the overall transportation time, a hierarchical Genetic Algorithm-Ant Colony Optimization (GA-ACO) algorithm is proposed to solve the problem.

After the task assignment is completed, the path planning of the robots can be completed by robot navigation. Traditional navigation methods such as road sign navigation, magnetic navigation, and two-dimensional code navigation [8] are greatly affected by the environment, and auxiliary equipment needs to be laid on the site, and the robot has no autonomous decision-making. At present, autonomous navigation solutions such as laser navigation and visual navigation solutions have become mainstream navigation solutions. The simultaneous localization and mapping (SLAM) [9] and navigation scheme based on laser [10] is not affected by ambient light, and it has better effect and performance as a multi-robot [11] navigation scheme.

The rest of the paper is organized as follows: Sect. 2 details the objective model and hierarchical GA-ACO to minimize the total transit time, as well as the robot navigation method. We present our experimental results in Sect. 3 and conclude the paper in Sect. 4.

2   Methods

2.1   Problem Description

In the warehousing environment, a batch of orders has P tasks, and there are N mobile robots of the same type. The mobile robots start from the starting point, go to the task point to load the goods, return to the starting point to unload the goods after executing the assigned task point, and then execute the next pick-up task. Robots stay at each task point and start point for t seconds to pick up or unload the goods, and the load cannot exceed its maximum capacity. Assuming that the robot has a constant velocity of v, the acceleration and deceleration of the robot during the start-stop phase and the power problem of the robot are not considered. It is required to design a reasonable route to achieve the goal of minimum overall transportation time. In our work, the robot with the longest transportation time when its tasks are finished is actually the shortest transportation time consumption of the overall scheduling. Therefore, we have the objective function:

$$f = \min\{\max_i \sum_{j=1}^{L_i} [[\sum_{k=1}^{n_{ij}} d_{r_{ij(k-1)}r_{ijk}}/v + d_{r_{ij0}r_{ijn_{ij}}}\delta(n_{ij})/v] + (n_{ij}+1)t\delta(n_{ij})]\}$$

(1)

Subject to:

$$\delta(n_{ij}) = \begin{cases} 1, & n_{ij} \geq 1 \\ 0, otherwise \end{cases}$$

(2)

$$\sum_{k=1}^{n_{ij}} w_{r_{ijk}} \leq W_i$$

(3)

$$\sum_{i=1}^{N}\sum_{j=1}^{L_i} n_{ij} = P$$

(4)

$$R_{ij} \in \{r_{ijk}|r_{ijk} \in [1,2,\cdots,P], k = 1,2,\cdots,n_{ij}\}$$

(5)

$$R_{ij(1)} \cap R_{ij(2)} = \emptyset, \forall ij(1) \neq ij(2)$$

(6)

where n_{ij} represents the number of pick-up task points for the j-th transportation of robot i, L_i is the round-trip transportation times of robot i, r_{ijk} represents the k-th pick-up task point of robot i for the j-th transportation, r_{ij0} is the starting point of the robot, and the distance between the task points a, b is the Euclidean distance $d_{ab} = \sqrt{(x_a - x_b)^2 + (y_a - y_b)^2}$. The robot transport time calculated by the objective function includes two components: the path travel time between task points and the dwell time at each task point.

Equation (2) describes that when the pick-up task point of the j-th transportation of robot i is greater than or equal to 1, i.e., $n_{ij} \geq 1$, indicates this transportation is required, then $\delta(n_{ij}) = 1$, When the task point is less than 1, i.e., $n_{ij} < 1$, indicating that this transportation is not required, then $\delta(n_{ij}) = 0$; Eq. (3) means that the total weight of the goods at the task point on each path shall not exceed the load of the mobile robot, where W_i is the maximum load of each mobile robot. The weight of the cargo at the k-th pick-up task point of the j-th transport of robot i is $w_{r_{ijk}}$; Eq. (4) indicates that each pick-up task point can be completed, P is the number of tasks, and N is the number of mobile robots; Eq. (5) represents the task composition of each path, and R_{ij} represents the j-th transportation path of robot i; Eq. (6) shows that there are no identical tasks in any two robot paths. It restricts that each task point can only be transported once by one mobile robot.

2.2 Hierarchical Structure of GA-ACO

The hierarchical Genetic Algorithm-Ant Colony Optimization (GA-ACO) is divided into two layers. As shown in Fig. 1, The gray squares represent robot 1 and the gray circles are the corresponding assigned tasks, while the white squares represent robot 2 and the white circles are its assigned tasks. Connecting lines indicate the path of the robot. In the upper layer, the Genetic Algorithm is used to calculate the solution space to be searched, i.e., the task point set of each mobile robot is determined. In the lower layer, the Ant Colony Optimization algorithm is used to search each solution space for the optimal solution, i.e., the optimal task route of each robot is searched. By repeating the upper and lower operations, the optimal solution of the entire system can be obtained, which is a reasonable allocation of multi-robots and multi-tasks. The flow chart of the algorithm is shown in Fig. 2.

2.2.1 Genetic Algorithm to Solve the Upper Layer

Code: First, the task set is encoded with natural numbers, and the encoding length is the number of task points. The task set breakpoints assigned to each robot are encoded with natural numbers. The encoding length is the number of robots minus one. If there are 10 task points (numbered 4-13), 3 mobile robots (numbered 1-3), the task set code is 11-4-8-12-10-6-9-13-5-7, the task set breakpoint code is 3-8, decode Later, it indicates that the No. 1 robot task set is "11, 4, 8", the No. 2 robot task set is "12, 10, 6, 9, 13", and the No. 3 robot task set is "5, 7".

Fitness: The fitness function of designing a Genetic Algorithm can be mapped by the objective function as follows:

$$F = \frac{1}{t_{\max}} \tag{7}$$

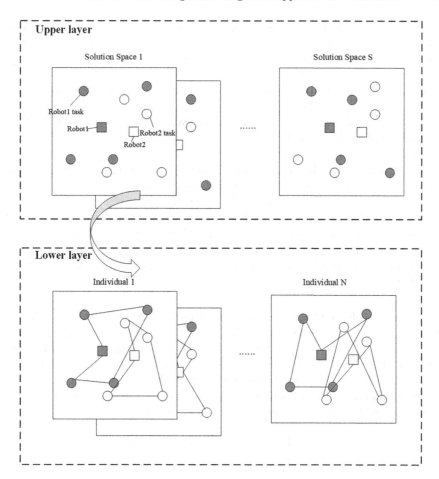

Fig. 1. Conceptual diagram of hierarchical structure of GA-ACO

$$t_{\max} = \max_i \sum_{j=1}^{L_i} [[\sum_{k=1}^{n_{ij}} d_{r_{ij(k-1)}r_{ijk}}/v + d_{r_{ij0}r_{ijn_{ij}}}\delta(n_{ij})/v] + (n_{ij}+1)t\delta(n_{ij})]$$

(8)

where $t_{\max}$ is the longest robot transportation time when all tasks are completed, and the shortest task set path of each robot is obtained. The path length and robot speed are known, and the time consumption of the path is obtained, plus the pickup and starting point of each task point. The unloading time can be used to obtain the time consumed by all robots to perform tasks, and the transportation time of the robot with the longest time can be obtained, which is the total transportation time. Genetic algorithms need to generate the next generation population through selection, crossover and mutation.

Selection: This algorithm adopts the method of tournament selection. The tournament selects the elite individuals in the population with shorter consumption time as the parent and retains their breakpoint genes.

Crossover: The sequential crossover method (OX) was used. For this task assignment problem, it is to exchange the position of task points in the code of the total task set, and the method of random generation within the rules is adopted for the code of task set breakpoints.

Variation: Randomly select individuals to exchange two random positions of the coding chromosome to obtain a new generation of chromosomes. The task points in the two robot task sets are randomly exchanged.

After the selection, crossover and mutation operation, a new population will be generated, and then the ant colony optimization algorithm is repeated to solve the lower layer, and the genetic algorithm calculates the operation until the maximum number of iterations is reached, the optimal task allocation scheme will be obtained.

2.2.2 Ant Colony Optimization Algorithm to Solve the Lower Layer

After decoding the Genetic Algorithm, the task point set of each robot is obtained, and the Ant Colony Optimization algorithm is used to solve the most task path of each robot. First, generate a path table for each ant, put the starting point of the robot into the starting point of the path table, and calculate the selection probability of going to other task points according to the transition probability. At time t, the probability of ant moving from city i to city j is:

$$
p_i^k(t) = \begin{cases} \dfrac{[\tau_{ij}(t)]^\alpha [\eta_{ij}(t)]^\beta}{\sum\limits_{s \in J_k(i)} [\tau_{is}(t)]^\alpha [\eta_{is}(t)]^\beta}, j \in J_k(i) \\ 0, j \notin J_k(i) \end{cases} \tag{9}
$$

$$
J_k(i) = \{1, 2, \cdots, n\} - tabu_k, \eta_{ij} = 1/d_{ij} \tag{10}
$$

where $\tau_{ij}(t)$ is the pheromone between task points i and j at time t, $\eta_{ij}(t)$ is the heuristic factor between task points i and j at time t, α and β represent the relative importance of the pheromone and the heuristic factor respectively, d_{ij} is the distance between task points i and j, $tabu_k$ is the tabu table for storing all task points, and n is the number of task concentration points of the robot. Select the next task point from the robot's set of task points according to the transfer probability using the roulette wheel method, determine whether the accumulated weight exceeds the robot's load when adding the weight of the next task point's cargo, and if not, add this task point to the path table, otherwise add the starting point to the path table (return to the starting point). Repeat until all task points in the task set are added to the path table, and finally add the starting point (back to the starting point). Calculate the path lengths in all ant path tables, select the shortest path to store, and update the pheromone on the path:

$$
\tau_{ij}(t+1) = (1-\rho)\tau_{ij}(t) + \Delta\tau_{ij} \tag{11}
$$

$$\Delta\tau_{ij} = \sum_{k=1}^{m}\Delta\tau_{ij}^{k}, \Delta\tau_{ij}^{k} = \begin{cases} \frac{Q}{L_{K}}, if\ ant\ k\ passes\ ij\ during\ this\ tour \\ 0, otherwise \end{cases} \quad (12)$$

where $\rho(0 < \rho < 1)$ represents the evaporation coefficient of the pheromone on the path, Q is a constant, and L_k represents the length of the path that the k-th ant traveled in this tour.

Repeat the above steps to reach the number of iterations of the ant colony algorithm, and obtain the shortest task set path of each robot.

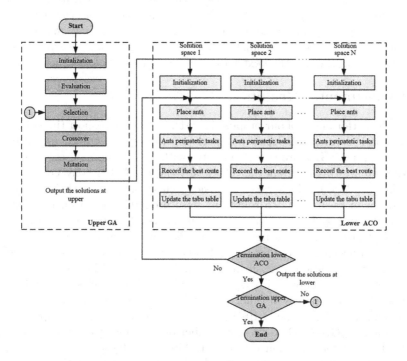

Fig. 2. Hierarchical GA-ACO algorithm flow chart

2.3 ROS-Based Multi-robot System

We first build a multi-robot system based on ROS by using a host computer and multiple robot industrial computers as slaves. By giving robots different namespaces, we can manage and schedule multiple robots.

The host directs a mobile robot to use the Cartographer algorithm [12] to build an environmental grid map through the carried radar, and obtain the shelf coordinate points in the map and the starting point of each mobile robot in the map. A message sending node is set up. When the order comes, the path of each robot task set is calculated by the hierarchical GA-ACO algorithm. We define the task point coordinates as a differently named array message type, which is sent to the node manager of the multi-robot system in the form of a topic.

Each robot subscribes to the path message of its own task set, obtains the order of task points and their coordinates, calculates the number of task points in the message, and starts to execute the navigation function according to the order. The navigation of the robot to the task point first calculates the global path to the target point, and performs local path planning and obstacle avoidance when encountering obstacles during the path execution process. The Monte Carlo positioning algorithm is used to realize the positioning of the robot's pose. After the robot reaches a task point, it stays at the task point for a fixed period of time, waits for the completion of the pickup task, and then goes to the next task point in the list. Repeat the above operations until the picking and unloading tasks at all task points are completed, and the robot stays at the starting point of the robot. Waiting for the next batch of order tasks to be assigned.

3   Experimental Results

To test the performance of our proposed shortest overall transport time objective function and hierarchical GA-ACO algorithm, we implemented it in MATLAB (R2020a) on a 2.9 GHz processor computer with 16G random access memory.

We compare this with another classical K-means-Ant Colony Optimization algorithm (K-ACO), which takes a K-mean clustering approach in the upper layer to assign task points to robots in close proximity based on the distance between the task point and the robot starting point, and uses the same Ant Colony Optimization algorithm as ours in the lower layer to solve for the task point path of a single robot.Our algorithm is similar in structure to the K-ACO, but the genetic iteration mechanism used makes our algorithm superior.

We set up two robots with a load of 90kg and a travel speed of 0.1m/s. There are 20 cargo points of different weights to be completed, and the robot picking and unloading time is 10s.We set the Genetic Algorithm population of the hierarchical GA-ACO to 40, the number of iterations to 100, the crossover rate to 0.9, and the variation rate to 0.1. The upper layer of the K-ACO uses the principle of proximity assignment. The Ant Colony Optimization algorithms of the two algorithms are set as follows: the number of ant colonies is 20, the number of iterations is 20, the pheromone importance factor α is 1, the heuristic factor importance factor β is 5, and the pheromone volatility factor ρ is 0.5.

The solution path of hierarchical GA-ACO is shown in Fig. 3, which shows that this algorithm can effectively solve the multi-robot multi-task assignment problem. Additionally, 10 independent runs are calculated and the statistic results are shown in Fig. 4. Furthermore,a comparison experiment is conducted with different combinations of 2, 3, 5 robots and 20, 30, 50 task points as shown in Fig. 5. The parameters of the two algorithms are set as in the two-robot experiment in the previous paragraph. It is observed from Figs. 4 and 5 that the hierarchical GA-ACO solves the overall transportation time significantly shorter than the K-ACO.

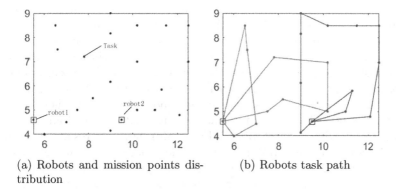

(a) Robots and mission points distribution

(b) Robots task path

Fig. 3. The solution result of hierarchical GA-ACO

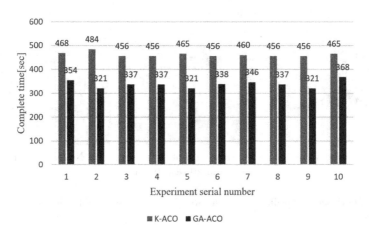

Fig. 4. Comparison of the overall transportation time by K-ACO and hierarchical GA-ACO

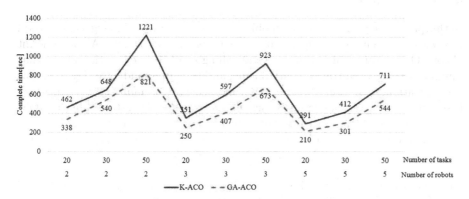

Fig. 5. Comparison of the completion times of K-ACO and hierarchical GA-ACO algorithms for different number of robots and number of tasks

Finally, our approach is validated through a two-robot system, and it successfully controlled two robots to complete their individual tasks. In Fig. 6, we controlled two robots to complete 20 tasks in a real scene. Figure 6(a) shows the actual scene that simulates a warehouse environment, in which the black rectangles are obstacles. Figure 6(b) shows the visualization map by rviz after building a map, in which the black squares denote the obstacles in Fig. 6(a), while the white areas represent the accessible areas of robots, and the blue marked points are 20 task points. The red square is the starting point of robot 1 and the yellow square is the starting point of robot 2. Figure 6(c) shows the results of rviz when two robots are performing task point navigation. Both robot 1 in the upper left corner and robot 2 in the lower right corner are performing navigation of their respective task points. The robots travel from the current locations to their individual blue dots at the end of the green lines, each representing the obtained global path planning, and the lavender area on the map is the map expansion layer for local obstacle avoidance.

 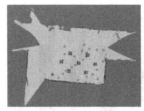 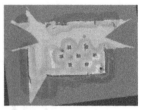

(a) Testing scene (b) Map, robot starting points and 20 task points (c) Navigation in rviz

Fig. 6. Experimental verification of a two-robot dispatch navigation system (Color figure online)

4   Conclusions

In this paper, we propose a multi-mobile robot scheduling and navigation method in a warehouse environment, and propose a hierarchical Genetic Algorithm-Ant Colony Optimization algorithm by establishing an objective function model that minimizes the overall transportation time, and the upper and lower layers repeat the operation to obtain the shortest time scheme for multi-robot multi-task assignment. The task assignment scheme is executed by establishing a ROS-based multi-robot system to control multi-robot navigation. The results show that our proposed method can effectively reduce the overall transportation time and achieve a multi-robot navigation execution task set.

Acknowledgment. This work was supported by National Natural Science Foundation of China (No. 61876024), and partly by the higher education colleges in Jiangsu province (No. 21KJA510003), and Suzhou municipal science and technology plan project (No. SYG202129).

References

1. Li, J., Dong, T., Li, Y.: Research on task allocation in multiple logistics robots based on an improved ant colony algorithm. In: 2016 International Conference on Robotics and Automation Engineering (ICRAE), pp. 17–20. IEEE (2016)
2. Li, D., Zhua, J., Xu, B., Lu, M., Li, M.: An ant-based filtering random-finite-set approach to simultaneous localization and mapping. Int. J. Appl. Math. Comput. Sci. **28**(3), 505–519 (2018)
3. Singh, A., Baghel, A.S.: A new grouping genetic algorithm approach to the multiple traveling salesperson problem. Soft. Comput. **13**, 95–101 (2009). https://doi.org/10.1007/s00500-008-0312-1
4. Wei, C., Ji, Z., Cai, B.: Particle swarm optimization for cooperative multi-robot task allocation: a multi-objective approach. IEEE Robot. Autom. Lett. **5**(2), 2530–2537 (2020)
5. Kong, X., Gao, Y., Wang, T., Liu, J., Xu, W.: Multi-robot task allocation strategy based on particle swarm optimization and greedy algorithm. In: 2019 IEEE 8th Joint International Information Technology and Artificial Intelligence Conference (ITAIC), pp. 1643–1646. IEEE (2019)
6. Tao, Q., Sang, H., Guo, H., Han, Y.: Research on multi-AGVs scheduling based on genetic particle swarm optimization algorithm. In: 2021 40th Chinese Control Conference (CCC), pp. 1814–1819. IEEE (2021)
7. Wu, S., Liu, X., Wang, X., Zhou, X., Sun, M.: Multi-robot dynamic task allocation based on improved auction algorithm. In: 2021 6th International Conference on Automation, Control and Robotics Engineering (CACRE), pp. 57–61. IEEE (2021)
8. Song, Z., Wu, X., Xu, T., Sun, J., Gao, Q., He, Y.: A new method of AGV navigation based on Kalman filter and a magnetic nail localization. In: 2016 IEEE International Conference on Robotics and Biomimetics (ROBIO), pp. 952–957. IEEE (2016)
9. Bailey, T., Durrant-Whyte, H.: Simultaneous localization and mapping (SLAM): part II. IEEE Robot. Autom. Mag. **13**(3), 108–117 (2006)
10. Balasuriya, B.L.E.A., et al.: Outdoor robot navigation using Gmapping based SLAM algorithm. In: 2016 Moratuwa Engineering Research Conference (MERCon), pp. 403–408. IEEE (2016)
11. Sun, S., Xu, B.: Online map fusion system based on sparse point-cloud. Int. J. Autom. Control **15**(4–5), 585–610 (2021)
12. Hess, W., Kohler, D., Rapp, H., Andor, D.: Real-time loop closure in 2D LIDAR SLAM. In: 2016 IEEE International Conference on Robotics and Automation (ICRA), pp. 1271–1278. IEEE (2016)

Self-organizing Mobile Robots Swarm Movement Control Simulation

E. V. Larkin[1]([✉]), T. A. Akimenko[1]([✉]), and A. V. Bogomolov[2]([✉])

[1] Tula State University, Tula 300012, Russia
elarkin@mail.ru, tantan72@mail.ru
[2] A.I. Burnasyan Federal Medical Biophysical Center FMBA, Moscow, Russia
avbogomolov@gmail.com

Abstract. The control principle of mobile robots, self-organized in convoy structure, is considered. It is shown that in this case a high autonomy of swarm operation may be achieved, due to such parameters of motion, as swarm azimuth angle and velocity of movement may be set only for the master robot, while slave robots follow master one at the pre-determined distance. The flowchart of the swarm control system is worked out, according which all slave robots measure their own deviation from the direction on previous swarm unit and distance till it, and actuate mechanics of the robot for zeroing azimuth angles difference and setting pre-determined distance. Mathematical models of the swarm, as the united object under control, and of the distributed Von Neumann controller, which closes control system, are worked out. It is shown, that characteristic equation of closed system have complex exponent at the left side, which is due to time delays, born by Von Neumann type controllers. Method of evaluation of time delays, based on semi-Markov simulation of control algorithm, is proposed. Theoretical results are confirmed by modeling the motion control of a convoy, including pair mobile robots.

Keywords: Swarm · Mobile robot · Convoy · Von Neumann · Control algorithm · Closed control system · Transfer function · Delay · Semi-Markov process

1 Introduction

Unmanned vehicles, operating autonomously on terrestrial surface, below referred as mobile robots (MR), presently are widely used in various domains of human activity: such as on transport, during environmental monitoring, in the defense sector, etc. [1–3]. Effectiveness of MR utilization increases significantly, when they are gathered in swarms, aligned in those or that order [4]. Managing by MR swarm is a much more difficult problem, than control by a single robot due to the fact, that it is necessary not only to provide the specified movement parameters of each robot, but also to maintain the swarm structure [4, 5]. In this case the soft runtime of Von-Neumann type controllers are superimposed on the inertial properties of swarm units during maneuvers in the

Y. Tan et al. (Eds.): ICSI 2022, LNCS 13345, pp. 56–65, 2022.
https://doi.org/10.1007/978-3-031-09726-3_6

space, and both interfering factors deteriorate swarm performance, and make it difficult to maintain its structure.

One of possible application of terrestrial MR, which makes the control problem easier, is the ordering swarm in master-slave convoy structure, in which control system of master robot determines velocity and azimuth angle of movement the swarm as a whole, while control systems of slaves MRs maintain distance till previous unit, for example using a rangefinder [6, 7], and direction on previous robot, using information about deviation from direction, obtaining from sensor, installed [8] on the slave robots itself. Changes in speed and direction due to obstacles that arise at a trace both the master, and a slave robots are compensated by their control systems. To organize the movement of the swarm according to master-slave convoy principle, an adequate model of the swarm as an object under of vector control, and an integrated control system, including Von Neumann controllers distributed among MRs, is required [5]. Such a control principle has a high degree of self-organization, since when one of the moving objects leaves the convoy, the moving object following it begins to maintain the distance to the previous one, i.e. of the object that the retired object followed before the elimination.

However, there is a certain difficulty when control swarm with such structure, due to the fact that the Von Neumann controllers born delays in closed loop system, whose effect is enhanced by cross links between control contours, which emerging according chosen principle of self-organizing and control [9–11].

Methods for assessing the quality of control, which would take into account both the complexity of the structure of the MR column and the delay in the feedback contours, are not sufficiently developed in engineering practice, which explains the urgency and relevance of the study.

2   Model of the MR Convoy Swarm as the Object Under Control

MR master-slave convoy swarm, operating under Von Neumann type controllers, distributed onto vehicles, is shown on the Fig. 1 [12]. Self-organizing convoy swarm includes master robot MR_0 and K slave robots $MR_1,..., MR_k,..., MR_K$. Distributed digital control system, consists of digital controllers, nominated as $W_{c,k}(s)/Int_k$, where $0 \leq k \leq K$, s is the Laplace differentiation operator. Parameters of swarm movement are set with vectors $F_k(s)$, which are inputted into controllers $W_{c,k}(s)$ from outer source (on the Fig. 1 is not shown). The vector $F_0(s) = \begin{bmatrix} F_{0,1}(s), & F_{0,2}(s) \end{bmatrix}^\theta$, where θ is the transposition operation sign, defines parameters of swarm movement in common, namely, azimuth angle $F_{0,1}(s)$ and swarm velocity $F_{0,2}(s)$. Vectors $F_k(s) = \begin{bmatrix} 0, & F_{k,2}(s) \end{bmatrix}^\theta$, $1 \leq k \leq K$, define distance $x_{k-1,k}(s)$ from MR_k till previous MR_{k-1}. Except $F_k(s)$, feedback vector signal $V_{b,k}(s) = \begin{bmatrix} V_{b,k,1}(s), & V_{b,k,1}(s) \end{bmatrix}^\theta$ is inputted through Int_k into k-th digital controller, which, in turn, computes action vector $U_{c,k}(s) = \begin{bmatrix} U_{c,k,1}(s), & U_{c,k,1}(s) \end{bmatrix}^\theta$, actuated MR_k. The computed action vector $U_{c,k}(s)$ is outputted through the Int_k, and as analogue vector signal $U_k(s) = \begin{bmatrix} U_{k,1}(s), & U_{k,1}(s) \end{bmatrix}$ is applied to actuators of MR_k. Except the control signal action $U_k(s)$ on the MR_k affects the physical resistance force $R_k(s) = \begin{bmatrix} R_{k,1}(s), & R_{k,2}(s) \end{bmatrix}^\theta$, and MR_k mechanics generate state vector

$V_k(s) = \begin{bmatrix} V_{k,1}(s), & V_{k,2}(s) \end{bmatrix}^\theta$, in which $V_{k,1}(s)$ is longitudinal movement velocity, $V_{k,2}(s)$ is azimuth angle change velocity.

Feed-forward control by robots is described as

$$V_k(s) = W_{u,k}(s)U_k(s) + W_{r,k}(s)R_k(s), 0 \le k \le K, \qquad (1)$$

where

$$W_{u,k}(s) = \begin{bmatrix} W_{u,k,11}(s) & W_{u,k,21}(s) \\ W_{u,k,12}(s) & W_{u,k,22}(s) \end{bmatrix}; \quad W_{r,k}(s) = \begin{bmatrix} W_{r,k,11}(s) & W_{r,k,21}(s) \\ W_{r,k,12}(s) & W_{r,k,22}(s) \end{bmatrix}.$$

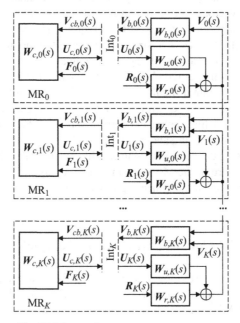

Fig. 1. Master-slave convoy swarm structure

Feedback in the MR convoy swarm is provided with blocks $W_{b,k}(s)$ $0 \le k \le K$, which measure state $V_k(s)$ of MR_k and return to k-th digital controller feedback vector signal $V_{b,k}(s)$. In the master robot MR_0, sensor subsystem measures velocity with speed sensor, and azimuth angle with, f.e. inertial measurement system. So matrix transfer function, which converts $V_0(s)$ into $V_{b,0}(s)$, is as follows:

$$V_{b,0}(s) = W_{b,0}(s) \cdot V_0(s) = \begin{bmatrix} \frac{\kappa_{b,0,1}}{s} & 0 \\ 0 & \kappa_{b,0,2} \end{bmatrix} \cdot V_0(s), \qquad (2)$$

where $\kappa_{b,0,1}$ is transmission ratio of master robot MR_0 azimuth angle sensor; $\kappa_{b,0,2}$ is transmission ratio of MR_0 speed sensor.

In slave robots, $MR_1 \div MR_K$, sensor subsystem measures distance till previous robot [6, 7] and the difference between azimuth angles [8] of MR_k and MR_{k-1}, and converts it into the vector signal $V_{b,k}(s) = [V_{b,k,1}(s), V_{b,k,2}(s)]^{\theta}$. The measuring principles is explained with Fig. 2, where $x_{k-1,k}$ is the distance from MR_k till MR_{k-1}; v_{k-1} and v_k are vectors of MR_{k-1} and MR_k velocities, correspondingly; $\psi_{k-1,k}$ is the angle between v_{k-1} and v_k; $\overline{AC}$ is the MR_k axial line, which coincides with direction of angle sensor sight mark; $\overline{AB}$ is the direction on the measured mark, situated on the MR_{k-1}; $\overline{BC}$ is the transverse component of vector v_{k-1}, respectfully to vector v_k; $\varphi_{k-,k}$ is the measured angle between MR_k axial line and the direction on the measured mark, which tends to zero, when angle $\psi_{k-1,k}$ tends to zero too, that is sufficiently to control MR_k movement direction.

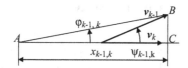

Fig. 2. The feedback measuring principle

According the flowchart, shown on the Fig. 1, and feedback measurement principle, shown on the Fig. 2, feedback signal $V_{b,k}(s) = [V_{b,k,1}(s), V_{b,k,2}(s)]^{\theta}$ is formed as follows:

$$\begin{bmatrix} V_{b,k,1}(s) \\ V_{b,k,2}(s) \end{bmatrix} = W_{b,k}(s) \begin{bmatrix} V_{k-1,2}(s) \\ V_{k,1}(s) \\ V_{k,2}(s) \end{bmatrix}, \tag{3}$$

where $V_{k-1}(s)$ is the second component of the vector $V_{k-1}(s)$; $W_{b,k}(s)$ is the 2×3 matrix

$$W_{b,k}(s) = \begin{bmatrix} \frac{\kappa_{b,k-1,10}}{s} & \frac{\kappa_{b,k,11}}{s} & \frac{\kappa_{b,k,12}}{s} \\ \frac{\kappa_{b,k-1,20}}{s} & 0 & \frac{\kappa_{b,k,22}}{s} \end{bmatrix}; \tag{4}$$

$\kappa_{b,k-1,10}, \kappa_{b,k,11}, \kappa_{b,k,12}, \kappa_{b,k-1,20}, \kappa_{b,k,22}$ are transmission ratios.

As it follows from the (3), measured angle between MR_k axial line and the direction on the measured mark depend on the rotation angle of MR_{k-1} respectfully to MR_k, and distance between MR_{k-1} and MR_k, which in turn is the product of difference velocities integration.

Vectors $U_k(s), V_{b,k}(s), R_k(s)$ and matrices $W_{u,k}(s), W_{r,k}(s)\, W_{b,k}(s)$ may be ordered into common feed-forward matrix equation, describing MR swarm convoy as united object under control:

$$V(s) = W_b(s) \cdot [W(s) \cdot U(s) + W_r(s) \cdot R(s)], \tag{5}$$

where $U(s) = [U_0(s), ..., U_k(s), ..., U_K(s)]^{\theta}$ is the united action vector; $R(s) = [R_0(s), ..., R_k(s), ..., R_K(s)]^{\theta}$ is the united resistance vector; $W(s)$ and $W_r(s)$ are united matrices, describing mechanics of MRs under control action and under resistance forces, respectively, of cellular diagonal type; $W_b(s)$ is cellular matrix, which null row consists of matrix $W_{b,0}(s)$ and K 2×2 zero matrices, while rows from the one till the K include 2×2 matrices $W'_{b,k-1}(s)$ and $W''_{b,k}(s)$, situated on the $(k-1)$-th and the k-th place, correspondingly, and 2×2 zero matrices, situated on all other places; $W'_{b,k-1}(s)$ is 2×2 matrix, in which the first column is the zero one, and second column is the null column of matrix $W_{b,k}(s)$; $W''_{b,k}(s)$ is 2×2 matrix, including first and second matrix $W_{b,k}(s)$ columns.

Self-organizing of convoy swarm at the physical level is provided through descending master-slave interconnections between previous and next MR, according to which MR_k maintain direction on MR_{k-1} and distance till it.

3   Model of Distributed Control System

The swarm control system has a distributed structure, shown on the Fig. 1 at the left. Movement parameters $V_{b,0}(s) \div V_{b,K}(s)$ through interfaces $Int_0 - Int_K$ are inputted to Von Neumann type controllers $W_{c,0}(s) \div W_{c,K}(s)$ in discrete form $V_{cb,0}(s) \div V_{cb,K}(s)$, processed by their software as discrete data, and control discrete vectors $U_{c,0}(s) \div U_{c,K}(s)$ through corresponding interface are outputted to MR's mechanical parts as analogue control actions $U_0(s) \div U_K(s)$ [13–15]. If in the software principle of linear processing is laid, then computation of united data vector is produced according matrix equation

$$U_c(s) = W_{c,f}(s) \cdot F(s) + W_{c,v}(s) \cdot V_{\tilde{n}b}(s), \tag{6}$$

where $U_c(s) = \left[U_{c,0}(s), \ ..., \ U_{c,k}(s), \ ..., \ U_{c,K}(s) \right]^\theta$, $F(s) = [F_0(s), \ ..., \ F_k(s), \ ..., \ F_K(s)]^\theta$ and $V_{cb}(s) = \left[V_{cb,0}(s), \ ..., \ V_{cb,k}(s), \ ..., \ V_{cb,K}(s) \right]^\theta$ are column vectors of size $2(K+1)$; $W_{c,f}(s)$ and $W_{c,u}(s)$ are cellular diagonal type matrix of transfer functions of processing $U_c(s)$ and $F(s)$, correspondingly, in which on the main diagonal 2×2 matrices $W_{c,f,k}(s)$, $W_{c,v,k}(s)$ are situated, and all other 2×2 sells are the zero ones.

Von Neumann controllers interpret processing algorithm sequentially, operator-by-operator, so in k-th controller there is delay between inputting data $V_{cb,k}(s)$, and outputting data $U_{c,k}(s)$. Let control algorithm sequentially inputting data $V_{cb,k,1}(s)$ into controller, that in time domain form the starting point of cycle. After that, with delay τ_0, data $V_{cb,k,2}(s)$ are inputted into controller [16–19]. A computation of action takes some time. So signals $U_{c,k,1}(s)$ and $U_{c,k,2}(s)$ are outputted with delays τ_1 and τ_2 respectively to the starting point. With taking into account delays, matrix $W_{c,v,k}(s)$ takes the form

$$W_{c,v,k}(s) = \begin{bmatrix} W_{c,v,k,11}(s) \cdot \exp(-\tau_{k,1}s) & W_{c,v,k,21}(s) \cdot \exp(-\tau_{k,1}s + \tau_{k,0}s) \\ W_{c,v,k,12}(s) \cdot \exp(-\tau_{k,2}s) & W_{c,v,k,22}(s) \cdot \exp(-\tau_{k,2}s + \tau_{k,0}s) \end{bmatrix}, \tag{7}$$

where $W_{c,v,k,11}(s) \div W_{c,v,k,22}(s)$ are transfer functions, realizing in control algorithm; $\exp(-\tau_{k,1}s)$ is Laplace transform of shifted Dirac function.

Desired parameters of swarm movement are inputted into controllers in advance, in such a way, delays in processing may be omitted from consideration and matrix $W_{c,v,k}(s)$ is as follows:

$$W_{c,f,k}(s) = \begin{bmatrix} W_{c,f,k,11}(s) & W_{c,f,k,21}(s) \\ W_{c,f,k,12}(s) & W_{c,f,k,22}(s) \end{bmatrix}; \tag{8}$$

where $W_{cf,k,11}(s) \div W_{cf,k,22}(s)$ are transfer functions, realizing in control algorithm.

To estimate time delays between transactions in polling algorithm of arbitrary complexity the ergodic semi-Markov chain should be formed [17, 20, 21]. Structure of semi-Markov process coincides with the structure of algorithm and in most common case is represented by complete graph without loops (Fig. 3 a).

Process of transactions generation is described with semi-Markov matrix

$$h(t) = \lfloor h_{i,j}(t) \rfloor = \lfloor g_{i,j}(t) \rfloor \otimes \lfloor p_{i,j}(t) \rfloor, \tag{9}$$

where $h(t)$ is the $N \times N$ semi-Markov matrix; t is the physical time; $p_{i,j}(t)$ is the probability of direct switching from the state i to the state j; $g_{i,j}(t)$ is the time of sojourn at the state i, when a priori is known, that next switching will be to the state j; $\otimes$ is the direct multiplying sign; N is the number of states in semi-Markov process, equal to number of operators in polling algorithm.

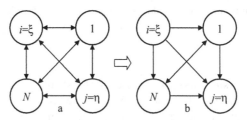

Fig. 3. Structure of polling control algorithm (a), semi-Markov process for delays estimation (b)

Contribute to time interval $\tilde{\tau}$ between f.e., inputting of $V_{cb,k,1}(s)$ ($i = \xi$) and outputting of $U_{c,k,1}(s)$ ($j = \eta$) both the direct switching, and the wandering through the semi-Markov chain from ξ to η. For estimation of time delay $\tilde{\tau}$ the semi-Markov matrix (13) should be transformed as follows (Fig. 3 b):

$$h(t) \rightarrow h'(t) = \lfloor g'_{i,j}(t) \cdot p'_{i,j} \rfloor, \tag{10}$$

where $g'_{\eta,j}(t) = 0$ and $p'_{\eta,j} = 0$ for all $1 \leq j \leq N$; $g'_{i,\xi}(t) = 0$ and $p'_{i,\xi} = 0$ for all $1 \leq i \leq N$; $p'_{i,j} = \frac{p_{i,j}}{1-p_{i,\xi}}$ for all columns except η-th and for columns, except ξ-th.

The time density of wandering from ξ to η is as follows:

$$\tilde{g}_{\xi,\eta}(t) = I_{R,\xi} \cdot L^{-1} \left[\sum_{\kappa=1}^{\infty} \{L[h'(t)]\}^{\kappa} \right] \cdot I_{C,\eta} \tag{11}$$

where $I_{R,\xi}$ is the row vector of size N, all elements of which, except ξ-th, which is equal to 1, are equal to zero; $I_{C,\eta}$ is the column vector of size N, all elements of which, except η-th, which is equal to 1, are equal to zero.

According to «three sigma rule» [22, 23] delay $\tilde{\tau}$ may be estimated as follows:

$$\tilde{\tau} = \tilde{T} + 3\sqrt{\tilde{D}}, \tag{12}$$

where $\tilde{T} = \int\limits_{0}^{\infty} \tilde{g}_{\xi,\eta}(t)t\,dt; \tilde{D} = \int\limits_{0}^{\infty} \left(t - \tilde{T}\right)^2 \tilde{g}_{\xi,\eta}(t)(t)\,dt.$

Matrix Eqs. (5), (6) should be used for obtaining of closed loops system matrix description.

$$V(s) = \left[E - W_b(s) \cdot W(s) \cdot W_{c,v}(s)\right]^{-1} \cdot W_b(s) \cdot W(s) \cdot W_{c,f}(s) \cdot F(s)$$
$$+ \left[E - W_b(s) \cdot W(s) \cdot W_{c,v}(s)\right]^{-1} \cdot W_b(s) \cdot W_r(s) \cdot R(s), \tag{13}$$

where E is the identity matrix.

It is necessary to admit, that performance of the MR swarm, as a whole, is determined by following characteristic equation:

$$\left|E - W_b(s) \cdot W(s) \cdot W_{c,v}(s)\right| = 0. \tag{14}$$

Equation (14) contains complex exponent on the left side, which brings the system closer to stability boundary, or even takes it beyond the boundary [24, 25].

4   Example

Confirm the approach to simulation of swarm control system with example in which convoy consists of master and one slave robot, having acceleration characteristics, described with first order differential equation with time constant, equal to 0,1 s. Responses of the system on the standard Heaviside function, when delay in feedback contour is equal to 0 s, 0,01 s, 0,02 s and 0,04 s in dimensionless form is shown on the Fig. 4 a), b), c), d), correspondingly.

From the curves picture it is clear, that overshooting and transient time of swarm, as a whole, directly depends on a soft runtime of Von-Neumann type controller due to the fact, that both sampling intervals and delays in feedback contours are determined by polling, embedded into the structure of control algorithm, and time of operators interpreting. An overshooting in the slave MR is more, than in master robot due to a resonant phenomenon, which emerging in the slave closed control system, when it has just the same characteristic equation, as master closed control system. It should also be noted rather uneven curve, according which distance between robots is set, that is linked with the same resonance, but already in speed control contours.

To minimize resonant phenomena one should to vary performances of slave robots with respect to previous master/slave robots, for example by changing their polling algorithm.

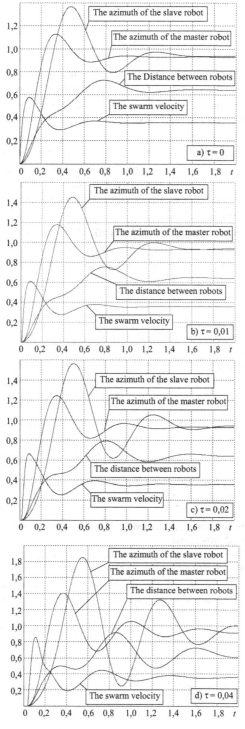

Fig. 4. Performance of the swarm at different time delay

5   Conclusion

As a result, the model of managing a swarm, organized in the convoy structure according to the "master/slave" principle, is proposed. It is shown that the main factor, determining the result of management is a time, spent by swarm units Von Neumann controllers on "thinking over" the control action, based on the processing of information from sensors. Delays in the feedback contours lead to a significant overshooting and an increase of performance time. This factor should be taken into account when designing unmanned mobile robots and their control systems. The method of simulation proposed may be recommended to hard- and software designers for utilization in a wide engineering practice of working out principles of organizing and control by swarms.

Further research in this area can be directed to the development of optimal polling algorithms that ensure, with minimal computational complexity, the control quality is not worse than the specified one.

This article was written within the framework of project RSF 22-29-00808: Fundamental interdisciplinary research on the development of "smart" personal protective equipment against infectious diseases transmitted by airborne droplets.

References

1. Tzafestas, S.G.: Introduction to Mobile Robot Control, p. 750. Elsevier (2014)
2. Siciliano, B., Khatib, O. (eds.): Springer Handbook of Robotics. Springer, Cham (2016). https://doi.org/10.1007/978-3-319-32552-1
3. Godwin, M.F., Spry, S.C., Hedrick, J.K.: A Distributed System for Collaboration and Control of UAV Groups: Experiments and Analysis. Center for the Collaborative Control of Unmanned Vehicles University of California, Berkeley, p. 224 (2007)
4. Leng, Y., Yu, C., Zhang, W., Zhang, Y., He, X., Zhou, W.: Hierarchical self-organization for task-oriented swarm robotics. In: Tan, Y., Shi, Y., Buarque, F., Gelbukh, A., Das, S., Engelbrecht, A. (eds.) ICSI 2015. LNCS, vol. 9140, pp. 543–550. Springer, Cham (2015). https://doi.org/10.1007/978-3-319-20466-6_57
5. Yu, C.Y., Wu, M.H., He, X.S.: Vehicle swarm motion coordination through independent local-reactive agents. In: Wu, Y. (ed.) Advanced Materials Research, vol. 108–111, pp. 619–624 (2010)
6. Morcom, J.: Optical distance measurement. US Patent No. 6753950. US CI. 356/4.01. Int. CI. G01S 17/00; G01C 3/08 (2004)
7. Bosch, T.: Laser ranging: a critical review of usual techniques for distance measurement. Opt. Eng. 40(1), 10 (2001). https://doi.org/10.1117/1.1330700
8. Gan, L., Zhang, H.: Optical circumferential scanning azimuth measurement method. In: IOP Conference Series. Materials Science and Engineering, vol. 428, p. 012013 (2018). https://doi.org/10.1088/1757-899X/428/1/012013
9. Landau, I.D., Zito, G.: Digital Control Systems, Design, Identification, p. 484. Springer, Heidelberg (2006)
10. Aström, J., Wittenmark, B.: Computer Controlled Systems: Theory and Design, p. 557. Tsinghua University Press, Prentice Hall (2002)
11. Fadali, M.S., Visioli, A.: Digital Control Engineering: Analysis and Design, pp. 239–272. Elsevier Inc. (2013)

12. Chen, I.-M., Yang, G., Huat, S.: Automatic modeling for modular reconfigurable robotic system: theory and practice. In: Cubero, S. (ed.) Yeo Industrial Robotics: Theory, Modelling and Control, pp. 43–82. IntechOpen (2006)

13. Yeh, Y.-C., Chu, Y., Chiou, C.W.: Improving the sampling resolution of periodic signals by using controlled sampling interval method. Comput. Electr. Eng. **40**(4), 1064–1071 (2014)

14. Pavlov, A.V.: About the equality of the transform of Laplace to the transform of Fourier. Issues Anal. **23**(1), 21–30 (2016). https://doi.org/10.15393/j3.art.2016.3211

15. Li, J., Farquharson, C.G., Hu, X.: Three effective inverse Laplace transform algorithms for computing time-domain electromagnetic responses. Geophysics **81**(2), E75–E90 (2015)

16. Arnold, K.A.: Timing analysis in embedded systems. In: Ganssler, J., Arnold, K., et al. (eds.) Embedded Hardware, pp. 239–272. Elsevier Inc. (2008)

17. Larkin, E.V., Ivutin, A.N.: Estimation of latency in embedded real-time systems. In: 3rd Meditteranean Conference on Embedded Computing (MECO 2014), pp. 236–239 (2014)

18. Fridman, E., Shaked, U.: A descriptor system approach to H control of linear time-delay systems. IEEE Trans. Autom. Control **47**(2), 253–270 (2002)

19. Zhang, X.M., Min, W.U., Yong, H.E.: Delay dependent robust control for linear systems with multiple time-varying delays and uncertainties. Control Decis. **19**(5), 496–500 (2004)

20. Janssen, J., Manca, R.: Applied Semi-Markov Processes, p. 310. Springer, Cham (2006)

21. Jiang, Q., Xi, H.-S., Yin, B.-Q.: Event-driven semi-Markov switching state-space control processes. IET Control Theory Appl. **6**(12), 1861–1869 (2012)

22. Kobayashi, H., Marl, B.L., Turin, W.: Probability, Random Processes and Statistical Analysis. Cambridge University Press, p. 812 (2012)

23. Pukelsheim, F.: The three sigma rule. Am. Stat. **48**(2), 88–91 (1994)

24. Li, D., Chen, G.: Impulses-induced p-exponential input-to-state stability for a class of stochastic delayed partial differential equations. Int. J. Control **92**(8), 1805–1814 (2019)

25. Hamann, A., Racu, R., Ernst, R.: Multi-dimensional robustness optimization in heterogeneous distributed embedded systems. In: Proceedings of the 13th IEEE Real Time and Embedded Technology and Applications Symposium, RTAS 2007, pp. 269–280. IEEE Computer Society, Washington, DC (2007)

Multi-Objective Optimization Robot Navigation Through a Graph-Driven PSO Mechanism

Timothy Sellers[1], Tingjun Lei[1], Gene Eu Jan[2], Ying Wang[3], and Chaomin Luo[1(✉)]

[1] Department of Electrical and Computer Engineering, Mississippi State University, Mississippi State, MS 39762, USA
Chaomin.Luo@ece.msstate.edu
[2] Department of Electrical Engineering, National Taipei University, and Tainan National University of the Arts, Tainan, Taiwan
[3] Department of Robotics and Mechatronics Engineering, Kennesaw State University, Marietta, GA 30060, USA

Abstract. In real-world robot applications such as service robots, mining mobile robots, and rescue robots, an autonomous mobile robot is required to visit multiple waypoints that it achieves multiple-objective optimizations. Such multiple-objective optimizations include robot travelling distance minimization, time minimization, turning minimization, etc. In this paper, a particle swarm optimization (PSO) algorithm incorporated with a Generalized Voronoi diagram (GVD) method is proposed for a robot to reach multiple waypoints with minimized total distance. Firstly, a GVD is used to form a Voronoi diagram in an obstacle populated environment to construct safety-conscious routes. Secondly, the sequence of multiple waypoints is created by the PSO algorithm to minimize the total travel cost. Thirdly, while the robot attempts to visit multiple waypoints, it traverses along the edges of the GVD to form a collision-free trajectory. The regional path locally from waypoints to nearest nodes or edges needs to be created to join the trajectory. A Node Selection Algorithm (NSA) is developed in this paper to implement such a protocol to build up regional path from waypoints to nearest nodes or edges on GVD. Finally, a histogram-based local reactive navigator is adopted for moving obstacle avoidance. Simulation and comparison studies validate the effectiveness and robustness of the proposed model.

1 Introduction

With the increasing demands and limited on board resources of autonomous robot, an autonomous robot requires the ability to visit several targets in one mission to optimize multiple objectives, such objectives include time, robot travel distance minimization, and spatial optimization [3,12,15,26,28,30]. Developing an autonomous robot multi-waypoint navigation system is crucial to effectively deploy robotics in various real-world fields [4,6,10,13]. Many algorithms have

© Springer Nature Switzerland AG 2022
Y. Tan et al. (Eds.): ICSI 2022, LNCS 13345, pp. 66–77, 2022.
https://doi.org/10.1007/978-3-031-09726-3_7

been proposed to solve autonomous robot navigation issue, such as graph-based model [5], ant colony optimization (ACO) [7,25], bat-pigeon algorithm (BPA) [8], neural networks [14,16], fuzzy logic [18], artificial potential field (APF) [21], sampling-based strategy [24], etc. Lei *et al.* produced a hybrid graph based algorithm associated with ant colony optimization (ACO) method used to optimize the trajectory of the global path generated [5]. Lei *et al.* [7] proposed an ACO model integrated with a variable speed module, which can decrease the robot motion speed in vicinity of the obstacles. In order to effectively control the speed modulation of autonomous vehicles, Lei *et al.* developed a bat-pigeon algorithm with the ability to adjust the speed navigation of autonomous vehicles [8]. Luo and Yang [14] set out to develop a complete coverage path planning algorithm for real-time map building and robot navigation in unknown environments. Luo *et al.* [16] then extended the model for multiple robots complete coverage navigation while using a bio-inspired neural network to dynamically avoid obstacles. Sangeetha *et al.* [18] developed a fuzzy gain-based navigation model to plan smooth and collision-free path in real time. Shin *et al.* [21] used a combination of potential risk fields to generate a hybrid directional flow, which can guide an autonomous vehicle with obstacle avoidance. Wang *et al.* [24] presented a sampling-based strategy with an efficient branch pruning model to avoid unnecessary growth of the tree and rapidly search a collision-free trajectory for the robot [11].

Autonomous robot multi-waypoint navigation as a special topic of robot path planning has also been studied for years [9]. For instance, Brunch *et al.* [2] proposed a model for real-world waypoint navigation using a variety of sensors for accurate environmental analysis. Nakamura and Kobayashi [17] brought about a multi-waypoint navigation system based on unmanned systems controller. The safety-conscious road model is developed utilizing the Generalized Voronoi diagram (GVD) approach. Once the safety-conscious roads are defined, a particle swarm optimization (PSO) algorithm-based multi-waypoint path planning algorithm is proposed to visit each waypoint in an explicated sequence while simultaneously avoiding obstacle. The Node Selection Algorithm (NSA) is developed in this paper to select the closest nodes on the safety-conscious roads to generate the final collision-free trajectories with minimal distance. Furthermore, within our hybrid algorithm, we utilize a reactive local navigator to avoid dynamic and unknown obstacles within the workspace.

2   Safety-Conscious Model

Voronoi diagrams (VD) model is an elemental data structure used as a minimizing diagram of a finite set of continuous functions [19]. The VD model decomposes the workspace into several regions, each of which consists of the points near a given object than to the others. Let $\mathcal{P} = \{p_1 \cdots, p_n\}$ be a set of points of $\mathbb{R}^d$ to each p_i associated with a specific Voronoi region $V(p_i)$. This is expressed by

$$V(p_i) = \{x \in \mathbb{R}^d : \|x - p_i\| \le \|x - p_j\|, \forall j \le n\}. \tag{1}$$

The intersection of $n-1$ half spaces can be denoted by the region $V(p_i)$. Within each half space holds a point p_i along with other point of $\mathcal{P}$. The regions $V(p_i)$ are convex polyhedron due to the bisectors acting as hyperplanes between each region. Using the Generalized Voronoi diagram (GVD) method we can establish the most obstacle-free path. The GVD model represents the workspace as a graph comprised of nodes, edges and vertices [22].

GVD nodes are the Euclidean distance between two or more obstacles, while the edges are the junction of two nodes that represent the distance between each neighboring nodes to another. Using these features from the GVD model, we can effectively create an obstacle-free path for our safety-conscious model. The safety-conscious roads are the clearest path between obstacles that occupies the available space in the map, this can be seen in Fig. 1.

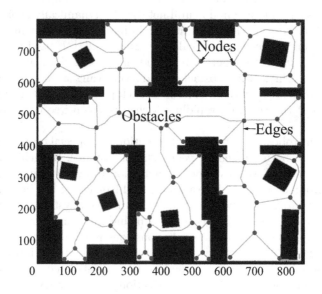

Fig. 1. Illustration of the safety-conscious model by creation of the GVD. The generated safety-conscious roads are represented by the yellow line, which are the edges in the GVD graph. (Color figure online)

3   PSO Based Multi-waypoint Navigation

The PSO algorithm is an optimization algorithm that uses an iterative methodology to optimize random initialized particles to define a path from the initial position to the target. This heuristic based algorithm can be utilized to resolve different types of optimization issues often confronted in engineering, such as multi-waypoint navigation problems.

3.1 Multi-waypoint Visiting Sequence

In the real-world applications of multi-waypoint navigation and mapping, multiple waypoints are provided by GPS coordinates. These waypoints are a series of targets needed to be reached and the associated cost of traveling is based on the distance between each waypoint. The goal is to search for the route of all waypoints at once and find the minimal cost of overall trajectories. The proposed model first uses the PSO algorithm to find waypoint order for the shortest path in the environment with only the information of the waypoints' coordinates. PSO algorithm uses randomized particle to find the best sequence to visit waypoints. In the system the local best position is denoted as *pbest*, and the global best positions is defined as *gbest*. The randomized particles use a fitness value to attract the particles to the local and global best positions. The velocities of the particles are updated as follows [20]:

$$v_p(\tau + 1) = v_p(\tau) + c_1 r_1 [pbest_p(\tau) - x_p(\tau)] + c_2 r_2 [gbest_p(\tau) - x_p(\tau)], \quad (2)$$

where $v_p(\tau)$ represents the velocity of particle p at instant τ, $x_p(\tau)$ is the position of particle p at instant τ, c_1 and c_2 are the positive acceleration constants used to scale the contribution of cognitive and social components. r_1 and r_2 are the uniform random number between 0 and 1. $pbest_p(\tau)$ is the best position the particle p achieved up to instant τ at current iteration, $gbest_p(\tau)$ is the best position that any of p's neighbors has achieved up to instant τ. Using this method, we can optimize the order in which each waypoint is visited.

The algorithm for the implemented PSO for finding multi-waypoint visiting sequence is explained in Algorithm 1. The objective of the algorithm is to minimize the total path length of Cartesian coordinates (X_n, Y_n) of waypoints given.

Algorithm 1: PSO Algorithm: Finding Order for Multi-waypoint

Initialize a population of particles
while *a stop criterion is not satisfied* **do**
    for *each particle p with position* x_p **do**
        if x_p *is better than* $pbest_p$ **then**
        | $pbest_p \leftarrow x_p$
        end
    end
    Define $gbest_p$ **as the best position**
    for *each particle p* **do**
        $v_p \leftarrow Compute\_velocity(x_p, pbest_p, gbest_p)$
        $x_p \leftarrow update\_position(x_p, v_p)$
    end
end

3.2 Safety-Conscious PSO Multi-waypoint Path Planning

By taking advantage of the pre-established information from the safety-conscious model we can construct our obstacle-free paths as seen in Fig. 2. A new Node Selection Algorithm (NSA) is developed as shown in Algorithm 2 for connecting each waypoint to a node in the graph. The NSA finds this entry point by employing two different methods. The first method constructs a path in the free-space between the waypoint and the node, while the other, constructs a path around obstacles that lie between the waypoint and node. In the first case, we calculate the distance from the nodes to the waypoint using the Euclidean distance. Then the node with the shortest distance from the waypoint is used as the entry point in the graph, but in case such as in Fig. 2, it is not that simple.

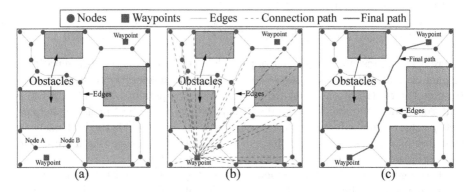

Fig. 2. Illustration of the NSA. (a) The workspace with nodes, edges and waypoints. (b) The connection path from the waypoints to the nodes. (c) The final generated path.

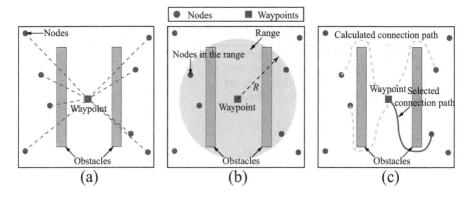

Fig. 3. Illustration of NSA method within a more specific sense, where an obstacle obstructing the connection path. (a) The multiple connection paths have been obstructed by the obstacles. (b) Selected the nodes in the defined range. (c) Conducted PSO point-to-point algorithm to achieve the optimal path to the selected node.

In Fig. 2 at the low left corer, the left node (Node A) is within proximity to the waypoint and would make the more obvious selection, but given a second opinion that the right node (Node B) is the most optimal node and they would be correct due to the path generated by using node A to reach the next waypoint being significantly greater than the path created by Node B. Thus, establishing a general method for node selection that is necessary to ensure that the minimal distance is always obtained. The NSA algorithm employs a second method for developing a path around obstacles that are between the waypoint and the node.

The algorithm constructs a path by utilizing nodes within a given range and selecting one of those nodes as an entry point, while constructing an obstacle-free path. The NSA defines the range by calculating the average length of all edges within graph $\mathcal{R}$, which is then used to create a field around the waypoint. Nodes that lie within that field will be utilized as an entry point in to the graph. Given the case in Fig. 3, where there are no direct path to any nodes we conduct point-to-point navigation using the PSO algorithm to obtain an obstacle-free path. $\mathcal{R}$ is utilized to reduce the number of nodes we have to conduct point-to-point navigation for. This enables us to find an optimal path from the waypoint to the node that avoids all obstacles. Utilizing these two methods in our NSA model we can construct an optimal path from each waypoint to its entry point.

Algorithm 2: Node Selection Algorithm for Waypoint Access Point

---

Initialize Parameters
$\mathcal{N}$ ←nodes within the graph
$\mathcal{R} \leftarrow Average\_Lenght(\mathcal{E})$
$\mathcal{PATH}$ ←contains all paths to all possible nodes
$\mathcal{WP} \leftarrow$ the position of the current waypoint
while $\mathcal{PATH} \in \mathcal{O}_i$ **do**
    if $N \in \mathcal{R}$ **then**
    | $\mathcal{NWR} \leftarrow \mathcal{N}$
    end
    for $N \in NWR$ **do**
    | $\mathcal{NP} \leftarrow PTP\_Path(\mathcal{WP}, \mathcal{NWR})$
    end
    $\mathcal{PATH} \leftarrow min(\mathcal{NP})$
end

---

4 Reactive Local Navigation

We use the Vector Field Histogram (VFH) model as a local navigator for obstacle avoidance. The local navigator determines the velocity commands that enable the autonomous mobile robot to move towards its waypoints. When applying VFH to a sequence of markers, the global trajectory can be broken down into a sequence of segments, which makes the model more efficient for workspaces that are heavily populated with obstacles. The VFH 2D cell-based map is filled with equally sized cells that are classified as either free or containing an obstacle [23]. The map is simultaneously being constructed as the mobile robot moves throughout the workspace as shown in Fig. 4. Concurrent map building and navigation are essential in autonomous obstacle avoidance [29]. A precise estimate of the robot pose (X, Y, Yaw) is demanded by map building so that precise registration of the local map on the global map is capable of being carried out. This map building aims to construct a occupancy-cell-based map. When the VFH model is used in conjunction with the GVD and PSO algorithm we can successfully navigate through our map and avoid all obstacles.

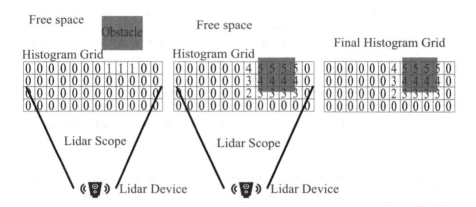

Fig. 4. Illustration of how the VHF uses a probability along with histogram based grid to detect and build a map simultaneously.

5   Simulation and Comparison Studies

In this section, simulations are reported to validate the effectiveness and robustness of the proposed model.

5.1   Comparison Study with Benchmark Datasets

We evaluate our PSO based waypoint order model through a comparison study in conjunction with well-known TSP test data sets and heuristic based algorithms, in which multi-target is implemented by various traveling salesman problem (TSP) algorithms. The chosen data sets are the 150-city problem by Chur Ritz (ch150), 200-city problem A, by Krolik/Felts/Nelson (kroA200), 299-city problem by Patberg/Rinaldi (pr-299), and 561-city problem by Kleinschmidt. The chosen data sets are well known previously used to verify other efficient TSP models [1].

The heuristic algorithms chosen for the comparison study were Ant Colony Optimization (ACO) algorithm, Genetic Algorithms (GA), Simulated Annealing (SA) algorithm, and Grey Wolf Optimization (GWO) algorithm. Due to the heuristic nature of the PSO algorithm, we were able to use the same set of parameters for each model. In our TSP comparison study, we focused on six key factors being min length (m), average length (m), length standard deviation (STD), min time (s), average time (s), and time standard deviation (STD). By evaluating parameters we can show a distinct difference between each model. Table 1 illustrates the effectiveness of the proposed model and its ability to generate the shortest path to connect waypoints. In respect to the time parameter, the proposed model does not achieve the shortest run time. However compared to the paths established by other models, the path generated was significantly larger than the proposed model. The STD parameter also showed the significance of our proposed model. From the obtained results, we can more than show the validly of our PSO based model in obtaining an optimal waypoint order.

Table 1. Comparison of minimum path length, average path length, STD of path length, minimum time, average time and STD of time with other models. The parameter for the test each model where: 100 initialized particles, 10 run per data set, and a maximum of 10 min per run

Test data set	Model	Min length (m)	Average length (m)	Length STD (m)	Min time (s)	Average time (s)	Time STD (s)
Ch150	Proposed model	1.67E+04	1.77E+04	5.99E+02	1.75E+04	1.31E+01	6.09E−02
	ACO	1.84E+04	2.23E+04	4.40E+03	1.67E+04	1.76E+02	1.32E+00
	GA	4.22E+04	5.04E+04	5.38E+03	1.79E+04	1.39E−02	1.25E−03
	SA	2.47E+04	3.05E+04	4.11E+03	1.75E+04	6.51E+00	7.25E−01
	GWO	3.23E+04	3.66E+04	2.64E+03	5.78E−01	9.39E−01	1.73E−01
KroA200	Proposed model	1.09E+05	1.17E+05	5.50E+03	2.15E+01	2.15E+01	9.73E−02
	ACO	1.30E+05	2.81E+05	4.09E+05	2.44E+02	5.17E+02	8.36E+02
	GA	2.39E+05	2.57E+05	1.15E+04	1.21E−02	1.54E−02	4.45E−03
	SA	1.96E+05	2.13E+05	1.18E+04	6.27E+00	6.62E+00	8.70E−01
	GWO	2.17E+05	2.40E+05	1.24E+04	1.12E+00	1.45E+00	3.68E−01
PR299	Proposed model	2.77E+05	2.89E+05	5.92E+03	4.28E+01	4.29E+01	6.06E−02
	ACO	3.37E+05	4.33E+05	4.82E+04	2.09E+02	2.37E+02	1.41E+01
	GA	3.19E+05	3.44E+05	1.17E+04	3.24E−02	3.38E−02	8.43E−04
	SA	5.26E+05	5.59E+05	1.94E+04	6.27E+00	6.52E+00	6.68E−01
	GWO	2.90E+05	3.90E+05	8.15E+04	6.71E+01	8.00E+01	7.51E+00
PA561	Proposed model	1.11E+05	1.14E+05	1.60E+03	1.41E+02	1.42E+02	3.65E−01
	ACO	–	–	–	–	–	–
	GA	1.37E+05	1.93E+05	6.30E+04	9.36E−02	9.57E−02	2.19E−03
	SA	1.88E+05	1.92E+05	2.07E+03	6.27E+00	6.42E+00	4.29E−01
	GWO	3.02E+05	4.21E+05	5.94E+04	7.64E+01	8.26E+01	2.21E+00

5.2 Model Comparison Study

We also conducted a model analysis using Zhang et al. [27] model proposed in. They developed a model to address the issues with multi-waypoint navigation within an indoor environment. They implemented an improved A* algorithm with the dynamic window approach (DWA) to handle multi-waypoint dynamic path planning. We selected this paper based on map configuration and overall efficiency in solving the multi-waypoint navigation problem. Our comparison study analyzes the number of nodes, the paths generated, and the total time to achieve the fastest route.

We first observe at the waypoint order and paths obtained by each model in an obstacle-free environment, as depicted in Fig. 5(a). The length generated by the Zhang's model was 240.84 m while the proposed model generated a shorter path of 219.99 m. We can see that this is due to the established waypoint orders in the environment. As seen from Table 2, we can see that the proposed model creates more nodes and has an increased length is 1.09% greater than Zhang's model, but the overall time spent by the proposed model was 6.1% faster than Zhang's model. Another advantage over the compared model is that there is no need for a node selection algorithm. We analyze the model before the critical node selection method is initialized and found the number of nodes and path generated were vastly greater than our proposed model. This further validates the proposed model ability to surpass and outperform compared model. The proposed model require no an addition method to achieve measurable results, which gives the proposed model significant advantage over Zhang's model.

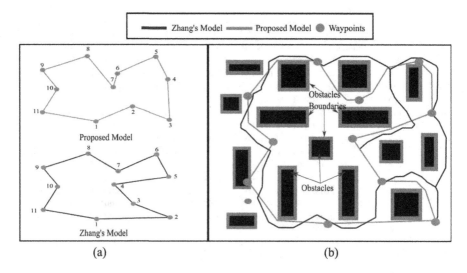

Fig. 5. Illustration of the waypoint sequencing (a) and how both models traverse through the map (b). The blue path indicates Zhang's method and red indicating the proposed method. The waypoints illustrated by the orange circles. (Color figure online)

Table 2. An illustration of the number of nodes, distance, and time spent traversing the map to each waypoint.

Model	Nodes	Distance	Time spent (s)
Zhang's model before node reduction	242	271.1	2.25
Zhang's model after node reduction	24	253.4	0.66
Proposed model	38	277.7	0.40

6 Conclusion

We proposed the safety-conscious multi-waypoint navigation method by a PSO and GVD model. We use a PSO-based multi-waypoint algorithm to define an order for waypoint navigation. Then NSA algorithm is developed to establish connections between the waypoints and the safety-conciseness roads to reach multi-objective optimization. We proved the feasibility and effectiveness of our model by performing a benchmark test and model comparison analysis.

References

1. Al-Khatib, R.M., et al.: MGA-TSP: modernised genetic algorithm for the travelling salesman problem. Int. J. Reason.-Based Intell. Syst. **11**(3), 215–226 (2019)
2. Brunch, M.H., Gilbreath, G., Muelhauser, J., Lum, J.: Accurate waypoint navigation using non-differential GPS. Technical report, Space and Naval Warfare Systems Center, San Diego, CA (2002)
3. Chen, J., Luo, C., Krishnan, M., Paulik, M., Tang, Y.: An enhanced dynamic Delaunay triangulation-based path planning algorithm for autonomous mobile robot navigation. In: Intelligent Robots and Computer Vision XXVII: Algorithms and Techniques, vol. 7539, pp. 253–264. SPIE (2010)
4. Jayaraman, E., Lei, T., Rahimi, S., Cheng, S., Luo, C.: Immune system algorithms to environmental exploration of robot navigation and mapping. In: Tan, Y., Shi, Y. (eds.) ICSI 2021. LNCS, vol. 12690, pp. 73–84. Springer, Cham (2021). https://doi.org/10.1007/978-3-030-78811-7_7
5. Lei, T., Luo, C., Ball, J.E., Rahimi, S.: A graph-based ant-like approach to optimal path planning. In: 2020 IEEE Congress on Evolutionary Computation (CEC), pp. 1–6. IEEE (2020)
6. Lei, T., Luo, C., Jan, G.E., Bi, Z.: Deep learning-based complete coverage path planning with re-joint and obstacle fusion paradigm. Front. Robot. AI **9**, 843816 (2022). https://doi.org/10.3389/frobt.2022.843816
7. Lei, T., Luo, C., Jan, G.E., Fung, K.: Variable speed robot navigation by an ACO approach. In: International Conference on Swarm Intelligence, pp. 232–242 (2019)
8. Lei, T., Luo, C., Sellers, T., Rahimi, S.: A bat-pigeon algorithm to crack detection-enabled autonomous vehicle navigation and mapping. Intell. Syst. Appl. **12**, 200053 (2021)
9. Lei, T., Luo, C., Sellers, T., Wang, Y., Liu, L.: Multi-task allocation framework with spatial dislocation collision avoidance for multiple aerial robots. IEEE Trans. Aerosp. Electron. Syst. (2022). https://doi.org/10.1109/TAES.2022.3167652

10. Lei, T., Sellers, T., Rahimi, S., Cheng, S., Luo, C.: A nature-inspired algorithm to adaptively safe navigation of a Covid-19 disinfection robot. In: Liu, X.-J., Nie, Z., Yu, J., Xie, F., Song, R. (eds.) ICIRA 2021. LNCS (LNAI), vol. 13015, pp. 123–134. Springer, Cham (2021). https://doi.org/10.1007/978-3-030-89134-3_12

11. Li, X., Luo, C., Xu, Y., Li, P.: A fuzzy PID controller applied in AGV control system. In: International Conference on Advanced Robotics and Mechatronics (ICARM), pp. 555–560 (2016)

12. Liu, L., Luo, C., Shen, F.: Multi-agent formation control with target tracking and navigation. In: IEEE International Conference on Information and Automation (ICIA), pp. 98–103 (2017)

13. Luo, C., Anjos, M.F., Vannelli, A.: Large-scale fixed-outline floorplanning design using convex optimization techniques. In: 2008 Asia and South Pacific Design Automation Conference, pp. 198–203. IEEE (2008)

14. Luo, C., Yang, S.X.: A bioinspired neural network for real-time concurrent map building and complete coverage robot navigation in unknown environments. IEEE Trans. Neural Netw. **19**(7), 1279–1298 (2008)

15. Luo, C., Yang, S.X., Krishnan, M., Paulik, M.: An effective vector-driven biologically-motivated neural network algorithm to real-time autonomous robot navigation. In: IEEE International Conference on Robotics and Automation (ICRA), pp. 4094–4099 (2014)

16. Luo, C., Yang, S.X., Li, X., Meng, M.Q.-H.: Neural-dynamics-driven complete area coverage navigation through cooperation of multiple mobile robots. IEEE Trans. Ind. Electron. **64**(1), 750–760 (2016)

17. Nakamura, R., Kobayashi, K.: A remote control system for waypoint navigation based mobile robot using JAUS. In: 2018 57th Annual Conference of the Society of Instrument and Control Engineers of Japan (SICE), pp. 1751–1756. IEEE (2018)

18. Sangeetha, V., et al.: A fuzzy gain-based dynamic ant colony optimization for path planning in dynamic environments. Symmetry **13**(2), 280 (2021)

19. Shao, M., Lee, J.Y.: Development of autonomous navigation method for nonholonomic mobile robots based on the generalized Voronoi diagram. In: Internation Conference on Control, Automation and Systems, pp. 309–313 (2010)

20. Shi, Y., Eberhart, R.: A modified particle swarm optimizer. In: 1998 IEEE International Conference on Evolutionary Computation Proceedings, pp. 69–73 (1998)

21. Shin, Y., Kim, E.: Hybrid path planning using positioning risk and artificial potential fields. Aerosp. Sci. Technol. **112**, 106–640 (2021)

22. Torpelund-Bruin, C., Lee, I.: Generalized Voronoi diagrams with obstacles for use in geospatial market analysis and strategy decisions. In: 2008 International Workshop on Geoscience and Remote Sensing, vol. 2, pp. 287–290 (2008)

23. Ulrich, I., Borenstein, J.: VFH+: reliable obstacle avoidance for fast mobile robots. In: Proceedings. In: 1998 IEEE International Conference on Robotics and Automation, vol. 2, pp. 1572–1577 (1998)

24. Wang, J., Li, B., Meng, M.Q.H.: Kinematic constrained bi-directional RRT with efficient branch pruning for robot path planning. Expert Syst. Appl. **170**, 114541 (2021)

25. Wang, L., Luo, C., Li, M., Cai, J.: Trajectory planning of an autonomous mobile robot by evolving ant colony system. Int. J. Robot. Autom. **32**(4), 406–413 (2017)

26. Yang, Y., Deng, Q., Shen, F., Zhao, J., Luo, C.: A shapelet learning method for time series classification. In: IEEE 28th International Conference on Tools with Artificial Intelligence (ICTAI), pp. 423–430 (2016)

27. Zhang, B., Jin, W., Gao, X., Chen, W.: A multi-goal global dynamic path planning method for indoor mobile robot. In: 2021 3rd International Symposium on Robotics Intelligent Manufacturing Technology (ISRIMT), pp. 97–103 (2021)

28. Zhao, W., et al.: A privacy-aware Kinect-based system for healthcare professionals. In: IEEE International Conference on Electro Information Technology (EIT), pp. 0205–0210 (2016)

29. Zhao, W., et al.: LiftingDoneRight: a privacy-aware human motion tracking system for healthcare professionals. Int. J. Handheld Comput. Res. (IJHCR) **7**(3), 1–15 (2016)

30. Zhu, D., Tian, C., Jiang, X., Luo, C.: Multi-AUVs cooperative complete coverage path planning based on GBNN algorithm. In: 29th Chinese Control and Decision Conference (CCDC), pp. 6761–6766 (2017)

Deep Neural Networks

Training Quantized Deep Neural Networks via Cooperative Coevolution

Fu Peng, Shengcai Liu$^{(\boxtimes)}$, Ning Lu, and Ke Tang

Guangdong Key Laboratory of Brain-Inspired Intelligent Computation,
Department of Computer Science and Engineering, Southern University of Science
and Technology, Shenzhen 518055, China
liusc3@sustech.edu.cn

Abstract. This work considers a challenging Deep Neural Network
(DNN) quantization task that seeks to train quantized DNNs without involving *any* full-precision operations. Most previous quantization
approaches are not applicable to this task since they rely on full-precision
gradients to update network weights. To fill this gap, in this work we
advocate using Evolutionary Algorithms (EAs) to search for the optimal
low-bits weights of DNNs. To efficiently solve the induced large-scale discrete problem, we propose a novel EA based on cooperative coevolution
that repeatedly groups the network weights based on the confidence in
their values and focuses on optimizing the ones with the least confidence.
To the best of our knowledge, this is the first work that applies EAs to
train quantized DNNs. Experiments show that our approach surpasses
previous quantization approaches and can train a 4-bit ResNet-20 on the
Cifar-10 dataset with the same test accuracy as its full-precision counterpart.

Keywords: Cooperative coevolution · Evolutionary algorithm ·
Large-scale discrete optimization · Neural network quantization

1 Introduction

Deep Neural Networks (DNNs) are powerful and have a wide range of applications in several fields such as image recognition [9], object detection [31], visual
segmentation [8], text classification [15], etc. However, DNNs generally require a
lot of computational resources. For example, the size of the well-known VGG-16
model built by Caffe is over 500MB and it consumes 16 GFLOPs, which makes
it impractical to be deployed on low-end devices. Hence, over the past few years,
many methods have been proposed to reduce the computational complexity
of DNNs, such as pruning [7], low-rank decomposition [41], knowledge distil-
lation [38], and quantization [6,7,11,30,36,37,42,43]. Specifically, DNN quantization maps the network weights from high bits to low bits, significantly reducing
memory usage, speeding up inference, and enabling the deployment of networks
on mobile devices with dedicated chips [12].

© Springer Nature Switzerland AG 2022
Y. Tan et al. (Eds.): ICSI 2022, LNCS 13345, pp. 81–93, 2022.
https://doi.org/10.1007/978-3-031-09726-3_8

Although existing quantization approaches have achieved notable success, most of them rely on full-precision gradients to update the network weights [6, 11,30,43], hindering their practical usages. In real-world applications, one may need to quantize a pre-trained full-precision DNN on different low-end devices for better adaptability, and the quantization procedure that is conducted on the device cannot involve any full-precision operations [36]. On the other hand, as a powerful search framework, EAs do not use any gradient information [18,20], which is naturally suitable for this scenario. Therefore, in this work we advocate using EAs to search for the low-bits weights of quantized DNNs.

Specifically, we first formulate DNN quantization as a large-scale discrete optimization problem. Since this problem involves a huge number of variables (network weights), e.g., ResNet-20 has 269722 parameters, we propose a novel EA based on cooperative coevolution to solve it. Given a pre-trained full-precision DNN, our algorithm first quantizes it to obtain an initial solution and then leverages estimation of distribution algorithm (EDA) to optimize the low-bits weights. To improve search efficiency, the algorithm repeatedly groups the network weights according to the confidence in their values and focuses on optimizing the ones with the least confidence. Finally, we compare our algorithm with exiting quantization approaches by applying them to train a 4-bit ResNet-20 on the Cifar-10 dataset, without involving any full-precision operations. The results show that our algorithm performs better and the quantized DNN obtains the same test accuracy as its full-precision counterpart, i.e., quantization without loss of accuracy. In summary, we make the following contributions in this paper:

1. We propose a novel EA based on cooperative coevolution to train quantized DNNs. To the best of our knowledge, this is the first work that applies EAs to search for the optimal low-bits weights of DNNs.
2. We conduct experiments to verify the effectiveness of the proposed algorithm. Notably, it can train a 4-bit ResNet-20 without accuracy degradation compared to the full-precision DNN, which indicates the great potential of EAs in DNN quantization.

2   Related Work

This section presents a brief literature review on the field of DNN quantization and cooperative coevolution.

2.1   DNN Quantization

DNN quantization is a popular research area, and researchers have proposed many quantization approaches [12], which can be classified into two categories: quantization-aware training (QAT) and post-training quantization (PTQ).

PTQ directly quantizes well-trained full-precision networks without retraining [37]. Two representative PTQ approaches are Outlier Channel Splitting

(OCS) [42] and Deep Compression [7]. The former deals with outliers during quantization by duplicating channels containing outliers and halving the channel values. The latter introduces a three-stage pipeline: pruning, trained quantization, and Huffman coding, which work together to reduce the memory storage for DNNs.

Unlike PTQ, QAT quantizes and finetunes network parameters in the training process [25], which usually obtains better performance, thus attracting much more research interest. BinaryConnect [6] restricts the weights to two possible values, i.e., -1 or 1, but the activations are still full-precision. BNN [11] quantizes both weights and activations to -1 or 1. XNOR-Net [30] proposes filter-wise scaling factors for weights and activations to minimize the quantization error. To further accelerate the training of DNNs, some work also attempts to quantize gradients. DoReFa-Net [43] uses quantized gradients in the backward propagation, but the weights and gradients are stored with full precision when updating the weights as the same as previous works. To the best of our knowledge, WAGE [36] is currently the only work that updates the quantized weights with discrete gradients.

2.2 Cooperative Coevolution

Cooperative coevolution is a powerful framework that leverages the "divide-and-conquer" idea to solve large-scale optimization problems. As first shown by Yang and Tang [5,14,22,39,40], the framework of cooperative coevolution consists of three parts: problem decomposition, subcomponent optimization, and subcomponents coadaptation. Among them, problem decomposition is the key step [29]. An effective decomposition can ease the optimization difficulty of a large-scale problem [33]. In contrast, an improper decomposition may lead the algorithms to local optimums [21,22,32].

There are three categories of problem decomposition approaches: static decomposition, random decomposition, and learning-based decomposition [21]. Static decomposition approaches do not take account into the subcomponents interactions and fixedly decompose the decision variables into subcomponents [3,4,28]. Conversely, selecting decision variables randomly for each subcomponent is the main idea of random decomposition approaches [13,27,35,39]. One of the most famous random decomposition methods is EACC-G proposed by Yang and Tang [39]. The main idea of this method is to divide the interdependent variables into the same subcomponent, but the dependencies between subcomponents should be as weak as possible. The learning-based approaches try to discover the interactions between variables [5,23,26].

3   Method

In this section, we first formulate DNN quantization as a large-scale discrete optimization problem and introduce our quantization functions. Then we detail the EDA applied to this problem. Finally, to further improve the algorithm performance, the cooperative coevolution framework is proposed.

3.1 Problem Formulation

In a DNN with L layers, let $\boldsymbol{w}_l$ represent the full-precision weights and $\hat{\boldsymbol{w}}_l$ represent the k bits quantized weights at layer l, which both are an n_l-dimension vector, i.e., there are n_l paremeters at layer l. We combine the quantized weights from all layers into $\hat{\boldsymbol{w}} = [\hat{\boldsymbol{w}}_1, \hat{\boldsymbol{w}}_2, \dots, \hat{\boldsymbol{w}}_L]$. The parameters in $\hat{\boldsymbol{w}}_l$ can only take one of 2^k possible discrete values, i.e., $\hat{\boldsymbol{w}}_l \in \{t_{l1}, t_{l2}, \dots, t_{l2^k}\}^{n_l}$. We formulate the DNN low-bit quantization problem as the following large-scale discrete optimization problem.

$$\max_{\hat{\boldsymbol{w}}} f(\hat{\boldsymbol{w}}) \quad s.t. \quad \hat{\boldsymbol{w}}_l \in \{t_{l1}, t_{l2}, \dots, t_{l2^k}\}^{n_l}, \quad l = 1, 2, ..., L, \tag{1}$$

where $f(\hat{\boldsymbol{w}})$ represents the accuracy of the quantized DNN. Since a DNN usually has a huge number of parameters, e.g., ResNet-152 has around 11 million paremeters, this is a large-scale discrete optimization problem.

3.2 Quantization Functions

To obtain a quantized DNN and construct the search space of our algorithm, we need to identify all the possible discrete values for each weight and activation. Moreover, the initial solution of our algorithm is obtained from a full-precision DNN. Based on the above two considerations, we design two linear quantization functions to map full-precision weights and activations to discrete ones separately.

For the weights, let the maximum and minimum values of each layer weights $\boldsymbol{w}_l$ be $[w_l^{min}, w_l^{max}]$, then the full-precision weights $\boldsymbol{w}_l$ at layer l are discretized with a uniform distance δ_l:

$$\delta_l(k) = \frac{w_l^{max} - w_l^{min}}{2^k - 1}, \tag{2}$$

where k is the number of bits. The quantization function for weights can be represented as:

$$Q(\boldsymbol{w}_l) = Clip\{round(\frac{\boldsymbol{w}_l}{\delta_l(k)}) \cdot \delta_l(k), w_l^{min}, w_l^{max}\}, \tag{3}$$

where the $Clip$ function is the saturation function, and the $round$ function maps continuous values to their nearest integers.

For the remaining parameters including activations and the parameters in batch normalization layers, we assume that the range of parameters is $[-1, 1]$ as WAGE [36]. The quantization function for activations $\boldsymbol{a}_l$ at layer l can be represented as:

$$Q(\boldsymbol{a}_l) = round(\frac{\boldsymbol{a}_l}{\delta_l(k)}) \cdot \delta_l(k). \tag{4}$$

3.3 Estimation of Distribution Algorithm

We propose to use the Estimation of Distribution Algorithm (EDA) to search for discrete weights. The overall framework of EDA for training quantized DNNs

---

Algorithm 1. Estimation of Distribution Algorithm

Require: the number of best individuals N_{best} to update the probabilistic model; updating step α ; generation number G; the size of population S
Output: best individual I_{best}
1: Initialize best individual $I_{best} = [\hat{w}_1 = a_1, \hat{w}_2 = a_2, \ldots, \hat{w}_n = a_n]$
2: Initialize probabilistic model P using σ-greedy strategy
3: **for** generation i from 0 to G **do**
4:     Generate S new individuals according to P
5:     Get the fitness values of the new individuals
6:     Rank the new individuals by fitness values in descending order
7:     Update the best individual I_{best}
8:     Select the first N_{best} best individuals
9:     Construct the probabilistic model P_{best} of the N_{best} best individuals
10:     Update the probabilistic model P: $P = (1 - \alpha)P + \alpha P_{best}$
11: **end for**

---

is summarized in Algorithm 1. We encode the quantized DNN weights into a fixed-length 1-dimensional array as the representation of our solution, i.e., $\hat{\boldsymbol{w}} = [\hat{w}_1, \hat{w}_2, \ldots, \hat{w}_n]$, where n represents the total number of parameters in a DNN. Then we construct a probabilistic model over it. For simplicity, we assume that the weights of the neural network are all independent of each other like PBIL [2]. Specifically, for each weight $\hat{w}_i$, there are 2^k possible values. Each possible value corresponds to a probability p_j, where $j = 1, 2, \ldots, 2^k$, k is the bit length of weights, and $\sum_{j=1}^{2^k} p_j = 1$. After the initial quantized network is obtained (Line 1), i.e., $\hat{\boldsymbol{w}} = [\hat{w}_1 = a_1, \hat{w}_2 = a_2, \ldots, \hat{w}_n = a_n]$, we initialize the probabilistic model P of the weights using σ-greedy strategy (Line 2), which is shown as Eq. (5):

$$\begin{cases} P(\hat{w}_i = a_i) = \sigma, \\ P(\hat{w}_i = \text{one of the other possible values}) = \dfrac{1 - \sigma}{2^k - 1}. \end{cases} \quad (5)$$

That is, if $\hat{w}_i$ takes the value a_i, then $P(\hat{w}_i = a_i) = \sigma$. The probability of other possible values is $(1 - \sigma)/(2^k - 1)$, where $0 < \sigma < 1$. For each generation, we sample weights from the probabilistic model to generate new individuals (Line 4), get the fitness values of them (Line 5) and rank them by their fitness values in descending order (Line 6). To update the probabilistic model P, we calculate the probability of each possible value for w_i according to the first N_{best} new individuals and construct the probabilistic model P_{best} of them (Line 9). Finally, we update P using $(1 - \alpha)P + \alpha P_{best}$ (Line 10), where α is updating step.

3.4   Cooperative Coevolution

Since our optimization problem has a huge number of decision valuables, to further improve the search efficiency, we propose a novel cooperative coevolution algorithm based on EDA inspired by Yang and Tang [39], namely EDA+CC.

The most important part of the cooperative coevolution algorithm lies in the efficient grouping of variables. As the EDA searches, the probabilistic model P

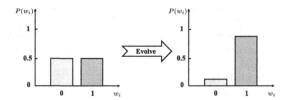

Fig. 1. $P(w_i)$ converges gradually as EDA evolving

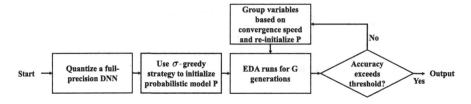

Fig. 2. EDA+CC framework

gradually converges. However, different decision variables have different convergence rates. Figure 1 gives a simple example of the convergence of the probabilistic model when applying EDA to a 0/1 optimization problem. Suppose the decision variables are encoded as $w = [w_1, w_2, \ldots, w_n]$ and variables are independent of each other. Initially, $P(w_i = 1) = 0.5$. As the evolution proceeds, $P(w_i = 1)$ gradually converges to 1. For w_i, if $P(w_i)$ converges quickly, it intuitively shows that EDA is confident about the value of w_i, which means w_i should not be changed in the subsequent searching process; conversely, if $P(w_i)$ converges slowly, then w_i should be further optimized.

Based on this intuition, we group the decision variables according to the confidence in their values, i.e., the speed of convergence. Specifically, we rank the decision variables according to the convergence speed of the probabilistic model in descending order during the EDA run. We divide the first $\beta \cdot n$ variables (which converge fast) into one group and the remaining variables (which converge slowly) into another group, where $\beta \in [0, 1]$ is a random number and n is the total number of weights in the network. For the former, we fix them. For the latter, we first perturb the probabilistic model of them with the σ-greedy strategy and then use EDA to optimize them. Figure 2 shows the framework of EDA+CC. Detials of EDA+CC are shown in Algorithm 2. Every G generations we regroup the variables, perturb the probabilistic model, and run EDA again until the network accuracy reaches the threshold.

4   Experiments

We use EDA+CC to train 4-bit quantized ResNet-20 on the Cifar-10 training set and test its performance on the test set. Firstly, we compare our EDA+CC algorithm with WAGE [36] (See Sect. 4.2). WAGE is the only work that can update

---

Algorithm 2. EDA+CC

Require: the number of best individuals N_{best} to update the probabilistic model; updating step α ; generation number G; the size of population S; accuracy threshold T; flag F; total number of weights n; random number β
Output: best individual I_{best}
1: Initialize best individual $I_{best} = [\hat{w}_1 = a_1, \hat{w}_2 = a_2, \ldots, \hat{w}_n = a_n]$
2: Get the fitness values of I_{best}
3: Initialize probabilistic model P using σ-greedy strategy
4: $F \leftarrow 0$
5: **while** fitness value of $I_{best} \leq T$ **do**
6:     **if** F=1 **then**
7:         Sort the weights by the convergence speed of P in descending order
8:         Divide the first $\beta \cdot n$ weights into group A
9:         Divide the remaining weights into group B
10:         Reinitialize the probabilistic model over group B using σ-greedy strategy
11:     **end if**
12:     **for** generation i from 0 to G **do**
13:         Generate S new individuals according to P
14:         Get the fitness values of the new individuals
15:         Rank the new individuals by fitness values in descending order
16:         Update the best individual I_{best}
17:         Select the first N_{best} best individuals
18:         Construct the probabilistic model P_{best} of the N_{best} best individuals
19:         Update the probabilistic model P: $P = (1 - \alpha)P + \alpha P_{best}$
20:         $F \leftarrow 1$
21:     **end for**
22: **end while**

---

discrete weights with quantized gradients. Secondly, we investigate the influence of different initial quantized DNNs by ablation study (See Sect. 4.3). Finally, besides EDA we also test the performance of Genetic Algorithm (GA) [24] and Local Search Algorithm (LS) [10] (See Sect. 4.4).

4.1 Experiment Settings

We implement EDA+CC based on TensorFlow 2.1.0 with python 3.6.9 and run the experiments on Nvidia RTX 2080ti. The settings of algorithms are as follows. The number of generations G is 500, the size of population S is 20, the number of best individuals N_{best} for each generation is 20, the updating step α is 0.1, and the parameter σ in the σ-greedy strategy is 0.95. To enforce the randomness of the algorithm, we set β as a random variable that obeys uniform distribution, $\beta \sim N(0.4, 0.6)$.

In the following, we show how to construct an initial quantized network, in which the range of weights of each layer is known. First, we use Eq. (3) and (4) to quantize the pre-trained full-precision ResNet-20. To increase the randomness of the initial quantized network we randomly select s% of all its parameters and perturb them to adjacent values ($s = 20, 30, 40$ separately). We denote the

Table 1. Accuracies of different initial quantized networks obtained by perturbing different proportional parameters. We also list the accuracy of the pre-trained full-precision network, namely ResNet20-Float.

Network	Training set	Test set
ResNet20-Float	98.65%	91.00%
ResNet20Q-Switch-20%	88.20%	85.38%
ResNet20Q-Switch-30%	46.65%	46.79%
ResNet20Q-Switch-40%	20.32%	19.60%

quantized network obtained by the above process as ResNetQ-Switch-s% and the pre-trained full-precision network as ResNet20-Float. Table 1 summarizes the accuracies of the different initial quantized DNNs.

Since ResNet-20 has 269722 parameters, to reduce the search space, we restrict weights to two possible values, i.e., the value before perturbation and the value next to it. Thus, the problem becomes a binary optimization problem and the search space size is 2^{269722}. Note that this is still a huge search space, it is 10^{81022} times larger than the Go search space.

4.2   Verifying the Effectiveness of EDA+CC

To verify the effectiveness of EDA+CC, we first compare EDA+CC with WAGE [36], a representative quantization method that quantizes gradients. The code of WAGE is available at [1]. To further examine the performance of the cooperative coevolution algorithm, we also compare EDA+CC with EDA w/o CC. EDA w/o CC re-initializes the probabilistic model using the σ-greedy strategy without grouping the decision valuables when EDA restarts. Both EDA+CC and EDA w/o CC use 150K fitness evaluations and take about 23.3 h separately, in which the time complexity is acceptable.

Table 2 shows the accuracies of the quantized DNNs obtained by different approaches. We use ResNet20Q-Switch-30% as our initial solution. The initial accuracy of ResNet20Q-Switch-30% is 46.50%. For EDA+CC, the training set accuracy only decreases by 0.15% and the test set accuracy increases by 0.4% compared to the full-precision network. In comparison, the accuracy of the network obtained by WAGE training is only about 43%, which is much worse than EDA+CC. We speculate the reason for the poor performance of WAGE might be that WAGE is designed for quantized DNNs with 2-bit weights and 8-bit activations, while our paper uses a more rigorous and hardware-friendly setting: 4-bit weights and 4-bit activations quantized DNNs. Comparing EDA+CC with EDA w/o CC, we can see the positive effect of cooperative coevolution. Applying cooperative coevolution increases the training accuracy from 98.05% to 98.50% and the testing accuracy from 89.40% to 91.40%. The effectiveness of cooperative coevolution is mainly shown in two aspects: improving the quality of the solution and accelerating the convergence. Figure 3 shows the training curves of

Table 2. Compare EDA+CC with WGAE and EDA w/o CC.

Algorithm	Training set	Test set
EDA+CC	**98.50%**	**91.40%**
EDA w/o CC	98.05%	89.40%
WAGE	43.44%	41.35%

Table 3. Results of different initial quantized networks.

Initial quantized DNN	Training set	Test set	No. of FEs
ResNet20Q-Switch-20	99.25%	91.50%	50K
ResNet20Q-Switch-30	98.50%	91.40%	150K
ResNet20Q-Switch-40	90.09%	82.75%	150K

EDA+CC and EDA w/o CC, i.e., the accuracy of the best individual in each generation. As Fig. 3 shows, after using the σ-greedy strategy to re-initialize the probabilistic model P and restarting EDA, EDA+CC can accelerate the convergence and help EDA find a better solution.

4.3 Ablation Study

We conduct more detailed studies on different initial quantized networks for EDA+CC. Table 3 shows the accuracies of the quantized networks obtained by EDA+CC with different initial networks. It can be seen that EDA+CC reaches 90.09% accuracy after 150K fitness evaluations for ResNet20Q-Switch-40. We estimate that it will take about 500K fitness evaluations(FEs) for EDA+CC to reach around 98% accuracy because each restart of EDA with σ-greedy strategy can improve the accuracy by about 0.9%. In summary, Table 3 illustrates that as the initial accuracy decreases, EDA+CC requires more FEs to train a quantized DNN without accuracy decay compared to the full-precision network.

4.4 Comparison of EDA, GA and LS

We compare three search algorithms, GA, LS, and EDA. We use ResNet20Q-Switch-50% as the initial quantized network. Each algorithm uses 100K fitness evaluations. Figure 4 shows the training curves of the three algorithms. It can be seen that EDA performs significantly better than LS and GA, which indicates that the distribution estimation mechanism is more suitable than the crossover and mutation mechanisms for the problem considered in this study. The crossover and mutation mechanisms might break some good patterns in the individuals imperceptibly, while the distribution estimation mechanism optimizes the individuals in a global way. It is worth noting that, theoretically, in the binary space, ResNet20Q-Switch-50% corresponds to random initialization, because half of the

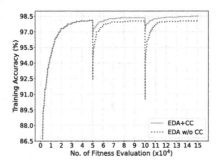

Fig. 3. Training curves of EDA+CC and EDA w/o CC.

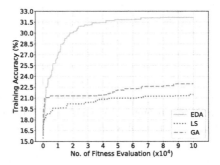

Fig. 4. Training curves of EDA, GA and LS.

parameters are randomly perturbed. All three algorithms can obtain better accuracy than ResNet20Q-Switch-50%, which illustrates the potential of search-based algorithms in training quantized DNNs.

5   Conclusion and Future Work

In this paper, we investigate search-based training approaches for quantized DNNs, focusing on exploring the application of cooperative coevolution to this problem. Unlike existing works, EDA+CC does not need gradient information. Considering the search space of this problem is extremely large (e.g., in our experiments it is 10^{81022} times larger than the Go search space), we propose to use cooperative coevolution to help solve this problem. The results show that our method can obtain quantized networks without accuracy decay compared to floating-point networks in our experiment setting.

Overall, this work is a proof of concept that EAs can be applied to train quantized DNNs. There are many subsequent lines of research to pursue, e.g., the effects of other variable grouping mechanisms. Moreover, the method of determining the ranges of discrete values should also be studied. Finally, based on the cooperative coevolution framework, it is interesting to investigate on solving different sub-problems by different algorithms [16,17,19,34], hopefully leading to better optimization performance.

Acknowledgments. This work was supported in part by the Shenzhen Peacock Plan (Grant No. KQTD2016112514355531), and the Guangdong Provincial Key Laboratory (Grant No. 2020B121201001).

References

1. Training and inference with integers in deep neural networks. https://github.com/boluoweifenda/WAGE
2. Baluja, S.: Population-based incremental learning. A method for integrating genetic search based function optimization and competitive learning. Technical report, Carnegie-Mellon University, Pittsburgh, PA, Department of Computer Science (1994)
3. Van den Bergh, F., Engelbrecht, A.P.: A cooperative approach to particle swarm optimization. IEEE Trans. Evol. Comput. **8**(3), 225–239 (2004)
4. Cao, Z., et al.: An effective cooperative coevolution framework integrating global and local search for large scale optimization problems. In: Proceedings of the 2015 IEEE Congress on Evolutionary Computation, CEC 2015, Sendai, Japan, pp. 1986–1993. IEEE, May 2015
5. Chen, W., Weise, T., Yang, Z., Tang, K.: Large-scale global optimization using cooperative coevolution with variable interaction learning. In: Schaefer, R., Cotta, C., Kołodziej, J., Rudolph, G. (eds.) PPSN 2010. LNCS, vol. 6239, pp. 300–309. Springer, Heidelberg (2010). https://doi.org/10.1007/978-3-642-15871-1_31
6. Courbariaux, M., Bengio, Y., David, J.P.: BinaryConnect: training deep neural networks with binary weights during propagations. In: Proceedings of the 28th Advances in Neural Information Processing Systems: Annual Conference on Neural Information Processing Systems, NeurIPS 2015, Montreal, Canada, pp. 3123–3131. Curran Associates Inc., December 2015
7. Han, S., Mao, H., Dally, W.J.: Deep compression: compressing deep neural networks with pruning, trained quantization and Huffman coding. arXiv preprint arXiv:1510.00149 (2015)
8. He, K., Gkioxari, G., Dollár, P., Girshick, R.: Mask R-CNN. In: Proceedings of the 2017 IEEE International Conference on Computer Vision, ICCV 2017, Venice, Italy, pp. 2961–2969. IEEE Computer Society, October 2017
9. He, K., Zhang, X., Ren, S., Sun, J.: Deep residual learning for image recognition. In: Proceedings of the 2016 IEEE Conference on Computer Vision and Pattern Recognition, CVPR 2016, Las Vegas, NV, pp. 770–778. IEEE Computer Society, June 2016
10. Hoos, H.H., Stützle, T.: Stochastic Local Search: Foundations and Applications. Morgan Kaufmann, San Francisco (2004)
11. Hubara, I., Courbariaux, M., Soudry, D., El-Yaniv, R., Bengio, Y.: Binarized neural networks. In: Proceedings of the 29th Advances in Neural Information Processing Systems: Annual Conference on Neural Information Processing Systems, NeurIPS 2016, Barcelona, Spain, pp. 4107–4115. Curran Associates Inc, December 2016
12. Krishnamoorthi, R.: Quantizing deep convolutional networks for efficient inference: a whitepaper. arXiv preprint arXiv:1806.08342 (2018)
13. Li, X., Yao, X.: Cooperatively coevolving particle swarms for large scale optimization. IEEE Trans. Evol. Comput. **16**(2), 210–224 (2012)

14. Liu, J., Tang, K.: Scaling up covariance matrix adaptation evolution strategy using cooperative coevolution. In: Yin, H., et al. (eds.) IDEAL 2013. LNCS, vol. 8206, pp. 350–357. Springer, Heidelberg (2013). https://doi.org/10.1007/978-3-642-41278-3_43

15. Liu, S., Lu, N., Chen, C., Tang, K.: Efficient combinatorial optimization for word-level adversarial textual attack. IEEE/ACM Trans. Audio Speech Lang. Process. **30**, 98–111 (2022)

16. Liu, S., Tang, K., Lei, Y., Yao, X.: On performance estimation in automatic algorithm configuration. In: Proceedings of the 34th AAAI Conference on Artificial Intelligence, AAAI 2020, New York, NY, pp. 2384–2391. AAAI Press, February 2020

17. Liu, S., Tang, K., Yao, X.: Automatic construction of parallel portfolios via explicit instance grouping. In: Proceedings of the 33rd AAAI Conference on Artificial Intelligence, AAAI 2019, Honolulu, HI, pp. 1560–1567. AAAI Press, January 2019

18. Liu, S., Tang, K., Yao, X.: Memetic search for vehicle routing with simultaneous pickup-delivery and time windows. Swarm Evol. Comput. **66**, 100927 (2021)

19. Liu, S., Tang, K., Yao, X.: Generative adversarial construction of parallel portfolios. IEEE Trans. Cybern. **52**(2), 784–795 (2022)

20. Liu, S., Wei, Y., Tang, K., Qin, A.K., Yao, X.: QoS-aware long-term based service composition in cloud computing. In: Proceedings of the 2015 IEEE Congress on Evolutionary Computation, CEC 2015, Sendai, Japan, pp. 3362–3369. IEEE, May 2015

21. Ma, X., Huang, Z., Li, X., Wang, L., Qi, Y., Zhu, Z.: Merged differential grouping for large-scale global optimization. IEEE Trans. Evol. Comput. 1 (2022)

22. Ma, X., et al.: A survey on cooperative co-evolutionary algorithms. IEEE Trans. Evol. Comput. **23**(3), 421–441 (2019)

23. Mei, Y., Omidvar, M.N., Li, X., Yao, X.: A competitive divide-and-conquer algorithm for unconstrained large-scale black-box optimization. ACM Trans. Math. Softw. **42**(2), 13:1–13:24 (2016)

24. Mitchell, M., Holland, J.H., Forrest, S.: The royal road for genetic algorithms: fitness landscapes and GA performance. Technical report, Los Alamos National Lab., NM, USA (1991)

25. Nahshan, Y., et al.: Loss aware post-training quantization. Mach. Learn. **110**(11), 3245–3262 (2021)

26. Omidvar, M.N., Li, X., Mei, Y., Yao, X.: Cooperative co-evolution with differential grouping for large scale optimization. IEEE Trans. Evol. Comput. **18**(3), 378–393 (2014)

27. Omidvar, M.N., Li, X., Yang, Z., Yao, X.: Cooperative co-evolution for large scale optimization through more frequent random grouping. In: Proceedings of the 2010 IEEE Congress on Evolutionary Computation, CEC 2010, Barcelona, Spain, pp. 1–8. IEEE, July 2010

28. Potter, M.A., De Jong, K.A.: A cooperative coevolutionary approach to function optimization. In: Davidor, Y., Schwefel, H.-P., Männer, R. (eds.) PPSN 1994. LNCS, vol. 866, pp. 249–257. Springer, Heidelberg (1994). https://doi.org/10.1007/3-540-58484-6_269

29. Potter, M.A., Jong, K.A.D.: Cooperative coevolution: an architecture for evolving coadapted subcomponents. Evol. Comput. **8**(1), 1–29 (2000)

30. Rastegari, M., Ordonez, V., Redmon, J., Farhadi, A.: XNOR-Net: ImageNet classification using binary convolutional neural networks. In: Leibe, B., Matas, J., Sebe, N., Welling, M. (eds.) ECCV 2016. LNCS, vol. 9908, pp. 525–542. Springer, Cham (2016). https://doi.org/10.1007/978-3-319-46493-0_32

31. Ren, S., He, K., Girshick, R., Sun, J.: Faster R-CNN: towards real-time object detection with region proposal networks. In: Proceedings of the 28th Advances in Neural Information Processing Systems : Annual Conference on Neural Information Processing Systems, NeurIPS 2015, Montreal, Canada, pp. 91–99. Curran Associates Inc, December 2015

32. Son, Y.S., Baldick, R.: Hybrid coevolutionary programming for nash equilibrium search in games with local optima. IEEE Trans. Evol. Comput. **8**(4), 305–315 (2004)

33. Sun, Y., Kirley, M., Halgamuge, S.K.: A recursive decomposition method for large scale continuous optimization. IEEE Trans. Evol. Comput. **22**(5), 647–661 (2018)

34. Tang, K., Liu, S., Yang, P., Yao, X.: Few-shots parallel algorithm portfolio construction via co-evolution. IEEE Trans. Evol. Comput. **25**(3), 595–607 (2021)

35. Trunfio, G.A., Topa, P., Was, J.: A new algorithm for adapting the configuration of subcomponents in large-scale optimization with cooperative coevolution. Inf. Sci. **372**, 773–795 (2016)

36. Wu, S., Li, G., Chen, F., Shi, L.: Training and inference with integers in deep neural networks. In: Proceedings of the 6th International Conference on Learning Representations, ICLR 2018, Vancouver, Canada. OpenReview.net, April 2018

37. Wu, Y., et al.: Rotation consistent margin loss for efficient low-bit face recognition. In: Proceedings of the 2020 IEEE/CVF Conference on Computer Vision and Pattern Recognition, CVPR 2020, Seattle, WA, pp. 6865–6875. Computer Vision Foundation/IEEE, June 2020

38. Xu, Y., et al.: Positive-unlabeled compression on the cloud. In: Proceedings of the 32nd Annual Conference on Neural Information Processing Systems, NeurIPS 2019, Vancouver, Canada, pp. 2565–2574. Curran Associates Inc., December 2019

39. Yang, Z., Tang, K., Yao, X.: Large scale evolutionary optimization using cooperative coevolution. Inf. Sci. **178**(15), 2985–2999 (2008)

40. Yang, Z., Tang, K., Yao, X.: Multilevel cooperative coevolution for large scale optimization. In: Proceedings of the 2008 IEEE Congress on Evolutionary Computation, CEC 2008, Hong Kong, China, pp. 1663–1670. IEEE, June 2008

41. Yu, X., Liu, T., Wang, X., Tao, D.: On compressing deep models by low rank and sparse decomposition. In: Proceedings of the 2017 IEEE Conference on Computer Vision and Pattern Recognition, CVPR 2017, Honolulu, HI, pp. 67–76. IEEE Computer Society, July 2017

42. Zhao, R., Hu, Y., Dotzel, J., Sa, C.D., Zhang, Z.: Improving neural network quantization without retraining using outlier channel splitting. In: Proceedings of the 36th International Conference on Machine Learning, ICML 2019, Long Beach, CA, pp. 7543–7552. PMLR, June 2019

43. Zhou, S., Wu, Y., Ni, Z., Zhou, X., Wen, H., Zou, Y.: DoReFa-Net: training low bitwidth convolutional neural networks with low bitwidth gradients. arXiv preprint arXiv:1606.06160 (2016)

An Improved Convolutional LSTM Network with Directional Convolution Layer for Prediction of Water Quality with Spatial Information

Ziqi Zhao[1], Yuxin Geng[2], and Qingjian Ni[1]([✉])

[1] School of Computer Science and Engineering, Southeast University, Nanjing, Jiangsu, China
nqj@seu.edu.cn
[2] Chien-Shiung Wu College, Southeast University, Nanjing, Jiangsu, China

Abstract. The prediction of water quality indicators is an important topic in environmental protection work. For the prediction of water quality data with multi-site data, this paper proposes an improved model based on ConvLSTM, which achieves the introduction of multi-site spatial relationships in water quality indicators prediction. On the basis of ConvLSTM, a directional convolutional layer is introduced to deal with the spatial dependence of multiple information collection stations with upstream and downstream relationship of a river to improve the prediction accuracy. The model proposed in this paper is applied to a dataset from three data collection stations to predict several indicators. Experiments on real-world data sets and results demonstrate that the improvements proposed in this paper make the model perform better compared to both the original and other common models.

Keywords: Water quality prediction · Convolutional LSTM network · Spatial dependence

1 Introduction

The monitoring and prediction of water quality data is an important topic in environmental protection work. The establishment of water quality information collection stations allows researchers to obtain enough data and use machine learning methods to predict future water quality indicators. Although water quality will change with natural weather and human factors, itself has a certain periodicity. At the same time, the spatial dependence of water quality indicators on the upstream and downstream is distinctive, so it is necessary to introduce spatial dependence in the task of predicting water quality indicators. Additionally, by introducing water quality data from multiple stations for modeling, we can further locate the location of the triggering factors while finding the water quality indicators anomaly.

© Springer Nature Switzerland AG 2022
Y. Tan et al. (Eds.): ICSI 2022, LNCS 13345, pp. 94–105, 2022.
https://doi.org/10.1007/978-3-031-09726-3_9

At present, there have been some preliminary studies on water quality, which define it as a time series prediction problem. Existing methods for precipitation water quality can roughly be categorized into three kinds: statistical analysis methods, machine learning methods and neural network models. For the statistical analysis methods, the autoregressive model (AR), moving average model (MA) and autoregressive integrated moving average model (ARIMA) are commonly used [5]. The advantage of these approaches is that the model can be trained quickly. The drawback is that it does not perform well on water quality data that is subject to more human interference. For the machine learning methods, support vector machine (SVM), bayesian network (BN), random forest (RF), decision tree (DT) and other models are applied to water quality prediction problems [1,3,16,17]. In addition to the application of existing models to the field of water quality prediction, more and more researchers began to improve the model according to the characteristics of water quality data [2].

In essence, water quality indicator prediction is a time-series prediction problem. Past water quality indicators are used as input, and a fixed number of future water quality indicators are used as output. However, statistical or regression-based models in practice face a problem: it highly requires smooth data sets, so that the prediction results are too smooth, leading to a lack of practical significance in the predicted results. Recent advances in deep learning, especially gate recurrent unit (GRU) and long short-term memory (LSTM) models, provide some useful insights on how to tackle this problem. In order to improve the accuracy of water quality prediction, more and more neural networks and hybrid models have been proposed. Long short-term memory (LSTM) deep neural networks, usually used in time series prediction, is also applied in the prediction of water quality [10]. And the hybrid model LSTM-MA is also applied for water indicator prediction [14]. Similarly, another hybrid model to improve RNN in the prediction of water indicator data is proposed [6]. As can be seen, current approaches to water quality prediction are mainly based on traditional time-series prediction machine learning models or deep learning networks. These prediction methods tend to ignore the spatial dependence of several different data collection sites in the same water system.

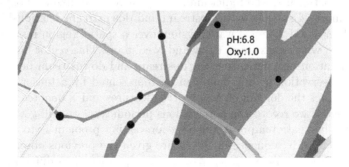

Fig. 1. An example of some sites in the same water system

Although there have been some studies that have introduced spatial information into time-series data predictions, the focus of their work wasn't on the relationships between individual sites [8]. In other fields, however, there are problems with the introduction of spatial dependencies in the analysis of time series forecasts. Examples include radar cloud prediction, traffic flow prediction, etc., and one of the representative methods is ConvLSTM [12]. So this paper proposes an improvement based on ConvLSTM, which is able to introduce spatial data in water quality indicators prediction and improve the prediction accuracy, having the following improvements and contributions:

1. Introducing spatial information. ConvLSTM is adopted for water quality prediction, taking the spatial relationship between different sites into consideration.
2. Improvement of ConvLSTM. ConvLSTM is usually used to solve problems where the spatial relationship has high uncertainty, but the river has a stable upstream and downstream relationship. So we introduce directional convolution layer to improve it.
3. Real evaluation. We evaluate our approach based on real-world data sets. Extensive experiments show the advantage that our method has a better performance.

This paper is organized as follows: in Sect. 2, this paper will introduce the research problems and related methods involved in our model; in Sect. 3, we will introduce the proposed model in detail; Sect. 4 is about experimental settings and results of experiments; the last section is the summary and prospects.

2 Basic Definitions

2.1 Problem Formulation

The goal of forecasting water indicators data is to use the previously observed data sequence to forecast the future water indicators of a fixed length in a water system (rivers or pools). In practical applications, the data are usually collected from the sites every 4 h. To predict 1-day ahead, this problem can be regarded as a spatiotemporal sequence forecasting problem.

In our experiments, we predict ammonia nitrogen and dissolved oxygen data from 3 monitoring stations with upstream and downstream relationships. Suppose we observe a dynamical water system over a spatial region represented by an $1 \times N$ grid which consists of 1 row and N columns. This vector implies a partial order relationship, indicating the upstream and downstream between sites. Thus, the observation at any time can be represented by a tensor $\chi \in R_{1 \times N}$ where R denotes the domain of the observed features and N denotes the domain of N stations. If we record the observations periodically, we will get a sequence of tensors. The spatiotemporal sequence forecasting problem is to predict the most likely length-K sequence in the future given the previous observations. In our experiments, we use $\chi_{t-w}, \chi_{t-w+1}, ..., \chi_t$ to predict χ_t in different stations, where w is the time window. Because in general, the influence of historical data on future data in time-series data forecasting decreases until it can be ignored.

2.2 Convolutional LSTM for Sequence Modeling with Spatial Information

ConvLSTM has been shown to be able to extract spatial relationships from temporal data. ConvLSTM not only inherits the advantages of the traditional LSTM, but due to its internal convolutional structure, it is well suited to spatial data and therefore can be applied to the prediction of the variation of various materials in a river with the movement of water [4,18]. The major innovation of ConvLSTM is that it extends FC-LSTM by using convolutional layers of variable sizes rather than fully connected layers to capture the motion patterns of objects with different velocities. This allows ConvLSTM to quickly capture spatial features in the data while discarding a certain degree of generality in the processing of time-series data. ConvLSTM may be seen as a universal version of FC-LSTM where the filter can take different values. When the filter size equals the size of input vector, it becomes FC-LSTM. In this paper, we follow the formulation of ConvLSTM as in paper [12]. The key equations are shown below.

$$
\begin{aligned}
i_t &= \sigma(W_{xi} * \chi_t + W_{hi} * \mathcal{H}_{t-1} + W_{ci}\mathcal{C}_{t-1} + b_i), \\
f_t &= \sigma\left(W_{xf} * \chi_t + W_{hf} * \mathcal{H}_{t-1} + W_{ci}\mathcal{C}_{t-1} + b_f\right), \\
o_t &= \sigma\left(W_{xo} * \chi_t + W_{ho} * \mathcal{H}_{t-1} + W_{co}\mathcal{C}_t + b_o\right), \\
\mathcal{C}_t &= f_t\mathcal{C}_{t-1} + i_t \tanh\left(W_{xc} * \mathcal{X}_t + W_{hc} * \mathcal{H}_{t-1} + b_c\right), \\
\mathcal{H} &= o_t \tanh\left(\mathcal{C}_t\right).
\end{aligned}
\tag{1}
$$

ConvLSTM can be combined with other deep learning models to form more complex structures. These structures have been applied to solve many real-world problems [7,9,11].

3 The Proposed Method

The proposed improvement structure is shown in figure (See Fig. 2). Although the ConvLSTM has been proved to be powerful at handling spatiotemporal data, it contains too much redundancy for water indicator data. To address this problem, we propose an extension of convolutional layer only in the input. By adding the layer that has fixed direction, we are able to build a network that can capture only the speed of data flow, not the direction of data flow. This means that the new structure will no longer apply to more general issues. At the same time, however, the new structure will perform better in water quality indicator predictions.

3.1 Reshaping Input Structure

The major drawback of ConvLSTM in handling water indicator data is its usage of convolutional layer, in which spatial information will be taken into account with all directions. In water quality data prediction tasks, however, the flow of data between sites is directed. This can be thought of as a specialised problem

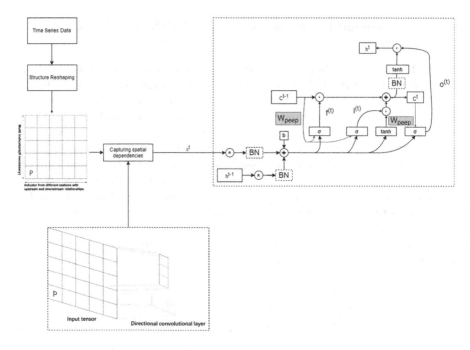

Fig. 2. The overview structure of the proposed model

for ordinary spatio-temporal data prediction. In this case, the individual stations form a partial order set with each other through upstream and downstream relationships. Their spatial dependence is unidirectional, i.e. from upstream to downstream. To overcome this problem, we have redesigned the data input form so that water quality data can contain both multiple indicators and spatial information in a two-dimensional tensor (See Fig. 3). In order to capture the potential linkages between multiple indicators and multiple sites, we reconstructed the data into a 2D tensor. One dimension of the matrix is the data originating from multiple sites. We will subsequently capture the spatial dependencies through a special convolution layer. The other dimension of the matrix is the multiple water quality indicators collected at the same site (if present). This provides the possibility of multiple indicators prediction.

3.2   Directional Convolution Layer

After reshaping the input structure, the convolutional layer used in ConvLSTM to extract spatial relations would no longer be applicable. Accordingly, they are replaced by two special convolutional layers in this paper instead. Their convolutional kernels are used to extract upstream-downstream relationships and relationships between multiple indicators at the same site, respectively (See Fig. 4).

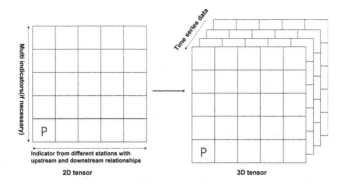

Fig. 3. Reshaping input structure

In proposed model, the directional kernel size is adjustable, which based on specific upstream-only relationships. The size of indicator kernel used to capture relationships between indicators is usually n, which is the number of indicators. Here is a special case, when predicting only one indicator, the input becomes a one-dimensional vector and the indicator kernel is no longer needed.

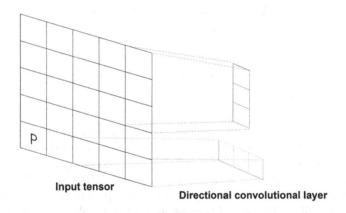

Fig. 4. Directional convolution layer proposed in this paper

4   Experiments and Results

We first compare our improved network with the LSTM network on data sets to gain some basic understanding of the performance of our model. To verify the effectiveness of our model, we also compared it with common models used for water quality indicator prediction. The results of the experiments lead to the following findings:

1. ConvLSTM with directional convolution layer performs better than pure LSTM and other time-series prediction models in handling water quality indicators data set.
2. Directional convolution layer can help capture the spatial relationships.
3. Directional convolution layer can speed up the training of the model to a great extent by simplifying convolutional layer operations.

4.1   Dataset and Preprocessing

In this paper, we used real-world water quality data from automatic monitoring stations in southern China. We conducted experiments on data from 3 monitoring stations with upstream and downstream relationships. The water quality indicators that we were committed to study is ammonia nitrogen and dissolved oxygen, which were important to pollution monitoring and ecological protection of rivers. The time range is from 2019/8/18 to 2020/8/18, and the time interval is 4 h. Other specific condition of data sets is in Table 1.

Our implementations of the models are in Python with the help of Keras. We run all the experiments on a computer with a single NVIDIA GTX2080Ti GPU.

Table 1. The condition of data set

Indicator	Max	Min
Ammonia nitrogen	17.77	0.13
Dissolved oxygen	12.29	0.09

In order to supplement the missing data, we mainly adopted the Seasonal and Trend decomposition using Loess (STL 2) to decompose the data into three parts so that we could obtain the cyclical and trend components of the data. According to this method, the sliding average of the time-series data (Y_t) can represent its trend. After subtracting the trend component (T_t) from the data, a fitting method allows the period component (S_t) to be obtained and finally the residuals (R_t).

$$Y_t = T_t + S_t + R_t, \quad t = 1, \cdots, N. \tag{2}$$

Here is an example of STL used to process dissolved oxygen data (See Fig. 5). The original data is decomposed into 3 parts: Trend, Seasonality and Residuals. From the 'Trend' sub-graph we can learn the variation of indicators over time. The 'Seasonality' sub-graph shows that the change of certain indicator. The 'Residual' sub-graph represents the amount of noise generated by irregular disturbance in the data.

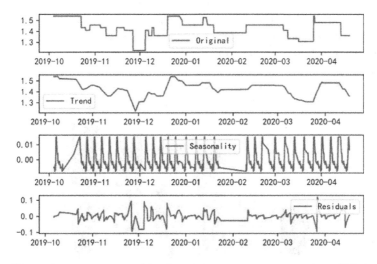

Fig. 5. Data decomposed by Trend decomposition using Loes (STL)

4.2 Baseline Methods for Comparision

In this paper, the proposed model is compared with some latest machine learning models in time series prediction.

1. Prophet: a procedure for forecasting time series data based on an additive model [13].
2. RNN-LSTM: LSTM recurrent neural network.
3. CNN: Convolutional neural network.
4. TCN: Temporal convolutional network.
5. Transformer: A network based solely on attention mechanisms, dispensing with recurrence and convolutions entirely [15].

4.3 Evaluation Metrics

In our datasets, since we have compared the proposed model with other models, we follow the metrics *RMSE*, which is defined as follow:

$$RMSE = \sqrt{\frac{\sum_{(i,t)\in\Omega_{test}}\left(Y_{it} - \hat{Y}_{it}\right)^2}{|\Omega_{test}|}}. \tag{3}$$

where Y and $\hat{Y}$ are true values and predicted values respectively. And the lower *RMSE* means better performance.

4.4 Results and Analysis

Since our model is based on the LSTM network framework and applies an improvement mechanism, the study is carried out to compare with the original model by introducing spatial dependency information. We conduct experiments in terms of *RMSE* and model training time. The comparison results are shown in figure (See Fig. 6 and Fig. 7). And raw data is presented in the table.

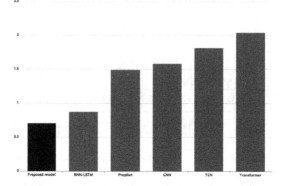

Fig. 6. Comparison of *RMSE* results between proposed model and baselines

The results (See Fig. 6, 9 and Table 2) show that our model can improve the accuracy of water quality indicators prediction by introducing spatial dependency information, which leads to an improved result and best between baselines.

Table 2. The experimental results of *RMSE*

Model	*RMSE*	Training time
Proposed model	0.70	33.91
RNN-LSTM	0.87	1105.64
Prophet	1.49	6.37
CNN	1.58	319.31
TCN	1.81	155.15
Transformer	2.04	734.89

At the same time, it can be seen in figure (See Fig. 8) that our directional convolution layer plays a role in assisting in the capture of spatial information. In fact the *RMSE* of results of Site 1 and Site 2 are a little higher than *RMSE* using the original model. It is because we make reduction in the number of network layers in the improved model, which is also the reason why our models can be trained and used in a short response time. However, when we tend to predict indicators of Site 3, the directional convolution layer can help us capture spatial relationships and improve prediction accuracy while having a small impact on the size of the model. The results in figure (See Fig. 7) show that our model's training time is only slightly higher than Prophet while at the highest levels of accuracy.

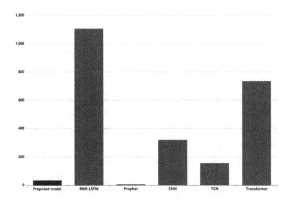

Fig. 7. Comparison of training time results between proposed model and baselines

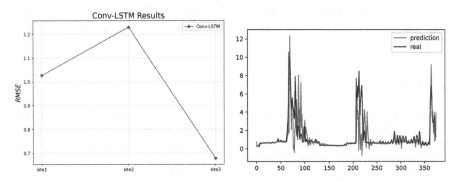

Fig. 8. *RMSE* of 3-site prediction

Fig. 9. Prediction results of ammonia nitrogen

5   Conclusion

In this paper, aiming at the prediction of time-series prediction of water quality indicators with multi-site, directional convolution layer is proposed, which is an improvement of convolution LSTM. Our model introduces two special convolution kernels before the input layer of original ConvLSTM for capturing spatial correlation between data. At the same time, it also provides a possibility to make multivariate predictions. Based on the real-water quality data, we predict water quality indicators on a 3-site data set. The experiments prove that the prediction accuracy of our model is significantly better than that of other models. What's more, by applying our improvement, the model can not only capture the spatial relationship but also make a significant reduction in model size, which leads to a short training time.

The experimental results show that the proposed improvement of ConvL-STM performs well in water indicator forecasting, through which, environmental authorities can analyse water quality more easily. Also, accurate prediction of water quality indicators for fixed stations can be an important basis for us to capture anomalous values and detect polluting discharges, which has a great effect on the monitoring of water quality.

In the future, the following problems deserve further study:

1. In the current study, this paper is more concerned with correctly predicting future data. In the field of environmental monitoring, however, we would also like to be able to assess existing data to determine when a particular body of water is unusually polluted. In future research, we expect to be able to construct a model for anomaly detection.
2. While this paper proposes a multivariate approach to time-space data prediction, it currently only introduces spatial relationships between multiple sites. In further experiments, we will try multivariate forecasting.
3. In the current study, this paper only predicts through water quality indicators as factors. However, water quality indicators are also influenced by other factors such as precipitation, temperature, air quality, etc. In future studies, we will consider introducing these factors to further improve the prediction accuracy.

References

1. Arora, S., Keshari, A.K.: ANFIS-ARIMA modelling for scheming re-aeration of hydrologically altered rivers. J. Hydrol. **601**, 126635 (2021)
2. Asadollah, S.B.H.S., Sharafati, A., Motta, D., Yaseen, Z.M.: River water quality index prediction and uncertainty analysis: a comparative study of machine learning models. J. Environ. Chem. Eng. **9**(1), 104599 (2021)
3. Chen, K., et al.: Comparative analysis of surface water quality prediction performance and identification of key water parameters using different machine learning models based on big data. Water Res. **171**, 115454 (2020)
4. Howard, A.G., et al.: MobileNets: efficient convolutional neural networks for mobile vision applications. arXiv preprint arXiv:1704.04861 (2017)
5. Katimon, A., Shahid, S., Mohsenipour, M.: Modeling water quality and hydrological variables using ARIMA: a case study of Johor river, Malaysia. Sustain. Water Resour. Manag. **4**(4), 991–998 (2018)
6. Li, L., Jiang, P., Xu, H., Lin, G., Guo, D., Wu, H.: Water quality prediction based on recurrent neural network and improved evidence theory: a case study of Qiantang river, China. Environ. Sci. Pollut. Res. **26**(19), 19879–19896 (2019)
7. Li, Z., Gavrilyuk, K., Gavves, E., Jain, M., Snoek, C.G.: VideoLSTM convolves, attends and flows for action recognition. Comput. Vis. Image Underst. **166**, 41–50 (2018)
8. Liang, Y., Ke, S., Zhang, J., Yi, X., Zheng, Y.: GeoMAN: multi-level attention networks for geo-sensory time series prediction. In: International Joint Conferences on Artificial Intelligence (IJCAI), vol. 2018, pp. 3428–3434 (2018)
9. Lin, Z., Li, M., Zheng, Z., Cheng, Y., Yuan, C.: Self-attention ConvLSTM for spatiotemporal prediction. In: Proceedings of the AAAI Conference on Artificial Intelligence, vol. 34, pp. 11531–11538 (2020)
10. Liu, P., Wang, J., Sangaiah, A.K., Xie, Y., Yin, X.: Analysis and prediction of water quality using LSTM deep neural networks in IoT environment. Sustainability **11**(7), 2058 (2019)

11. Majd, M., Safabakhsh, R.: A motion-aware ConvLSTM network for action recognition. Appl. Intell. **49**(7), 2515–2521 (2019). https://doi.org/10.1007/s10489-018-1395-8
12. Shi, X., Chen, Z., Wang, H., Yeung, D.Y., Wong, W.K., Woo, W.C.: Convolutional LSTM network: a machine learning approach for precipitation nowcasting. Neural Inf. Process. Syst. (NeurIPS) **28** (2015)
13. Taylor, S.J., Letham, B.: Forecasting at scale. Am. Stat. **72**(1), 37–45 (2018)
14. Than, N.H., Ly, C.D., Van Tat, P.: The performance of classification and forecasting Dong Nai river water quality for sustainable water resources management using neural network techniques. J. Hydrol. **596**, 126099 (2021)
15. Vaswani, A., et al.: Attention is all you need. Neural Inf. Process. Syst. (NeurIPS) **30** (2017)
16. Wang, H., Song, L.: Water level prediction of rainwater pipe network using an SVM-based machine learning method. Int. J. Pattern Recognit Artif Intell. **34**(02), 2051002 (2020)
17. Wang, J., Jiang, Z., Li, F., Chen, W.: The prediction of water level based on support vector machine under construction condition of steel sheet pile cofferdam. Concurr. Comput. Pract. Exp. **33**(5), e6003 (2021)
18. Zhang, L., Zhu, G., Mei, L., Shen, P., Shah, S.A.A., Bennamoun, M.: Attention in convolutional LSTM for gesture recognition. Adv. Neural. Inf. Process. Syst. **31** (2018)

COVID-19 Detection Method Based on Attentional Mechanism and LSTM

Wanpeng Zhu and Xiujuan Lei[✉]

School of Computer Science, Shaanxi Normal University, Xian, China
xjlei@snnu.edu.cn

Abstract. Since 2020, the Novel Coronavirus virus, which can cause upper respiratory and lung infections and even kill in severe cases, has been ravaging the globe. Rapid diagnostic tests have become one of the main challenges due to the severe shortage of test kits. This article proposes a model combining Long short-term Memory (LSTM) and Convolutional Block Attention Module for detection of COVID-19 from chest X-ray images. In this article, chest X-ray images from the COVID-19 radiology standard data set in the Kaggle repository were used to extract features by MobileNet, VGG19, VGG16 and ResNet50. CBAM and LSTM were used for classifcation detection. The simulation results showed that the experimental results showed that VGG16–CBAM–LSTM combination was the best combination to detect and classify COVID-19 from chest X-ray images. The classification accuracy of VGG-16-CBAM-LSTM combination was 95.80% for COVID-19, pneumonia and normal. The sensitivity and specificity of the combination were 96.54% and 98.21%. The F1 score was 94.11%. The CNN model proposed in this article contributes to automated screening of COVID-19 patients and reduces the burden on the healthcare delivery framework.

Keywords: Attentional mechanism · Long short-term memory · COVID-19 · Deep learning

1 Introduction

1.1 A Subsection Sample

In 2020, COVID-19 spread worldwide on March 11, 2020, COVID-19 (WHO) World Health Organization (WHO) announced that the new crown pneumonia became a global pandemic and a sudden health event, mainly caused by SARS-CoV-2. The novel coronavirus pneumonia has high transmission and is a new form of detection. It is mainly used to detect Reverse Transcription Polymerase Chain Reaction (RT-PCR) virus nucleic acid. It is an applicable form for detecting new crown pneumonia. However, RT-PCR has many disadvantages, such as high operation requirements, long time-consuming and low positive rate. Therefore, the development of reasonable and cost-effective diagnostic technology in

Supported by the National Natural Science Foundation of China (61972451, 61902230).

the early stage is of great significance to detect the spread of the virus [4]. Chest X-ray image, as a relatively easy to obtain medical image, is very suitable for diagnosis and detection in emergency. It is one of the effective ways to study the lung and its health status.

In recent years, deep learning has many applications in the infectious of disease diagnosis. The relevant research of Ghoshal *et al.* shows that the use of deep learning technology can make chest CT imaging applied to identify COVID-19 patients [5]. Wang *et al.* studied 453 CT images of COVID-19 confirmed cases, of which 217 images were used as the training set, and 73.1% accuracy was obtained using the initial model [12]. Hemdan *et al.* proposed a CNN based model, which is a variant of VGG-19. The model uses 50 images achieving 90% accuracy [6]. Ahsan *et al.* developed COVID-19 diagnostic model for mixed data (digital/classified data and image data) using multilayer perceptron and convolutional neural network (mlp-cnn) [1]. Apostolopoulos *et al.* used the transfer learning strategy and added it to CNN, so that CNN can automatically diagnose COVID-19 cases by learning the basic features in the chest film data set [2]. Wang *et al.* constructed a new database, chestx-ray8, containing 108948 X-ray frontal images of 32717 unique patients [13]. Song *et al.* used deep learning to detect COVID-19 on CT images, and the accuracy rate reached 90.8% [10].

2   Materials and Methods

The data processing model for COVID-19 detection proposed in this article is shown in Fig. 1. After the original X-ray image is preprocessed by size adjustment, transformation and normalization, the data is divided into training set and test set. Then we use the training set to train the cnn-cbam-lstm architecture we proposed. At the same time, the accuracy and loss of training are measured every certain time by using 50% cross verification, and the accuracy and loss of the whole system are verified. We used confusion matrix, accuracy, area under ROC curve (AUC), specificity, sensitivity and F1 score to evaluate the performance of the whole system.

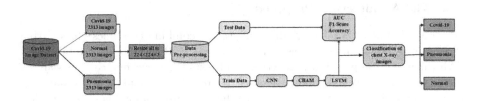

Fig. 1. Data processing model

2.1   The DataSet

COVID-19 data sets were organized into 3 folders (new crown pneumonia, pneumonia, normal). Since a large number of specific data sets could not be obtained, X-ray samples of COVID-19 were obtained from different sources. First, 1401 COVID-19 samples were collected using GitHub repository, radiopedia, Italian Society of Radiology, figshare data repository website. In addition, 912 enhanced images were collected from mendeley instead of explicitly using data enhancement technology. Finally, samples of 2313 normal and pneumonia cases were obtained from kaggle. A total of 6939 samples were used in the experiment, and 2313 samples were used in each case. We adjusted the size of the data set to 224 pixels, and divided it into 80% and 20% for training and testing, respectively. The data set used in this article are from https://data.mendeley.com/datasets/mxc6vb7svm and the division results of it is shown in Table 1.

2.2   Development of Network Models

The model structure proposed in this article is the combination of convolutional neural network (CNN) [8], attention mechanism [14] and long-term and short-term memory network (LSTM) [3] to learn kaggle dataset.

Table 1. Data set partition

Data/Cases	COVID-19	Normal	Pneumonia	Overall
Train (centered)	**1480**	1480	1480	4440
Test	**463**	463	462	1388
Val	**370**	370	371	1111
Overall	**2313**	2313	2313	6939

2.3   The Architecture Proposed

For the existing COVID-19 X-ray images, we combine combines the existing widely used pre training models VGG-16, VGG-19, ResNet50 and MobileNetv1 with the model proposed in this article, and tests the performance of classifying the images into COVID-19 positive cases, bacterial pneumonia and normal samples respectively. The following describes the structure combining different CNN with the model proposed in this article. In this article, the pneumonia X-ray image obtained from kaggle database is preprocessed and adjusted to 224 pixels. The convolution layer and CBAM are used to extract the features, and then it is input into LSTM. Finally, COVID-19 is classified through the full connection layer.

VGG-16 and VGG-19. In the current research, researchers have developed learning models VGG-16 and VGG-19 in the image recognition task of Imagenet [9]. We add these two pretraining models to the proposed CNN-CBAM-LSTM system, which can transfer the parameters of the original pretraining model instead of starting from scratch, so as to reduce its computational cost. VGG-16 model is a network composed of 16 layers network structure based on Imagenet database, which is mainly used for image recognition and classification. VGG-19 has a similar structure to VGG-16, except that VGG-19 has 16 convolution layers.

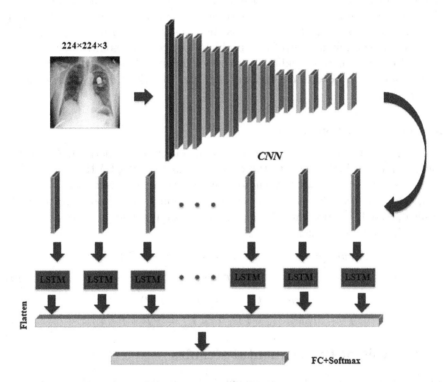

Fig. 2. CNN—CBAM—LSTM structural model

MobileNet. MobileNet is one of the representative networks of lightweight neural networks. With the application effect of deep learning network model getting better and better, deep learning also promotes the development of neural networks towards miniaturization . In this article, mobilenet is used as the convolution layer of CNN-CBAM-LSTM model system to extract features. The basic unit of mobilenet adopts depth level separable convolution. The deep convolution in the deep separable convolution is to turn the convolution kernel into a single channel and finally stacks it. The point by point convolution is the same as the traditional convolution layer, taking 1×1 Convolution kernel. This article preprocesses the X-ray images obtained from the kaggle database, adjusts them to 224 pixels, uses the convolution layer of MobileNetV1 and CBAM to extract features, then inputs them into LSTM to extract feature information, and finally realizes the classification of COVID-19 through the full connection layers.

ResNet. The residual network is obviously different from the proposed network. Its main idea is to add a direct channel in the whole network, which allows to retain the proportion and output of the previous network layer. The traditional convolutional network or fully connected network will more or less have problems such as information loss when transmitting information. At the same time, it will lead to the disappearance or explosion of gradient, which makes the deep network unable to train. ResNet retains the integrity of information after using the direct channel, which solves this problem to a certain extent. In this article, we uses ResNet50 as the backbone network of the model, adjusts the pictures of the data set to 224 pixels, and inputs them into the whole model to realize the classification.

2.4 The Evaluation Index

In this article, the Formula 1–4 were used to measure the performance of the proposed model: TP represents the correct prediction of COVID-19 cases, FP represents the normal cases or pneumonia cases that were systematically misclassified, TN represents the normal cases or pneumonia cases that were correctly classified, and FN represents the normal cases or pneumonia cases that were systematically misclassified.

$$Accuracy = (TP + TN)/(TN + FP + TP + FN) \qquad (1)$$

$$Sensitivity = TP/(TP + FN) \qquad (2)$$

$$Specificity = TN/(TN + FP) \qquad (3)$$

$$F1 - score = (2 * TP)/(2 * TP + FP + FN) \qquad (4)$$

3 Analysis of Experimental Results

This section describes the experimental device, data set and performance indicators, and then analyzes the performance of different combination using MobileNetV1, ResNet50, VGG-16 and VGG-19 respectively to find the best as the model structure proposed in this article.

3.1 The Experimental Device

In this article, the obtained data set, namely chest X-ray images, was converted into 224 × 224 × 3 pixel values, and then implemented using Keras and tensorflow2.0 on Intel I5, 3.30 GHZ processor. The overall dataset had 80% and 20% backed up for training and testing, respectively. The maximum number of iterations in this article was 100, and a fivefold cross validation approach was used to conduct the experiments.

3.2 Results Analysis

We conduct experimental analysis on the proposed different system models to verify their performance. Figure 3 shows the confusion matrix of the system model proposed in this article when VGG-19, VGG-16, MobileNetv1 and ResnR-Net50 are used as CNN respectively. Where a, b, c and d respectively represent the confusion matrix obtained by combining VGG-19, VGG-16, MobileNetv1 and ResNet50 with the system proposed in this article. Among 1380 images, 58 images in VGG-19-CBAM-LSTM system were misclassified, including 18 COVID-19 cases, 58 images in VGG-16-CBAM-LSTM system were misclassified, including 16 COVID-19 cases, and 58 images in MobileNetv1-CBAM-LSTM system were misclassified, including 18 COVID-19 cases. In the model with ResNet50 as the backbone network, 149 images were misclassified. In the above confusion matrix, we observed that VGG-16 has better true value and true negative value, and there are few misclassification cases.

Figure 4 shows the change of the accuracy value with the number of iterations of different systems in the training and verification stage,in which a, b, c and d represent the change in accuracy of the combination of the system proposed

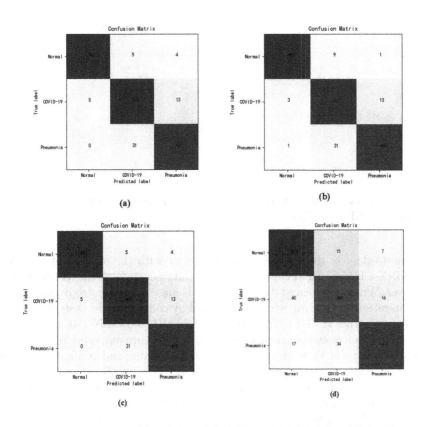

Fig. 3. Confusion matrix (a) VGG-19 (b) VGG-16 (c) MobileNet (d) ResNet 50

in this article and VGG-19, VGG-16, MobileNetV1 and ResNet50, respectively. In this article, 5-fold cross validation is adopted, and the number of iterations of training validation is 100. As shown in Fig. 4, when VGG-19 convolution is used for feature extraction, the accuracy of training and verification reaches 98.31% and 97.31%, respectively. When VGG-16 is used for feature extraction, its accuracy reaches 99.45% and 99.81% respectively. When MobileNetV1 is used for feature extraction, its accuracy reaches 98.20% and 98.88% respectively. When ResNet50 is used as the backbone network, the accuracy of its training set and verification set is only 85.71% and 90.39%.

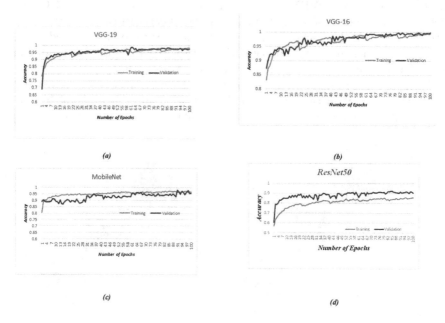

Fig. 4. Accuracy varies with the number of iterations (a) VGG-19 (b) VGG-16 (c) MobileNet (d) ResNet50

Figure 5 depicts the change of cross entropy with the number of iterations when CNN-CBAM-LSTM classifier uses different networks as the feature extraction layer. Where a, b, c and d represent the change in cross entropy as the number of iterations increasing which is obtained by combining the system proposed in this article and VGG-19, VGG-16, MobileNetV1 and ResNet50, respectively.

It can be seen from the image that in the architecture proposed in this article, the cross entropy of VGG-19 is 0.05 and 0.07 in training and verification respectively. The cross entropy of VGG-16 is 0.016 and 0.004 in training and verification ,respectively. The cross entropy of MobileNetV1 is 0.05 and 0.09 respectively. The cross entropy of ResNet50 is 0.38 and 0.50 in training and verification, respectively. According to the change of cross entropy, VGG16-CBAM-LSTM shows better performance.

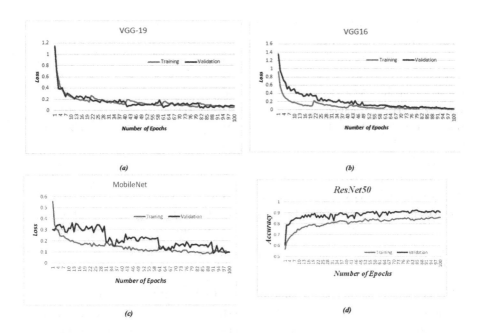

Fig. 5. Loss changes with the number of iterations (a) VGG-19 (b) VGG-16 (c) MobileNet (d) ResNet50

Table 2. Performance index of each method

CNN	Class	Acc (%)	Spe (%)	Sen (%)	F1-score(%)
VGG16-CBAM-LSTM	COVID-19	91.79	98.21	96.54	94.11
	Normal	99.11	98.92	99.80	98.45
	Pneumonia	96.85	96.58	93.07	94.92
VGG19-CBAM-LSTM	COVID-19	92.68	98.00	95.68	94.16
	Normal	98.06	99.03	99.56	98.80
	Pneumonia	96.82	96.67	92.21	94.46
MobileNetV1-CBAM-LSTM	COVID-19	91.60	98.00	94.17	92.86
	Normal	98.65	99.03	96.27	97.45
	Pneumonia	93.26	96.68	92.86	93.06
ResNet50-CBAM-LSTM	COVID-19	88.76	91.94	83.58	86.10
	Normal	84.90	94.47	95.16	89.74
	Pneumonia	94.70	95.16	88.96	91.74

Table 3. The existing method is compared with the method proposed in this article

Author	Model	Acc(%)	Acc(COVID-19)(%)	Dataset size
Wang et al. [12]	Inception	73.1	—	453 CT
Ghoshal et al. [5]	Bayes —CNN	89.92	—	5941 X-ray
Ahsan et al. [1]	MLP-CNN	95.4	—	168 X-ray
Hemdan et al. [6]	VGG19	90	—	50 X-ray
Wang et al. [11]	Tailored CNN	92.3	80.0	5941 X-ray
Bessinana et al. [2]	VGG19	93.48	—	1428 X-ray
Song et al. [10]	—	90.8	—	275 CT
Li et al. [7]	DenseNet	88.9	79.2	537 X-ray
Proposed method	VGG19-CBAM-LSTM	94.93	92.68	6939 X-ray
Proposed method	VGG16-CBAM-LSTM	95.80	91.79	6939 X-ray
Proposed method	MobileNet-CBAM-LSTM	94.42	91.60	6939 X-ray
-	SVM	82.89	9—	6939X-ray

Table 4. The existing method is compared with the method proposed in this article

Model	Accuracy(%)	Acc(COVID-19)(%)	Dataset size
VGG16	91.56	90.01	6939 X-ray images
VGG19	90.85	89.83	6939 X-ray images
MoBileNetV1	89.04	88.82	6939 X-ray images
SVM	82.89	80.0	6939 X-ray images
DenseNet	88.90	79.20	6939 X-ray images
VGG19–CBAM–LSTM	94.93	92.68	6939 X-ray images
VGG16–CBAM–LSTM	95.80	91.79	6939 X-ray images
MobileNet–CBAM–LSTM	94.42	91.60	6939 X-ray images

Table 2 summarizes the overall accuracy, specificity, sensitivity and F1 score of the proposed system under different networks. As shown in Table 2, the system proposed in this article shows better comprehensive performance under different neural networks, among which VGG-16 has the best performance. For COVID-19 cases, the specificity is 98.21%, the sensitivity is 96.54% and F1 score is 94.11%. The specificity for pneumonia cases was 96.85%, the sensitivity was 93.07%, and the F1 score was 94.92%. For normal cases, the specificity was 98.92%, the sensitivity was 99.80%, and the F1 score was 98.45%. Compared with other systems proposed in this article, the specificity and sensitivity of normal cases and F1 score are higher, while the sensitivity of pneumonia is lower.

In Fig. 6, VGG-16, VGG-19, MobiNetV1 and ResNet50 are successively added to the ROC curve (AUC) in the system model proposed in this article. It can be seen from the image that the area under the ROC curve exceeds 96% for COVID-19, normal and pneumonia, and the AUC value of COVID-19 of VGG-16 reaches 99.93%.

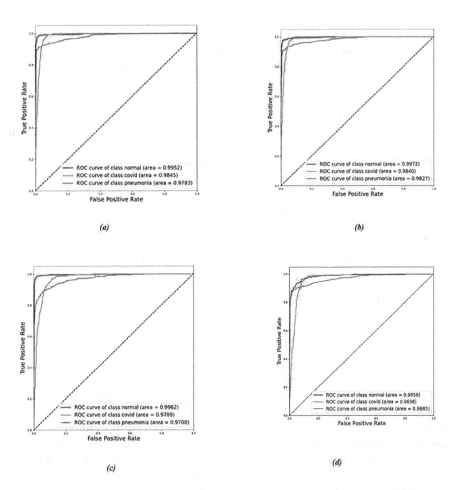

Fig. 6. ROC curve (a) VGG-16 (b) VGG-19 (c) MobileNet (d) Res Net50

The above experimental results show that the proposed CBAM and LSTM using CNN combined with attention mechanism have good performance. Among them, VGG-16 combined with the model proposed in this article obtains the best results. For COVID-19 cases, the specificity is 98.21%, the sensitivity is 96.54%, F1 score is 94.11%, and the AUC value reaches 99.93%. The main purpose of this experiment is to obtain a good effect in detecting COVID-19 cases, which is also confirmed in various indicators of the experimental results.

4   Discussion

As mentioned in Sect. 3, we use different backbone networks to conduct separate experiments on them, and presents the results in the form of performance indicators and confusion matrix. The model structure of VGG-16 combined with CBAM and LSTM obtained the highest accuracy, reaching 95.80%. The ROC curve in Fig. 6 also proves the advantages of VGG-16-CBAM-LSTM model compared with other networks tested in this article, and the model based on VGG-16 also shows better advantages in other performance indexes.

At the same time, this article compares the accuracy of the proposed system with some existing systems. The comparison results are shown in Table 4.

It can be seen from the chart that the proposed calculation model still has the disadvantage of low accuracy, and the data set used is relatively small. 6939 chest X-ray images are used in this article, which has the characteristics of rich and balanced. And in the architecture method proposed in this article, VGG-19, VGG-16, ResNet50 and MobileNetV1 are used as the convolution layer of feature extraction respectively. Their accuracy have reached more than 94% and have good scalability.

At the same time, we compare the system proposed in this article with the existing models under the unified data set, and the results are shown in Table 4. The VGG-16-CBAM-LSTM used in this article has a good accuracy of 95.80%, and has a good performance in the prediction of COVID-19.

5   Conclusion and Future Work

In this article, a system of CNN combining CBAM and LSTM in attention mechanism is proposed. VGG-19, VGG-16, ResNet50 and MobileNetV1 deep learning technologies are adopted on a large data set kaggle respectively. The 5-fold cross validation results show that VGG16-CBAM-LSTM combination is the best for detecting COVID-19 from chest X-ray images. The multi classification accuracy of this system is 95.80%, F1-score is 94.11%, sensitivity is 96.5% and specificity is 98.21%. The AUC value in ROC curve exceeds 0.98 for COVID-19, pneumonia and normal, and the whole network has good robustness. The system proposed in this article has good performance and can help medical practitioners reduce their workload. Under the current global epidemic, we hope to develop a method to reduce the pressure of doctors to fight the epidemic according to X-rays and contribute to the fight against the epidemic in terms of artificial intelligence.

References

1. Ahsan, M.M., Alam, T.E., Trafalis, T., Huebner, P.: Deep MLP-CNN model using mixed-data to distinguish between COVID-19 and non-COVID-19 patients. Symmetry **12**, 1526 (2020). https://doi.org/10.3390/sym12091526

2. Apostolopoulos, I.D., Mpesiana, T.A.: Covid-19: automatic detection from X-ray images utilizing transfer learning with convolutional neural networks. Phys. Eng. Sci. Med. **43**(2), 635–640 (2020). https://doi.org/10.1007/s13246-020-00865-4

3. Chen, G.: A gentle tutorial of recurrent neural network with error backpropagation (2016). https://doi.org/10.48550/arXiv.1610.02583

4. Esbin, M.N., Whitney, O.N., Chong, S., Maurer, A., Tjian, R.: Overcoming the bottleneck to widespread testing: a rapid review of nucleic acid testing approaches for covid-19 detection. RNA **26**(7), 771–783 (2020). https://doi.org/10.1261/rna.076232

5. Ghoshal, B., Tucker, A.: Estimating uncertainty and interpretability in deep learning for coronavirus (COVID-19) detection (2020). https://doi.org/10.48550/arXiv.2003.10769

6. Hemdan, E.D., Shouman, M.A., Karar, M.E.: Covidx-net: a framework of deep learning classifiers to diagnose COVID-19 in x-ray images (2020). https://doi.org/10.48550/arXiv.2003.11055

7. Li, X., Li, C., Zhu, D.: COVID-MobileXpert: On-device COVID-19 patient triage and follow-up using chest x-rays. arXiv e-prints (2020). https://doi.org/10.48550/arXiv.2004.03042

8. Lundervold, A.S., Lundervold, A.: An overview of deep learning in medical imaging focusing on MRI **29**(2), 102–127 (2019). https://doi.org/10.1016/j.zemedi.2018.11.002

9. Simonyan, K., Zisserman, A.: Very deep convolutional networks for large-scale image recognition. Computer Science (2014). https://doi.org/10.48550/arXiv.1409.1556

10. Song, Y., Zheng, S., Li, L., Zhang, X., Yang, Y.: Deep learning enables accurate diagnosis of novel coronavirus (COVID-19) with CT images. IEEE/ACM Trans. Comput. Biol. Bioinform. **PP**(99), 1 (2021). https://doi.org/10.1109/TCBB.2021.3065361

11. Wang, L., Wong, A.: Covid-net: a tailored deep convolutional neural network design for detection of COVID-19 cases from chest x-ray images (2020). https://doi.org/10.48550/arXiv.2003.09871

12. Wang, S., et al.: A deep learning algorithm using CT images to screen for Corona virus disease (COVID-19). Eur. Radiol. **31**(8), 6096–6104 (2021). https://doi.org/10.1007/s00330-021-07715-1

13. Wang, X., Peng, Y., Lu, L., Lu, Z., Bagheri, M., Summers, R.M.: Chestx-ray8: hospital-scale chest x-ray database and benchmarks on weakly-supervised classification and localization of common thorax diseases. IEEE, 3462–3471 (2017). https://doi.org/10.1109/CVPR.2017.369

14. Woo, S., Park, J., Lee, J.Y., Kweon, I.S.: CBAM: Convolutional Block Attention Module. Springer, Cham (2018). https://doi.org/10.1007/978-3-030-01234-2_1

Cooperative Positioning Enhancement for HDVs and CAVs Coexisting Environment Using Deep Neural Networks

Ao Zhang[1,2,3,4], Xuting Duan[1,2,3,4], Jianshan Zhou[1,2,3,4], Haiying Xia[5(✉)], Long Zhang[6], He Liu[7], and Wenjuan E[8,9]

[1] School of Transportation Science and Engineering, Beihang University, Beijing, China
[2] Beijing Advanced Innovation Center for Big Data and Brain Computing, Beijing, China
[3] Beijing Key Laboratory for Cooperative Vehicle Infrastructure Systems and Safety Control, Beijing, China
[4] National Engineering Laboratory for Comprehensive Transportation Big Data Application Technology (NEL-CTBD), Beijing, China
[5] Key Laboratory of Operation Safety Technology on Transport Vehicles, Ministry of Transport, PRC, No. 8, Xitucheng Road, Haidian District, Beijing 100088, China
hy.xia@rioh.cn
[6] National Key Laboratory of Science and Technology on Information System Security, Institute of Systems Engineering, AMS, PLA, Beijing, China
[7] Institute of Quartermaster Engineering and Technology, Institute of Systems Engineering, AMS, PLA, Beijing 100010, China
[8] School of Rail Transportation, Soochow University, Suzhou 215000, China
[9] Suzhou Automotive Research Institute, Tsinghua University, Suzhou 215000, China

Abstract. Accurate vehicle positioning is a key technology affecting traffic safety and travel efficiency. High precision positioning technology combined with the internet of vehicles (IoV) can improve the positioning accuracy of human-driving vehicles (HDVs), which is well suited for practical application requirements and resources saving. In this paper, a positioning error prediction model based on deep neural network (DNN) and positioning information sharing methods are proposed for traffic scenarios where connected and autonomous vehicles (CAVs) and HDVs with different positioning capabilities coexist. The CAVs with high precision positioning capability is utilized to share positioning information for

This research was supported in part by by the National Key Research and Development Program of China under Grant No. 2018YFB1600500, in part by the National Natural Science Foundation of China under Grant No. 62173012, U20A20155 and 52172339, in part by the Beijing Municipal Natural Science Foundation under Grant No. L191001, in part by the Newton Advanced Fellowship under Grant No. 62061130221, in part by the Project of HuNan Provinicial Science and Technology Department under Grant No. 2020SK2098 and 2020RC4048, in part by the CSUST Project under Grant No. 2019IC11.

Y. Tan et al. (Eds.): ICSI 2022, LNCS 13345, pp. 118–131, 2022.
https://doi.org/10.1007/978-3-031-09726-3_11

HDVs to enhance the cooperative positioning accuracy of vehicles with different positioning capabilities. Experimental results show the accuracy and timeliness of our proposal for enhancing vehicle positioning accuracy and sharing vehicle positioning information.

Keywords: Cooperative positioning · Deep neural network (DNN) · Accuracy enhancement · Positioning error

1 Introduction

Connected and autonomous vehicles (CAVs) and intelligent transportation systems (ITSs) have made breakthrough progress in order to solve the problems of efficiency, cost and safety in the transportation field. The accurate positioning technology of autonomous vehicles is the key technology to ensure the safety and travel efficiency of ITSs [1]. Global Navigation Satellite System (GNSS) is the most dominant technology for positioning and navigation signals can be used in most cases [2]. Nonetheless, despite the widespread use of GNSS positioning technology, accurate positioning in urban canyons and under heavy traffic remains challenging [3].

Under the influence of building occlusion and multi-path effect in urban areas [4], GNSS cannot meet the positioning accuracy requirements of intelligent vehicle infrastructure cooperative systems (IVICS) and automatic driving applications. The vehicular carrier-phase differential Global Navigation Satellite System (CDGNSS) can achieve better positioning performance in deep urban without the aid of inertial and odometry sensors [5], but it is still affected by the distance between the reference station and the user and the signal is unstable. In order to solve the problems of GNSS positioning, many researches are based on sensor technology [6], machine learning and IoV technology to improve the positioning accuracy of vehicles.

Autonomous vehicles can sense the environment through LiDAR and then match with the known 3D point clouds to estimate the position, which is a non-cooperative positioning method based on the vehicle's own sensor [7]. Based on vehicle-to-vehicle (V2V) and dedicated short-range communication (DSRC), the real-time relative position prediction method of adjacent vehicles is realized [8].

There are also some studies on vehicle cooperative positioning based on multi-agent reinforcement learning, such as the deep reinforcement learning, using decentralized scheduling algorithms to optimize a partially observable Markov decision process, can improve vehicle positioning accuracy [9]. El-Sheimy et al. proposed two multi-sensor information fusion schemes based on artificial neural networks (ANN) to reduce positioning system errors by utilizing multilayer feed-forward neural networks with a conjugate gradient training algorithm [10]. The cooperative positioning of CAVs with the same positioning ability was mainly studied, and the existing road measurement facilities were not fully utilized to form a systematic positioning system framework in the previous work. However, HDVs will still occupy a large proportion of the driving vehicles on the road.

In this paper, we focus on the cooperative positioning of vehicles with different positioning capabilities in the traffic scenario where CAVs and HDVs coexist. We first predict the positioning error of vehicles by constructing deep neural network (DNN) model, in which the vehicle motion state information and position information obtained by CAVs within the identification range of traffic signs are used as the training set of network model. We also propose three positioning error information sharing methods and compare their performance. Using the high accuracy positioning capability of CAVs, we provide positioning error information based on DNN for HDVs to enhance the cooperative positioning accuracy of vehicles with different positioning capabilities.

The rest of the paper is organized as follows: In Sect. 2, we show the system model and discusses assumptions that our work is based on. Following this positioning error prediction algorithm and positioning accuracy enhancement method are shown in Sect. 3, while Sect. 4 gives the numerical results of the simulation and discusses a lot. Finally, the conclusions are presented in Sect. 5.

2   System Model

In this section, we introduce the vehicle positioning scenario and the overall system architecture. At the same time, we analyze the feasibility of improving positioning accuracy by sharing vehicle positioning errors.

2.1   Positioning Scenarios

In this part, we give a detailed description of the vehicle positioning scenario and its road facilities.

- Road section description: Road sections with continuous characteristics are equipped with special traffic signs. Passing vehicles identify these special traffic signs through electronic signals (such as RFID) to determine the road and obtain the location of traffic signs.
- Vehicle description: There are two kinds of vehicles with different positioning capabilities in the vehicle positioning scenario, CAVs and HDVs. All vehicles are equipped with the same GNSS receiver. In addition to using GNSS to obtain positioning information, CAVs are equipped with other on-board sensors to assist in precise positioning. Therefore, CAVs can recognize the roadside traffic sign and obtain its coordinates, and then calculate its exact position through the relative distance and angle with sign. However, there is a large error in the positioning information obtained by GNSS for HDVs.
- Communication and computing capability: The vehicle can realize V2V and vehicle-to-infrastructure (V2I) communication through the DSRC communication module. Among them, CAVs have stronger computing power and larger data storage space while HDVs can only do some simple calculations. And All vehicles have access to the mobile edge computing nodes (MECNs) within communication range.

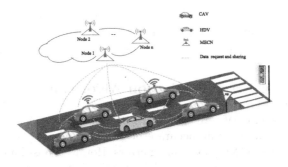

Fig. 1. System architecture of positioning error sharing.

Our overall system architecture is shown in Fig. 1. In this vehicle positioning scenario, we will solve the positioning error problem of two types of vehicles with different positioning capabilities respectively:

1) For CAVs, when the vehicle's turning is too large to recognize the traffic sign or the traffic sign is not continuous, the current position of the vehicle cannot be calculated by obtaining the coordinates of the traffic sign and the positioning error of the vehicle cannot be corrected.

2) For HDVs, large positioning errors always exist in the driving process because the vehicle positioning information is only obtained by GNSS.

3) For the positioning scene where vehicles with different positioning abilities coexist, the positioning scene is enhanced through the cooperative positioning method of high precision positioning vehicles and vehicles with positioning errors.

2.2 Feasibility Analysis of Sharing Positioning Error

Some researches divide error sources into systematic error and random error [11]. Then GNSS positioning error can be expressed as

$$E = E_s + E_r \tag{1}$$

where E_s is the systematic error and E_r is the random error. And the value of the random error is much less than the systematic error.

In the positioning scenario of this paper, the satellite clock difference and ionospheric delay which are the source of systematic error are basically the same under the condition that the satellite combination observed by different vehicles is basically the same in the similar time period and all vehicles are equipped with the same type of GNSS receiver. The position difference of two vehicles driving on the same road section is far less than the distance between vehicles and the satellite. For vehicle i and vehicle j that can share positioning error information, the systematic error between vehicles is considered to be approximately equal in this paper, that is, $E_{si} \approx E_{sj}$. We can get

$$\|\Delta E_{ij}\| \leq \|(E_{ri} - E_{rj})\| \tag{2}$$

In GNSS positioning error, the random error E_r is too small to participate in the calculation of positioning error compared with the systematic error E_s [12]. According to the above analysis, $E_i \approx E_j$ is considered to be established.

3   Positioning Error Prediction Algorithm

As mentioned above, CAVs are equipped with abundant sensors, which can recognize traffic signs and obtain accurate positioning information of the vehicles themselves to correct GNSS positioning errors. However, positioning errors need to be predicted for areas not covered by traffic signs or for HDVs that only obtain positioning information through GNSS.

In this section, we develop an algorithm for vehicle positioning error prediction with DNN. Furthermore, we use DNN algorithm to predict the positioning error information at the current time and discuss how to train the DNN.

3.1   Design of the DNN Algorithm

Deep learning models can not only simulate multilayer nonlinear mapping but also have high fault tolerance and adaptive ability [13]. In the vehicular positioning problem, there are many factors that affect the accuracy of positioning and the correlation of each factor is complex. Therefore, DNN with nonlinear problem-solving ability can be used to predict vehicular positioning error. The data obtained when CAV is within the identification range of traffic signs are used as DNN training data. The proposed DNN prediction algorithm runs on MECNs with powerful computing power and data storage space.

1) DNN structure: We use a fully connected neural network with multiple hidden layers and multiple hidden nodes to predict location errors. The number of nodes in each hidden layer is the same if the hidden layer is greater than 1. The DNN we designed as shown in Fig. 2.

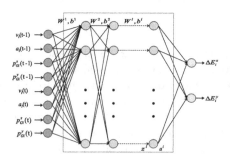

Fig. 2. Structure of DNN.

2) Input and output nodes: We mainly consider the motion state of the vehicle in the process of driving as the input node. There are 8 input nodes in

total, including the vehicle speed $v_i(t-1)$, acceleration $a_i(t-1)$, GNSS position $p_{Gi}^x(t-1)$ and $p_{Gi}^y(t-1)$ at the last moment; The vehicle speed $v_i(t)$, acceleration $a_i(t)$, GNSS position $p_{Gi}^x(t)$ and $p_{Gi}^y(t)$ at the current moment.

The output nodes are the positioning errors we need to predict, which are the positioning error in the X direction ΔE_i^x and the positioning error in the Y direction ΔE_i^y respectively.

3) Hidden layers and nodes: The number of hidden layers can affect the accuracy of network training while too many layers will also lead to time-consuming problems. Therefore, we need to balance network accuracy and time cost and the value of hidden layer is between the number of input nodes and the number of output nodes [14].

The number of neurons in each hidden layer is an important parameter affecting network performance. An appropriate number of neurons will enable the network to have excellent information processing ability without reducing training efficiency or even falling into local optimal. In this paper, we consider the number of training set samples, input nodes and output nodes to define the number of hidden layers nodes of the network that does not lead to overfitting [15]:

$$N = \frac{S}{\lambda * (I + O)} \tag{3}$$

where S is the number of training set samples. The number of input nodes and output nodes is represented by I and O. The constant λ is a scaling factor satisfies $\lambda \in [2,10]$.

As shown in Fig. 2, z^l represents the input vector of l'th layer, a^l represents the output vector of l'th layer and b^l represents the bias vector of l'th layer. Therefore, the relationship between them can be expressed as

$$z^l = W^l * a^{l-1} + b^l \tag{4}$$

where W is the weight of output and input between two hidden layers. By defining the activation function $\sigma(\cdot)$, we can get

$$\begin{aligned} a^l &= \sigma(z^l) \\ &= \sigma(W^l * a^{l-1} + b^l) \end{aligned} \tag{5}$$

3.2   Training Phase

In this section, the parameter learning and model training process of DNN are introduced. In order to reduce the dependency between parameters and prevent over-fitting, Linear rectification function (ReLU) was selected as the activation function of the hidden layer

$$\sigma(z) = \begin{cases} z & z \geq 0 \\ 0 & z < 0 \end{cases} \tag{6}$$

In the process of training, the cost function is constantly minimized. In order to be applicable to numerical prediction problems, we choose the mean square error (MSE) of output value and target value as the cost function, that is

$$J(W, b, x, y) = \frac{1}{2n} \sum_{i=1}^{n} (y_x - a_x^L)^2 \tag{7}$$

where y_x and a_x^L are the target value and output value of each sample respectively. There are L hidden layers in the network. W and b are the coefficient matrix and bias vector respectively. x and y are input vectors and output vectors respectively. According to (5), we can get the gradient of the output layer

$$\frac{\partial J}{\partial W^L} = \frac{\partial J}{\partial a^L} \frac{\partial a^L}{\partial z^L} \frac{\partial z^L}{\partial W^L} = (a^L - y) \circ \sigma'(z^L)(a^{L-1})^T \tag{8}$$

and

$$\frac{\partial J}{\partial b^L} = \frac{\partial J}{\partial a^L} \frac{\partial a^L}{\partial z^L} \frac{\partial z^L}{\partial b^L} = (a^L - y) \circ \sigma'(z^L) \tag{9}$$

where $\circ$ represents Hadamard product.

According to the forward propagation (4), we can obtain the gradient of the l-th

$$\frac{\partial J}{\partial W^l} = \delta^l (a^{l-1})^T \tag{10}$$

and

$$\frac{\partial J}{\partial b^l} = \delta^l \tag{11}$$

Combining (5) and considering the relationship between z^l and z^{l+1}, we can obtain

$$\frac{\partial J}{\partial b^l} = (W^{l+1})^T \delta^{l+1} \circ \sigma'(z^l) \tag{12}$$

and

$$\frac{\partial J}{\partial W^l} = (W^{l+1})^T \delta^{l+1} \circ \sigma'(z^l)(a^{l-1})^T \tag{13}$$

Therefore, the adjustment rules of W^l and b^l are as

$$W^l = W^l - \alpha * \delta^l (a^{l-1})^T \tag{14}$$

and

$$b^l = b^l - \alpha * \delta^l \tag{15}$$

where α is learning rate.

3.3   Positioning Accuracy Enhancement Method

In this paper, CAVs and HDVs use the positioning information of CAVs in the identification range of traffic signs as training sets to predict the current positioning error of vehicles. The positioning accuracy of HDVs is so low that it needs to be enhanced by positioning error information provided by CAVs. The following three methods are proposed to enhance the positioning accuracy of vehicles using the precise positioning information provided by CAVs, as shown in Fig. 3. The workflow of the proposed positioning accuracy enhancement method is as follows:

- Sharing current positioning error (SPE) : For the vehicle i, the positioning error of other J vehicles at the current moment is directly used and the average value is taken as itself current positioning error. Thus, The positioning error E_i can be expressed as $\Delta E_i = \frac{1}{J} \sum_{i=1}^{J} \Delta E_j$.
- Sharing positioning error data (SPD) : The positioning information data of CAVs are uploaded to edge nodes as a training set to train DNN model to predict positioning errors. The HDVs send requests to the MECNs within the communication range to obtain the trained DNN model and predict vehicle positioning errors by combining their own GNSS positioning and vehicle motion state information.
- Sharing error prediction model (SPM) : The model parameters and positioning errors of J vehicles were averaged on the MECN and a hidden layer was added to the DNN model to continue the training with the transfer learning strategy, so as to obtain new error evolution to predict the positioning errors of other vehicles. HDVs can send a request to the nearby MECN, and then can use the trained DNN prediction model to predict the vehicle positioning error combined with their own GNSS positioning and vehicle motion state information.

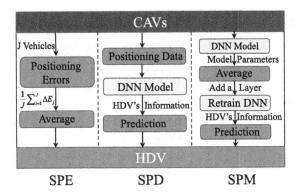

Fig. 3. The methods of positioning accuracy enhancement.

4 Simulation and Results

In this section, a series of experiments are carried out to verify the feasibility and superiority of our proposed positioning error prediction algorithm and sharing method. Firstly, the accuracy of the proposed DNN-based error prediction algorithm is evaluated. Then, the improvement of positioning accuracy of a single CAV through the above three sharing methods is compared and the influence of the number of shared vehicles on positioning results is evaluated. Finally, the robustness and timeliness of the proposed method are discussed.

We use data from the publicly available NGSIM data set for our experiments [16]. The data set includes the speed, acceleration and position information of the vehicle when driving on the road. The data in the data set is collected and recorded by 8 synchronous cameras with relatively accurate location information, which can be regarded as the data obtained by CAVs after accurate positioning by recognizing traffic signs. By reference [17,18], this paper sets the GNSS positioning system error E_s as $10m$, and the random error follows the Gaussian distribution $E_r \sim N(0,4)$. According to this rule, we will add system error and random error to the original data, and the processed data will be used as the positioning information of HDVs.

The vehicle trajectory data is selected from the data set as the positioning error information to share and its parameter requirements meet the requirements in Table 1. According to Sect. 3.1, the number of hidden layers is between 2 and 8. The number of neurons in each hidden layer is calculated according to (3). By comparing with the simulation experiments, it is found that the convergence and function loss are better when the number of hidden layers is set to 4 and the number of hidden layers is 5 after adding a hidden layer in SPM method in this paper. The specific parameter settings of DNN are shown in Table 2.

Based on the DNN model constructed in Sect. 3.1, we predict the vehicle positioning error and use the vehicle positioning error sharing method proposed

Table 1. Data selected

Parameters	Values
Maximum speed (m/s)	30
Acceleration (m/s^2)	$[-4, 4]$
Sampling frequency (Hz)	10
Number of shared vehicles	$[1, 4]$

Table 2. DNN settings

Parameters	Values
Input nodes	8
Output nodes	2
Hidden layers	4, 5
Hidden layer nodes	50, 100, 150, 200
Scaling factor	2
Learning rate	0.005
Batch size	32
Loss function	MSE
Optimization	Stochastic
Algorithm	Gradient descent [19]

in Sect. 3.3 to correct the vehicle positioning error. The results of positioning error correction by three methods are discussed in the following when the number of shared vehicles is different.

For convenience of comparison, only the first 80 sampling points are extracted here. The positioning error result after the positioning accuracy enhancement by SPE is shown in Fig. 4. The three broken lines in Fig. 4 are the real positioning error of the vehicle, the vehicle positioning error corrected by using the shared positioning information of one vehicle and the vehicle positioning error corrected by using the shared positioning information of two vehicles. The positioning accuracy enhancement effect is not obvious, so we stopped the experiment when we added two vehicles. The reason for this is that the positioning accuracy enhancement method based on SPE directly uses the error information provided by other vehicles, but does not combine with the vehicle motion state to improve its own positioning accuracy, thus affecting the performance of positioning accuracy enhancement.

Compared with SPE, SPD and SPM re-estimate the positioning error prediction model based on the vehicle's own data and information shared by other vehicles, which is closer to the situation of the vehicle itself, as shown in Fig. 5 and Fig. 6. The four broken lines in Fig. 5 and Fig. 6 are the real positioning error of the vehicle and the vehicle positioning error corrected by using the shared positioning information of one to three vehicles. SPD and SPM have similar positioning accuracy enhancement performance. The positioning accuracy of vehicles with low positioning ability can be enhanced by sharing positioning error information with other vehicles and correcting positioning error. However, there are still differences between vehicles, leading to small positioning errors after enhanced positioning accuracy. It can be seen from the two figures that the fluctuation of broken line after the error correction of SPD method is larger than that of SPM method, indicating that the error correction effect of SPM method is more stable.

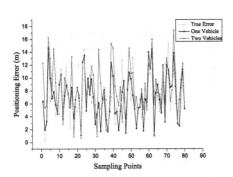

Fig. 4. The positioning accuracy enhancement of different shared vehicle numbers based on SPE.

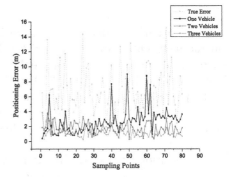

Fig. 5. The positioning accuracy enhancement of different shared vehicle numbers based on SPD.

The number of vehicles sharing positioning error information increases from 1 to 3. With the increase of the number of vehicles, the positioning error gradually decreases, that is, the performance of the positioning accuracy enhancement method becomes more obvious. In order to further compare SPD and SPM, we compared the timeliness of enhanced positioning accuracy under the same condition that 4 vehicles jointly provided positioning error information, as shown in Fig. 7.

It can be seen that the loss function value of SPM decreases faster than SPD, indicating that SPM has a higher training efficiency. This is because for SPD, the vehicle needs to retrain the DNN model based on the data shared by itself and other vehicles. For SPM, ordinary vehicles continue training after adding a hidden layer on the basis of other vehicle training network architecture. Since the previous network model has been trained and the network parameters have been determined, the loss function can be reduced more quickly to save time.

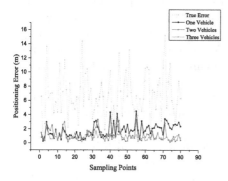

Fig. 6. The positioning accuracy enhancement of different shared vehicle numbers based on SPM.

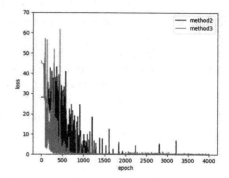

Fig. 7. The loss function decline curves of SPD and SPM.

In addition, in order to more intuitively compare the positioning accuracy enhancement effects of SPD and SPM, When the number of shared vehicles is set to 4, the positioning accuracy enhancement effect in the X and Y directions is shown in the Fig. 8 and Fig. 9. And the prediction results of positioning error in the process of positioning accuracy enhancement based on 4 shared vehicles are summarized in the in Table 3, which are the maximum, minimum, mean and MSE of the error between the real value and predicted value of SPD and SPM respectively.

The positioning errors in the X and Y directions become smaller after correction, that is, the positioning accuracy is significantly enhanced. As can be seen, the mean square error of each method is 0.9935 and 0.5305 respectively, indicating that positioning error of vehicles can be accurately predicted by SPM. The minimum value indicates that the predicted result is very similar to the

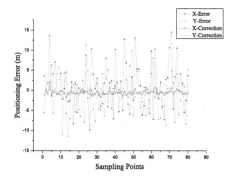 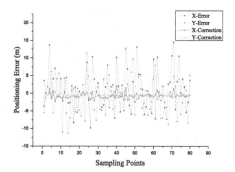

Fig. 8. The positioning accuracy enhancement of x direction and y direction based on SPD.

Fig. 9. The positioning accuracy enhancement of x direction and y direction based on SPM.

Table 3. Error prediction performant of different cases

	SPD	SPM
Maximum (m)	2.690	1.8571
Minimum (m)	0.2425	0.1686
Mean (m)	1.1883	0.8943
MSE (m^2)	0.9935	0.5305

actual error, which indicates that it is feasible to enhance the vehicle positioning accuracy by sharing the positioning error information.

The comparison of the three methods shows that the SPE does not perform well in enhancing the positioning accuracy because it does not combine the vehicle's own motion state and other information. SPD and SPM have similar positioning accuracy enhancement performance. When the number of shared vehicles is 4, the positioning error based on SPD can be reduced by 82.48% while that based on SPM can be reduced by 86.81%. Both of them can significantly reduce the positioning error of vehicles and enhance the positioning accuracy of vehicles. Among them, SPM is better than SPD in timeliness because it considers the transfer learning strategy and uses the trained model to continue training with its own data to make the network convergence faster. In addition, our shared information is the positioning error of the vehicle rather than the location information of the vehicle, which can protect the privacy of the vehicle and avoid some security problems.

5 Conclusions

This paper addressed the problem of enhancing the cooperative positioning accuracy of vehicles by combining the existing high-precision positioning technology and IoV technology in the traffic positioning scene where vehicles with different positioning abilities coexist in accordance with the actual situation. We build a vehicle positioning error prediction model based on DNN and proposed vehicle positioning error information sharing methods. Among them, SPD and SPM have better performance in enhancing positioning accuracy, while SPM has better convergence effect with the help of transfer learning strategy. Although highly preliminary, we believe that the proposal is promising in terms of enhanced positioning accuracy and error information sharing. The results provide many perspectives for future research on the enhancement of positioning accuracy of vehicles with different positioning abilities.

References

1. Liu, Q., Liang, P., Xia, J., Wang, T., Song, M., Xu, X.: A highly accurate positioning solution for C-V2X systems. Sensors **21**(4), 1175 (2021)
2. Nowak, A.: Dynamic GNSS mission planning using DTM for precise navigation of autonomous vehicles. J. Navig. **70**(3), 483C504 (2017)
3. Chen, S., Hu, J., Shi, Y., Zhao, L., Li, W.: A vision of C-V2X: technologies, field testing, and challenges with chinese development. IEEE Internet Things J. **7**(5), 3872–3881 (2020)
4. Rohani, M., Gingras, D., Gruyer, D.: A novel approach for improved vehicular positioning using cooperative map matching and dynamic base station DGPS concept. IEEE Trans. Intell. Transp. Syst. **17**(1), 230–239 (2016)
5. Humphreys, T., Murrian, M.J., Narula, L.: Deep-urban unaided precise global navigation satellite system vehicle positioning. IEEE Intell. Trans. Syst. Mag. **PP**(99), XX1 (2020)
6. Suhr, J.K., Jang, J., Min, D., Jung, H.G.: Sensor fusion-based low-cost vehicle localization system for complex urban environments. IEEE Trans. Intell. Trans. Syst. **18**(5), 1078–1086 (2017)
7. Javanmardi, E., Javanmardi, M., Gu, Y., Kamijo, S.: Autonomous vehicle self-localization based on multilayer 2D vector map and multi-channel LiDAR. In: 2017 IEEE Intelligent Vehicles Symposium (IV). IEEE (2017)
8. Ansari, K.: Cooperative position prediction: beyond vehicle-to-vehicle relative positioning. IEEE Trans. Intell. Transp. Syst. **21**(3), 1–10 (2019)
9. Peng, B., Seco-Granados, G., Steinmetz, E., Frohle, M., Wymeersch, H.: Decentralized scheduling for cooperative localization with deep reinforcement learning. IEEE Trans. Vehi. Technol. **PP**(5), 1 (2019)
10. El-Sheimy, N., Chiang, K.W., Noureldin, A.: The utilization of artificial neural networks for multisensor system integration in navigation and positioning instruments. IEEE Trans. Instrum. Meas. **55**(5), 1606–1615 (2006)
11. Grewal, M.S., Weill, L.R., Andrews, A.P.: Global Positioning Systems, Inertial Navigation, and Integration. Wiley Interdisciplinary Reviews Computational Statistics (2007)
12. Skog, I., Handel, P.: In-car positioning and navigation technologies-a survey. IEEE Trans. Intel. Transp. Syst. **10**(1), 4–21 (2009). https://doi.org/10.1109/TITS.2008.2011712
13. Wei, Z., Peng, G., Li, C., Chen, Y., Zhang, Z.: A new deep learning model for fault diagnosis with good anti-noise and domain adaptation ability on raw vibration signals. Sensors **17**(3), 425 (2017)

14. Goodfellow, I., Bengio, Y., Courville, A.: Deep Learning. MIT Press, Cambridge (2016). http://www.deeplearningbook.org
15. Hagan, M.T., Demuth, H.B., Beale, M.H., De Jesús, O.: Neural Network Design. Martin Hagan, Cambridge (2014)
16. Colyar, J., Halkias, J.: US highway 101 dataset. Federal Highway Administration (FHWA), Technical report FHWA-HRT-07-030 (2007)
17. Nam, S., Lee, D., Lee, J., Park, S.: CNVPS: cooperative neighboring vehicle positioning system based on vehicle-to-vehicle communication. IEEE Access 7, 16847–16857 (2019). https://doi.org/10.1109/ACCESS.2019.2894906
18. Soatti, G., Nicoli, M., Garcia, N., Denis, B., Raulefs, R., Wymeersch, H.: Implicit cooperative positioning in vehicular networks. IEEE Trans. Intell. Trans. Syst. **PP**(99), 1–17 (2017)
19. Bottou, L.: Stochastic gradient descent tricks. In: Montavon, G., Orr, G.B., Müller, K.-R. (eds.) Neural Networks: Tricks of the Trade. LNCS, vol. 7700, pp. 421–436. Springer, Heidelberg (2012). https://doi.org/10.1007/978-3-642-35289-8_25

Stability Analysis of Stochastic Neutral Hopfield Neural Networks with Multiple Time-Varying Delays

Yongcai Li[1], Shengbing Xu[2(✉)], and Jiqiang Feng[1]

[1] Department of Mathematics and Statistics, Shenzhen Institute of Computing Sciences, Shenzhen University, Shenzhen 518060, China
fengjq@szu.edu.cn
[2] School of Computer and Information, Dongguan City College, Dongguan 523401, China
1900201003@email.szu.edu.cn

Abstract. Neutral Hopfield neural networks have widely used in optimization problems while its stability analysis has received a great deal of attention. This paper investigates the problem of the global asymptotic stability of stochastic neutral Hopfield neural networks with multiple time-varying delays. Different form the previous reported results, the neural networks are affected by not only stochastic perturbations, but also the time delays including discrete, distributed and neutral types which are a variety of functions related to neural nodes. By constructing a novel Lyapunov functional and using stochastic analysis techniques, we derive the sufficient criteria of the global asymptotic stability of the networks. Finally, some numerical simulations are given to verify the effectiveness of the theoretical results.

Keywords: Hopfield neural networks · Stability · Multiple time-varying delays

1 Introduction

Neural network model was first proposed by scholars Culloch and Pitts imitating the structural characteristics and working principle of biological neurons [1]. Due to the needs of engineering application and theory, a variety of neural network models are designed, such as Hopfield neural networks [2], convolutional neural networks [3], cell neural networks [4], etc. Because of its powerful nonlinear mapping ability and parallel processing ability, neural networks are now widely used in intelligent optimization, image analysis, intelligent computing, machine learning and other fields [5–7]. Some engineering applications often involve the stability of neural networks. For example, when using neural networks to handle the optimization objective problem, the network states must converge to a constant solution independent of the initial conditions, which means that the unique equilibrium point of the network is required to be stable in a certain range [8]. Stability means the system to return to the equilibrium state after deviating from the equilibrium state, which reflect the system's ability to resist small external disturbance [9].

© Springer Nature Switzerland AG 2022
Y. Tan et al. (Eds.): ICSI 2022, LNCS 13345, pp. 132–142, 2022.
https://doi.org/10.1007/978-3-031-09726-3_12

When scholars analyze the stability of neural networks, they find that time delays play an important part. The signal propagation is not instantaneous due to the extension of neurons in space, which leads to time delays in the network and cause network instability or oscillation [10]. Moreover, the change of neural network state is not only related to the current state of neurons, but also depends on the previous state of neurons. If the neural network has both the current state derivative and the past state derivative of neurons, this type of network is called the neutral neural network [11]. Neutral neural networks with time delays have complete delay characteristics, which are more suitable for the analysis of practical problems. In addition, environmental noise inevitably exists in various systems, which may also lead to system instability. Recently, the stability analysis of stochastic neutral neural networks with time delays has attracted great attention [12].

At present, scholars have come up with a variety of neural networks, of which the typical representative is Hopfield neural networks. Hopfield neural network was first designed and implemented by Hopfield and John [2]. Nowadays, Hopfield neural network is widely used in various fields such as intelligent computing, optimization calculation, super-resolution recognition of remote sensing images, suppression of communication interference and so on [13]. For example, application of arbitrary order Hopfield neural networks to optimization problems was studied by doing experiments about solving Diophantine equation, Hamiltonian cycle and k-colorability problems [14]. Up to now, many scholars have put forward a large number of stability theories about Hopfield neural networks. Faydasicokd deduced the stability condition expressed by the parameters of the network model, which is completely independent of time delay and neutral time delay [9]. Arik studied the system and multiple delays, and obtained the stability conditions independent of delay parameters [11]. Xiao studied the stochastic stability of neutral systems with fractional Brownian motion and Markov jump [23]. It can refer to [13–23] for more details of stability achievement. However, previous studies are often based on linear matrix inequality (LMI) method, which cannot deal with multi delay problems related to nodes. In addition, some studies are limited to fixed constant delays for the most important time delays.

In this paper, the Hopfield neural networks are affected by not only stochastic per-turbations, but also the time-varying delays including discrete, distributed and neutral types related to neural nodes. By constructing a novel Lyapunov functional avoiding the use of LMI method, with the help of stochastic analysis techniques, we derive the stability criteria of the networks. Finally, some numerical simulations are given to verify the effectiveness of the theoretical results.

2 Stability Analysis

Throughout this paper, the superscript 'T' denotes transposition of a vector or a matrix, $C([-\Xi, 0], R)$ denotes all continuous functions from $[-\Xi, 0]$ to R. $\|\cdot\|$ represents 1-norm of vector or a matrix while $\|\cdot\|_2$ represents Euclidean norm. Let $N = \{1, 2, ..., N\}$ The model of stochastic neutral Hopfield neural network systems with multiple time-varying delays is mathematically expressed as:

$$
\begin{aligned}
d&\left(x_i(t) + \sum_{j=1}^{n} e_{ij}x_i\big(t - \tau_{ij}(t)\big)\right) \\
&= \left(\begin{array}{l} -c_ix_i(t) + \sum_{j=1}^{n} a_{ij}G_j\big(x_j(t)\big)+ \\ \sum_{j=1}^{n} b_{ij}G_j\big(x_j\big(t - \mu_{ij}(t)\big)\big) + \sum_{j=1}^{n} o_{ij} \int_{t-v_{ij}(t)}^{t} H_j\big(x_j(s)\big)ds \end{array}\right) dt \\
&\quad + \left(\alpha_i f_i(x_i(t - \mu_i(t))) + \sum_{j=1}^{n} \beta_{ij}M_{ij}\big(x_j(t)\big)\right)dB(t), i \in N
\end{aligned}
\tag{1}
$$

where $x_i(t)$ represents the state of the ith neuron. Constant c_i is a real positive number. The constant elements $a_{ij}, b_{ij}, o_{ij}, \alpha_i$ and β_{ij} represent the values of neuron interconnections, and e_{ij} represents neutral weight coefficient. Functions $G_j\big(x_j(t)\big), H_j\big(x_j(t)\big), f_i(x_i(t))$ and $M_{ij}\big(x_j(t)\big)$ are nonlinear functions determining the relationships between the states and the outputs of neurons. And $B(t)$ is 1-dimensional standard Brownian motion. Delay function $\tau_{ij}(t), \mu_{ij}(t)$, and $\mu_i(t)$ are time-varying functions satisfying specific conditions. Let $\Xi = \max\{\tau_{ij}(t), \mu_{ij}(t), v_{ij}(t), \mu_i(t)\}, i \in N$. The initial conditions of neutral type neural network (1) are described by $x_i(t) = \varphi_i(t)$ which are included in $C([-\Xi, 0], R)$.

For the stochastic neutral Hopfield neural network system (1), suppose that Hypotheses 1–3 are always satisfied.

Hypothesis 1. Functions $G_j\big(x_j(t)\big), H_j\big(x_j(t)\big), f_i(x_i(t))$ and $M_{ij}\big(x_j(t)\big)$ satisfy the Lipschitz condition, that is, for $\forall, i, j = 1, 2, \cdots, n$, there are some numbers g_i, h_i, κ_i and θ_{ij} such that

$$
G_i^2(x_i(t)) \le g_i^2 x_i^2(t), H_i^2(x_i(t)) \le h_i^2 x_i^2(t), f_i^2(x_i(t)) \le \kappa_i^2 x_i^2(t), M_{ij}^2(x_i(t)) \le \theta_{ij}^2 x_i^2(t)
$$

Hypothesis 2. There exist normal numbers μ_i, τ_{ij} and μ_{ij} such that

$$
\mu_i < \dot{\mu}_i(t) < 1, \tau_{ij} < \dot{\tau}_{ij}(t) < 1, \mu_{ij} < \dot{\mu}_{ij}(t) < 1, i, j = 1, 2, \cdots, n.
$$

Hypothesis 3. Network (1) has only one equilibrium point.

Based on above assumptions, we will analyze the global asymptotic stability of the network (1). In this paper, we use the following expressions for the sake of brevity:

$$
\zeta_i = \sum_{j=1}^{n} \left(|a_{ij}|+|b_{ij}|+|c_ie_{ij}|+|o_{ij}v_{ij}h_j|+|o_{ji}v_{ji}h_i| + \frac{|c_je_{ji}|}{1 - \tau_{ji}}\right),
$$

$$
\xi_{ijk} = |e_{ij}a_{ik}| + |e_{ij}b_{ik}| + |e_{ij}o_{ik}|, \Omega_i = \sum_{j=1}^{n}\sum_{k=1}^{n}\left(\frac{\xi_{jik}}{1 - \tau_{ji}} + \left|e_{ki}o_{ki}v_{ki}h_i^2\right|\right).
$$

Theorem 1. *The stochastic neutral Hopfield neural network system (1) is globally asymptotically stable if the parameters of system satisfy the following inequalities:*

$$A1.\ \rho_i = 2c_i - \zeta_i' - \Omega_i' \geq 0, \quad A2.\ \delta_i = 1 - \sum_{j=1}^{n} |e_{ji}| > 0, \ i = 1, 2, \cdots, n,$$

where $\zeta_i' = \zeta_i + \sum_{j=1}^{n}(\beta_{ji}^2\theta_{ji}^2 + |a_{ji}g_i^2| + \frac{|b_{ji}g_i^2|}{1-\mu_{ji}} + \frac{\alpha_i^2\kappa_i^2}{1-\mu_i})$,

$$\Omega_i' = \Omega_i + \sum_{j=1}^{n}\sum_{k=1}^{n}|e_{kj}a_{ki}g_i^2| + \sum_{j=1}^{n}\sum_{k=1}^{n}\frac{|e_{kj}b_{ki}g_i^2|}{1-\mu_{ki}}.$$

Proof: Constructing the following Lyapunov functional as

$$V_i(x_i(t), t) = (x_i(t) + \sum_{j=1}^{n}e_{ij}x_j(t - \tau_{ij}(t)))^2 + \sum_{j=1}^{n}\frac{|c_ie_{ij}|}{1-\tau_{ij}}\int_{t-\tau_{ij}(t)}^{t}x_j^2(s)ds$$

$$+ \sum_{j=1}^{n}\frac{|b_{ij}|}{1-\mu_{ij}}\int_{t-\mu_{ij}(t)}^{t}G_j^2(x_j(s))ds + \sum_{j=1}^{n}\sum_{k=1}^{n}\frac{\xi_{ijk}}{1-\tau_{ij}}\int_{t-\tau_{ij}(t)}^{t}x_j^2(s)ds$$

$$+ \sum_{j=1}^{n}\sum_{k=1}^{n}\frac{|e_{ij}b_{ik}|}{1-\mu_{ik}}\int_{t-\mu_{ik}(t)}^{t}G_k^2(x_k(s))ds + \frac{|\alpha_i^2|}{1-\mu_i}\int_{t-\mu_i(t)}^{t}f_i^2(x_i(s))ds \quad (2)$$

Taking the stochastic differential operator $\mathcal{L}V_i(x_i(t), t)$ in the Lyapunov functional (2) along the trajectories of neutral system (1) yields:

$$\mathcal{L}V_i(x_i(t), t)$$

$$= 2(x_i(t) + \sum_{j=1}^{n}e_{ij}x_j(t - \tau_{ij}(t)))(-c_ix_i(t)$$

$$+ \sum_{j=1}^{n}a_{ij}G_j(x_j(t)) + \sum_{j=1}^{n}b_{ij}G_j(x_j(t - \mu_{ij}(t)))$$

$$+ \sum_{j=1}^{n}o_{ji}\int_{t-v_{ij}(t)}^{t}H_j(x_j(s))ds) + \sum_{j=1}^{n}\frac{|c_ie_{ij}|}{1-\tau_{ij}}x_j^2(t)$$

$$- \sum_{j=1}^{n}\frac{|c_ie_{ij}|(1 - \dot{\tau}_{ij}(t))}{1-\tau_{ij}}x_j^2(t - \tau_{ij}(t))$$

$$+ \sum_{j=1}^{n}\frac{|b_{ij}|}{1-\mu_{ij}}G_j^2(x_j(t)) - \sum_{j=1}^{n}\frac{|b_{ij}|(1 - \dot{\mu}_{ij}(t))}{1-\mu_{ij}}G_j^2(x_j(t - \mu_{ij}(t)))$$

$$+ \sum_{j=1}^{n}\sum_{k=1}^{n}\frac{\xi_{ijk}}{1-\tau_{ij}}x_j^2(t) - \sum_{j=1}^{n}\sum_{k=1}^{n}\frac{\xi_{ijk}(1 - \dot{\tau}_{ij}(t))}{1-\tau_{ij}}x_j^2(t - \tau_{ij}(t))$$

$$+ \sum_{j=1}^{n}\sum_{k=1}^{n}\frac{|e_{ij}b_{ik}|}{1-\mu_{ik}}G_k^2(x_k(t)) - \sum_{j=1}^{n}\sum_{k=1}^{n}\frac{|e_{ij}b_{ik}|(1 - \dot{\mu}_{ik}(t))}{1-\mu_{ik}}G_k^2(x_k(t - \mu_{ik}(t)))$$

$$+ \frac{|\alpha_i^2|}{1-\mu_i} f_i^2(x_i(t)) - \frac{|\alpha_i^2|(1-\dot{\mu}_i(t))}{1-\mu_i} f_i^2(x_i(t-\mu_i(t))) + trace[(\alpha_i f_i(x_i(t$$

$$- \mu(t))) + \sum_{j=1}^{n} \beta_{ij} M_{ij}(x_j(t)))^T (\alpha_i f_i(x_i(t-\mu_i(t)))$$

$$+ \sum_{j=1}^{n} \beta_{ij} M_{ij}(x_j(t)))]$$

$$\leq -2c_i x_i^2(t) + 2x_i(t) \sum_{j=1}^{n} a_{ij} G_j(x_j(t)) + 2x_i(t) \sum_{j=1}^{n} b_{ij} G_j(x_j(t-\mu_{ij}(t))) +$$

$$2x_i \sum_{j=1}^{n} o_{ij} \int_{t-v_{ij}(t)}^{t} H_j(x_j(s))ds - 2c_i x_i(t) \sum_{j=1}^{n} e_{ij} x_j(t-\tau_{ij}(t)) +$$

$$2(\sum_{j=1}^{n} e_{ij} x_j(t-\tau_{ij}(t)))(\sum_{j=1}^{n} a_{ij} G_j(x_j(t))) + 2(\sum_{j=1}^{n} e_{ij} x_j(t-$$

$$\tau_{ij}(t)))(\sum_{j=1}^{n} b_{ij} G_j(x_j(t-\mu_{ij}(t)))) + 2(\sum_{j=1}^{n} e_{ij} x_j(t-$$

$$\tau_{ij}(t)))\left(\sum_{j=1}^{n} o_{ij} \int_{t-v_{ij}(t)}^{t} H_j(x_j(s))ds\right) + \sum_{j=1}^{n} \frac{|c_i e_{ij}|}{1-\tau_{ij}} x_j^2(t) - \sum_{j=1}^{n} |c_i e_{ij}| x_j^2(t-$$

$$\tau_{ij}(t)) + \sum_{j=1}^{n} \frac{|b_{ij}|}{1-\mu_{ij}} G_j^2(x_j(t)) - \sum_{j=1}^{n} |b_{ij}| G_j^2(x_j(t-\mu_{ij}(t))) +$$

$$\sum_{j=1}^{n} \sum_{k=1}^{n} \frac{\xi_{ijk}}{1-\tau_{ij}} x_j^2(t) - \sum_{j=1}^{n} \sum_{k=1}^{n} \xi_{ijk} x_j^2(t-\tau_{ij}(t)) +$$

$$\sum_{j=1}^{n} \sum_{k=1}^{n} \frac{|e_{ij} b_{ik}|}{1-\mu_{ik}} G_k^2(x_k(t)) - \sum_{j=1}^{n} \sum_{k=1}^{n} |e_{ij} b_{ik}| G_k^2(x_k(t-\mu_{ik}(t))) +$$

$$\frac{|\alpha_i^2|}{1-\mu_i} f_i^2(x_i(t)) - |\alpha_i^2| f_i^2(x_i(t-\mu_i(t))) + \alpha_i^2 f_i^2\left(x_i(t-\mu_i(t)) + \sum_{j=1}^{n} \beta_{ij}^2 M_{ij}^2(x_j(t))\right)$$

In addition, we can derive the following inequalities:

$$\sum_{i=1}^{n} 2x_i \sum_{j=1}^{n} o_{ij} \int_{t-v_{ij}(t)}^{t} H_j(x_j(s))ds \leq \sum_{i=1}^{n} 2 \sum_{j=1}^{n} |o_{ij} v_{ij} h_j| |x_i(t)x_j(t)|$$

$$\leq \sum_{i=1}^{n} \sum_{j=1}^{n} |o_{ij} v_{ij} h_j| x_i^2(t) + \sum_{i=1}^{n} \sum_{j=1}^{n} |o_{ij} v_{ij} h_j| x_j^2(t)$$

$$\leq \sum_{i=1}^{n} \sum_{j=1}^{n} |o_{ij} v_{ij} h_j| x_i^2(t) + \sum_{i=1}^{n} \sum_{j=1}^{n} |o_{ji} v_{ji} h_i| x_i^2(t) \qquad (3)$$

$$= \sum_{i=1}^{n} \sum_{j=1}^{n} (|o_{ij} v_{ij} h_j| + |o_{ji} v_{ji} h_i|) x_i^2(t)$$

$$\sum_{i=1}^{n} 2\left(\sum_{j=1}^{n} e_{ij} x_j(t-\tau_{ij}(t))\right)\left(\sum_{j=1}^{n} o_{ij} \int_{t-v_{ij}(t)}^{t} H_j(x_j(s))ds\right)$$

$$\leq \sum_{i=1}^{n} \sum_{j=1}^{n} \sum_{k=1}^{n} |e_{ij} o_{ik}| x_j^2(t-\tau_{ij}(t)) + \sum_{i=1}^{n} \sum_{j=1}^{n} \sum_{k=1}^{n} |e_{ik} o_{ik}| \int_{t-v_{ik}(t)}^{t} H_k^2(x_k(s))ds$$

$$\leq \sum_{i=1}^{n}\sum_{j=1}^{n}\sum_{k=1}^{n}|e_{ij}o_{ik}|x_j^2(t-\tau_{ij}(t)) + \sum_{i=1}^{n}\sum_{j=1}^{n}\sum_{k=1}^{n}\left|e_{ik}o_{ik}v_{ik}h_k^2\right|x_k^2(t)$$

$$\sum_{i=1}^{n}2\sum_{j=1}^{n}e_{ij}x_j(t-\tau_{ij}(t))\sum_{j=1}^{n}a_{ij}G_j(x_j(t)) \tag{4}$$

$$\leq 2\sum_{i=1}^{n}\sum_{j=1}^{n}\sum_{k=1}^{n}|e_{ij}a_{ik}x_j(t-\tau_{ij}(t))G_k(x_k(t))|$$

$$\leq \sum_{i=1}^{n}\sum_{j=1}^{n}\sum_{k=1}^{n}|e_{ij}a_{ik}|x_j^2(t-\tau_{ij}(t)) + \sum_{i=1}^{n}\sum_{j=1}^{n}\sum_{k=1}^{n}|e_{ij}a_{ik}|G_k^2(x_k(t)) \tag{5}$$

$$\sum_{i=1}^{n}2\sum_{j=1}^{n}e_{ij}x_j(t-\tau_{ij}(t))\sum_{j=1}^{n}b_{ij}G_j(x_j(t-\mu_{ij}(t)))$$

$$= \sum_{i=1}^{n}2\sum_{j=1}^{n}e_{ij}x_j(t-\tau_{ij}(t))\sum_{k=1}^{n}b_{ik}G_k(x_k(t-\mu_{ik}(t)))$$

$$\leq 2\sum_{i=1}^{n}\sum_{j=1}^{n}\sum_{k=1}^{n}|e_{ij}b_{ik}x_j(t-\tau_{ij}(t))G_k(x_k(t-\mu_{ik}(t)))|$$

$$\leq \sum_{i=1}^{n}\sum_{j=1}^{n}\sum_{k=1}^{n}|e_{ij}b_{ik}|x_j^2(t-\tau_{ij}(t)) + \sum_{i=1}^{n}\sum_{j=1}^{n}\sum_{k=1}^{n}|e_{ij}b_{ik}|G_k^2(x_k(t-\mu_{ik}(t))) \tag{6}$$

Thus, inserting (3)–(6) into $\mathcal{L}V_i(x_i(t), t)$ yields

$$\mathcal{L}V_i(x_i(t), t)$$

$$\leq -2c_i x_i^2(t) + \sum_{j=1}^{n}|a_{ij}|\left(x_i^2(t) + G_j^2(x_j(t))\right) + \sum_{j=1}^{n}|b_{ij}|x_i^2(t) + \sum_{j=1}^{n}|c_i e_{ij}|x_i^2(t)$$

$$+ \sum_{j=1}^{n}\sum_{k=1}^{n}|e_{ij}a_{ik}|G_k^2(x_k(t)) + \sum_{j=1}^{n}(|o_{ij}v_{ij}h_j|+|o_{ji}v_{ji}h_i|)x_i^2(t)$$

$$+ \sum_{j=1}^{n}\sum_{k=1}^{n}\left|e_{ki}o_{ki}v_{ki}h_i^2\right|x_i^2(t) + \sum_{j=1}^{n}\frac{|c_j e_{ji}|}{1-\tau_{ji}}x_i^2(t) + \sum_{j=1}^{n}\frac{|b_{ij}|}{1-\mu_{ij}}G_j^2(x_j(t)$$

$$+ \sum_{j=1}^{n}\sum_{k=1}^{n}\frac{\xi_{ijk}}{1-\tau_{ij}}x_i^2(t) + \sum_{j=1}^{n}\sum_{k=1}^{n}\frac{|e_{ij}b_{ik}|}{1-\mu_{ik}}G_k^2(x_k(t))$$

$$+ \frac{|\alpha_i^2|}{1-\mu_i}f_i^2(x_i(t)) + \sum_{j=1}^{n}|\beta_{ij}^2|M_{ij}^2(x_j(t))$$

$$= (-2c_i + \zeta_i + \Omega_i)x_i^2(t) + \sum_{j=1}^{n}|a_{ij}|G_j^2(x_j(t)) + \sum_{j=1}^{n}\sum_{k=1}^{n}|e_{ij}a_{ik}|G_k^2(x_k(t))$$

$$+ \sum_{j=1}^{n}\sum_{k=1}^{n}\frac{|e_{ij}b_{ik}|}{1-\mu_{ik}}G_k^2(x_k(t)) + \sum_{j=1}^{n}\frac{|b_{ij}|}{1-\mu_{ij}}G_j^2(x_j(t)$$

$$+ \frac{|\alpha_i^2|}{1 - \mu_i} f_i^2(x_i(t)) + \sum_{j=1}^{n} \beta_{ji}^2 \Big| M_{ji}^2(x_i(t)) \tag{7}$$

Utilizing Hypothesis 1, we have

$$\mathcal{L}V_i(x_i(t), t)$$

$$\leq (-2c_i + \zeta_i + \Omega_i)x_i^2(t) + \sum_{j=1}^{n} \Big|a_{ij}g_j^2\Big| x_j^2(t) + \sum_{j=1}^{n}\sum_{k=1}^{n} \Big|e_{ij}a_{ik}g_k^2\Big| x_k^2(t)$$

$$+ \sum_{j=1}^{n}\sum_{k=1}^{n} \frac{|e_{ij}b_{ik}g_k^2|}{1 - \mu_{ik}} x_k^2(t) + \sum_{j=1}^{n} \frac{|b_{ij}|}{1 - \mu_{ij}} G_j^2(x_j(t)) + \frac{|\alpha_i^2|}{1 - \mu_i} f_i^2(x_i(t))$$

$$+ \sum_{j=1}^{n} \Big|\beta_{ji}^2\Big| M_{ji}^2(x_i(t))$$

$$= (-2c_i + \zeta_i + \Omega_i)x_i^2(t) + \sum_{j=1}^{n} |a_{ji}g_i^2| x_i^2(t) + \sum_{j=1}^{n}\sum_{k=1}^{n} \Big|e_{kj}a_{ki}g_i^2\Big| x_i^2(t)$$

$$+ \sum_{j=1}^{n}\sum_{k=1}^{n} \frac{|e_{kj}b_{ki}g_i^2|}{1 - \mu_{ki}} x_i^2(t) + \sum_{j=1}^{n} \frac{|b_{ji}g_i^2|}{1 - \mu_{ji}} x_i^2(t) + \frac{|\alpha_i^2\kappa_i^2|}{1 - \mu_i} x_i^2(t) + \sum_{j=1}^{n} \Big|\beta_{ji}^2\theta_{ji}^2\Big| x_i^2(t)$$

$$= \left(-2c_i + \zeta_i' + \Omega_i'\right) x_i^2(t)$$

Therefore, according to condition A1, there is a positive constant λ such that

$$\mathcal{L}V(x(t), t) = \sum_{i=1}^{n} \mathcal{L}V(x(t), t) \leq \sum_{i=1}^{n} \left(-2c_i + \zeta_i' + \Omega_i'\right) x_i^2(t) \leq -\lambda \|x(t)\|_2^2.$$

Let $y_i(t) = x_i(t) + \sum_{j=1}^{n} e_{ij}x_j(t - \tau_{ij}(t))$, T be a sufficiently large positive constant such that $0 \leq t \leq T$, then

$$|x_i(t)| \leq |y_i(t)| + \sum_{j=1}^{n} |e_{ij}x_j(t - \tau_{ij}(t))|$$

$$\leq |y_i(t)| + \sum_{j=1}^{n} |e_{ij}| (\sup_{0 \leq t \leq T} |x_j(t)| + \sup_{-\Xi \leq t < 0} |x_j(t)|) \tag{8}$$

Formula (8) can be written as

$$\sum_{i=1}^{n} \sup_{0 \leq t \leq T} |x_i(t)| \leq \sum_{i=1}^{n} \sup_{0 \leq t \leq T} |y_i(t)|$$

$$+ \sum_{i=1}^{n}\sum_{j=1}^{n} |e_{ij}| (\sup_{0 \leq t \leq T} |x_j(t)| + |\sup_{-\Xi \leq t \leq 0} |x_j(t)|). \tag{9}$$

From (9), we have

$$\sum_{i=1}^{n}\left(1 - \sum_{j=1}^{n}|e_{ji}|\right)sup_{0\leq t\leq T}|x_i(t)| \leq \sum_{i=1}^{n}sup_{0\leq t\leq T}|y_i(t)| + \sum_{i=1}^{n}\sum_{j=1}^{n}|e_{ij}|sup_{-\Xi\leq t\leq 0}|x_j(t)|$$

(10)

According to condition A2, let

$$\alpha = min\{1 - \sum_{j=1}^{n}|e_{ji}|\} > 0$$

then

$$\alpha \sum_{i=1}^{n}sup_{0\leq t\leq T}|x_i(t)| \leq \sum_{i=1}^{n}sup_{0\leq t\leq T}|y_i(t)| + \sum_{i=1}^{n}\sum_{j=1}^{n}|e_{ij}|sup_{-\Xi\leq t\leq 0}|x_j(t)|$$ (11)

implying that

$$\alpha sup_{0\leq t\leq T}||x(t)|| \leq sup_{0\leq t\leq T}||y(t)|| + \sum_{i=1}^{n}\sum_{j=1}^{n}|e_{ij}|sup_{-\Xi\leq t\leq 0}|x_j(t)|.$$ (12)

Since $\sum_{i=1}^{n}\sum_{j=1}^{n}|e_{ij}|sup_{-\Xi\leq t\leq 0}|x_j(t)|$ is bounded, $||y(t)|| \to \infty$ as $||x(t)|| \to \infty$. Based on the expression of $V(x(t), t)$, it has $V(x(t), t) \geq \sum_{i=1}^{n}y_i^2(t) \geq \alpha||y(t)||_2^2$. In addition, since $||y_i(t)||_2^2 \geq \frac{1}{n}||y(t)||^2$, we have $V(x(t), t) \geq \frac{\alpha}{n}||y(t)||^2$. Thus, we can know that $V(x(t), t) \to \infty$ as $||y(t)|| \to \infty$. It is obtained that the system (1) is globally asymptotically stable. The proof is complete.

Remark 1. It is worth noting that the delays of the neural network studied in many literatures are limited to fix constants. For example, only one delay μ was used in literature [19]. In literature [20], the neural network involved multiple delays of μ_i and neutral delays τ_i. Literature [11, 21] extended the study of multiple time delays μ_{ij}. The delay studied in the literature [22, 23] is a function $\tau(t)$, but it is limited to a single delay. The delays studied in this paper are time-varying functions $\mu_{ij}(t)$ and $\tau_i(t)$ related to neuron nodes, which contain more rich delay information. Furthermore, above literature did not consider the distributed delay and stochastic factor. In sum, this paper is a further generalization of the theorem of above papers.

3 Numerical Example

In order to visually verify the validity of Theorem 1, a numerical example is given.

Example 1. Suppose the network (1) has the following network parameters:

$c = [443.8\ 4], \alpha = [0000], e_{ij} = 1/16,$

$$a_{ij} = 0.25 \times \begin{pmatrix} 1 & -1 & -2 & 1 \\ -1 & -2 & 1 & 1 \\ 2 & 0 & -1 & 1 \\ 1 & 1 & -1 & 2 \end{pmatrix}$$

$$b_{ij} = 0.2 \times \begin{pmatrix} -2 & 1 & -1 & 1 \\ -1 & -2 & 1 & 1 \\ 1 & 1 & -1 & 1 \\ -1 & 2 & -1 & 1 \end{pmatrix}$$

$$o_{ij} = 0.25 \times \begin{pmatrix} 2 & -1 & 1 & 1 \\ -1 & -1 & 2 & 1 \\ 1 & 2 & -1 & 1 \\ -1 & 1 & -1 & 2 \end{pmatrix}$$

Besides, we set functions $\mu_i(t) = 0.01|sint|, \tau_{ij}(t) = 0.01\sin^2(t), \mu_{ij}(t) = 0.01|cost|, \upsilon_{ij}(t) = 0.01|\cos t|$, which means $\mu_i = \mu_{ij} = -0.01, \upsilon_{ij} = 0.01$. Besides, we let $G_j(x) = \sin x, F_{ij}(x) = |x|, H_i(x) = \cos x, M_{ij}(x) = |\cos x|$. Thus, it is easy to calculate that $g_i = h_i = \kappa_i = \theta_{ij} = 1$, and we can get that $\delta_i = 0.75 > 0, \zeta' = [5.59\ 5.53\ 4.92\ 5.39], \Omega' = [2.06\ 2.06\ 2.00\ 2.06]$ and $\rho = [0.35\ 0.40\ 0.68\ 0.55]$. Therefore, it can be concluded that conditions of Theorem 1 are satisfied, which means the system is globally asymptotically stable. The neuronal state of system (1) is shown in Fig. 1.

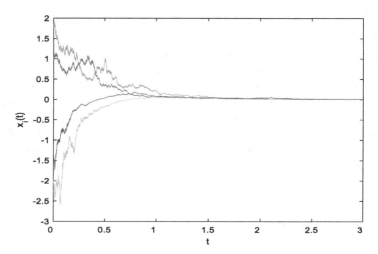

Fig. 1. Neuronal state of system (1).

4   Conclusions

In this paper, the stability of stochastic neutral Hopfield neural networks with multiple time-varying delays has been analyzed. Compared with previous studies, the time delays in this paper are time-varying and related to neuron nodes. By constructing a novel Lyapunov functional, the sufficient criteria of global asymptotic stability of the studied network are derived. Our theoretical results are the generalization of previous conclusions, which improves the applicability of the networks.

Acknowledgement. The authors really appreciate the valuable comments of the editors and reviewers. This work was supported by Social science and technology development project in Dongguan under grant 2020507151806 and Natural Science Foundation of Shenzhen under Grant 20200821143547001.

References

1. Culloch, W., Pitts, W.H.: A logical calculus of the ideas immanent in neural nets. Bull. Math. Biophys. **5**(4), 115–133 (1943)
2. Hopfield, J.J.: Neural networks and physical systems with emergent collective computational abilities. Proc. Natl. Acad. Sci. U.S.A. **79**(8), 2554–2558 (1982)
3. Shin, H.C., Roth, H.R., Gao, M., et al.: Deep convolutional neural networks for computer-aided detection: CNN architectures, dataset characteristics and transfer learning. IEEE Trans. Med. Imaging **35**(5), 1285–1298 (2016)
4. Zhang, Q., et al.: Delay-dependent exponential stability of cellular neural networks with time-varying delays. Chaos Solitons Fractals **23**, 1363–1369 (2005)
5. Niculescu, S.I.: Delay Effects on Stability: A Robust Control Approach. Springer, Berlin (2001). https://doi.org/10.1007/1-84628-553-4
6. Kolmanovskii, V.B., Nosov, V.R.: Stability of Functional Differential Equations. Academic Press, London (1986)
7. Kuang, Y.: Delay Differential Equations with Applications in Population Dynamics. Academic Press, Boston (1993)
8. Yang, Y., Cao, J.: A feedback neural network for solving convex constraint optimization problems. Appl. Math. Comput. **201**(1–2), 340–350 (2008)
9. Faydasicok, O.: Stability analysis of neutral-type Hopfield neural networks with multiple delays. Neural Netw. **129**, 288–297 (2020)
10. Li, S., Sun, H., Li, W.: Stochastic hybrid multi-links networks with mixed delays: stabilization analysis via aperiodically adaptive intermittent control. Int. J. Syst. Sci. **51**, 852–877 (2020)
11. Arik, S.: A modified Lyapunov functional with application to stability of neutral-type neural networks with time delays. J. Franklin Inst. **356**(1), 276–291 (2019)
12. Zhu, Q., Cao, J.: Stability analysis for stochastic neural networks of neutral type with both Markovian jump parameters and mixed time delays. Neurocomputing **73**(13–15), 2671–2680 (2010)
13. Cheng, K.S., Lin, J.S.: The application of competitive Hopfield neural network to medical image segmentation. IEEE Trans. Med. Imaging **15**(4), 560–567 (1996)
14. Joya, G., Atencia, M., Sandoval, F.: Hopfield neural networks for optimization: study of the different dynamics. Neurocomputing **43**, 219–237 (2002)
15. Zheng, C.D., Gu, Y., Liang, W., et al.: Novel delay-dependent stability criteria for switched Hopfield neural networks of neutral type. Neurocomputing **158**, 117–126 (2015)

16. Aouiti, C., Miaadi, F.: Finite-time stabilization of neutral Hopfield neural networks with mixed delays. Neural Process. Lett. **48**(3), 1645–1669 (2018). https://doi.org/10.1007/s11063-018-9791-y
17. Du, F., Lu, J.G.: New criteria on finite-time stability of fractional-order Hopfield neural networks with time delays. IEEE Trans. Neural Netw. Learn. Syst. **99**, 1–9 (2020)
18. Kobayashi, M.: Stability conditions of bicomplex-valued Hopfield neural networks. Neural Comput. **33**, 552–562 (2021)
19. Liu, Y., Wang, Z., Liu, X.: Global exponential stability of generalized recurrent neural networks with discrete and distributed delays. Neural Netw. **19**(5), 667–675 (2006)
20. Wu, F.: Global asymptotic stability for neutral-type Hopfield neural networks with multiple delays. J. Phys. Conf. Ser. **1883**, 012089 (2021)
21. Arik, S.: New criteria for stability of neutral-type neural networks with multiple time delays. IEEE Trans. Neural Netw. Learn. Syst. **99**, 1–10 (2019)
22. Ding, K., Zhu, Q., Li, H.: A generalized system approach to intermittent nonfragile control of stochastic neutral time-varying delay systems. IEEE Trans. Syst. Man Cybern.: Syst. **99**, 1–10 (2020)
23. Xiao, X.A., Li, W.A., Zd, B., et al.: Stochastic stabilization of Markovian jump neutral systems with fractional Brownian motion and quantized controller. J. Franklin Inst. **358**, 9449–9466 (2021)

Task Offloading Decision Algorithm for Vehicular Edge Network Based on Multi-dimensional Information Deep Learning

Xi Hu[(✉)], Yang Huang, Yicheng Zhao, Chen Zhu, Zhibo Su, and Rui Wang

Northeastern University at Qinhuangdao, Qinhuangdao 066004, China
hux@neuq.edu.cn

Abstract. Traditional vehicle edge network task offloading decision is based on the nature of tasks and network status, and each vehicular node makes a distributed independent decision. Nevertheless, the network state considered in the decision is single and lacks global information, which is not conducive to the overall optimization of the system. Therefore, this paper proposes a task offloading decision algorithm for vehicular edge network based on deep learning of multi-dimensional information. With the optimization goal of minimizing system overhead, the algorithm uses hybrid neural networks to deeply learn the state information of multi-dimensional networks and constructs the central task offloading decision model. A large number of simulation experiments show that the task offloading decision model trained by the hybrid neural network in this paper has high validity and accuracy when making the offloading decision and can significantly reduce system overhead and task computing delay.

Keywords: Vehicular edge network · Task offloading · Deep learning · Multi-dimensional information · System overhead

1 Introduction

With the rapid application of 5G communication technology, the Internet of Vehicles (IoV) has not only a wider range of applications [1] but also more and more types of applications carried, which puts forward higher requirements for the computing and transmission capabilities of the IoV [2]. Due to the limited capabilities of the vehicle itself, task offloading methods can be used to complete tasks' calculations from the remote cloud on the IoV. However, cloud computing has problems such as large tasks' calculations delays, high-energy consumption and low data security. The shortcomings of cloud computing will become more prominent as the number of data increases so some scholars have proposed mobile edge computing (MEC) technology to solve these problems and improve the efficiency of offloading in recent years [3]. Mobile edge computing provides users with fast and powerful computing capabilities and efficient offloading efficiency, etc. It allows users to enjoy a high-quality network experience

© Springer Nature Switzerland AG 2022
Y. Tan et al. (Eds.): ICSI 2022, LNCS 13345, pp. 143–154, 2022.
https://doi.org/10.1007/978-3-031-09726-3_13

while having features such as ultra-low latency, ultra-high bandwidth, strong computing power and firm real-time performance [4].

The offloading decision is one of the core problems of mobile edge computing. In the Internet of Vehicles environment, the main solution is whether and where vehicular tasks need to be offloaded [5]. The main goals of the offloading decision are the shortest time delay, the smallest energy consumption, the smallest trade-off value between delay and energy consumption, the largest system utility and so on. The problem of time delay minimization under mobile edge computing and designed an efficient one-dimensional search algorithm to find the optimal task offloading decision is proposed in [6]. In [7], Zhang proposed energy consumption problem of the offloading system under mobile edge computing. It designs a computing offloading scheme to jointly optimize offloading and wireless resource allocation, and obtain the minimum energy consumption under relevant constraints. All of the above are single optimization goals and the comprehensive influence of multiple goals on task offloading is ignored. A priority based joint computing offloading algorithm is proposed to solve the offloading decision problem of minimizing time delay and energy consumption in [8]. The offloading methods of vehicle tasks are mainly divided into local computing, full offloading, partial offloading, cloud servers, etc. In [9], each overall task follows the binary calculation offloading decision (local computing and full offloading). S. Bi proposed a joint optimization method based on ADMM to solve this problem. In [10], considering that MEC is unavailable or not enough, the surrounding vehicles are used as offloading points for calculation and a distributed computing offloading decision based on DQN is proposed to find the best offloading method. M. Yuqing proposes to implement offloading decision by drawing on the idea of the simulated annealing algorithm, improving the particle swarm algorithm and offloading tasks to local vehicles, MEC servers and nearby vehicles for decision in [11]. J. Long proposed a solution for computing offloading through moving vehicles in the edge cloud network of the Internet of Things in [12]. The device generates tasks and sends the tasks to vehicle and vehicle decides to calculate the task in the local vehicle, the MEC server or the cloud center. With the diversification of offloading scenarios, deep learning and deep reinforcement learning as a new algorithm are gradually used in offloading decision making. In [13], a deep learning based shunting decision algorithm for single server mobile edge computing network is proposed for computing shunting. However, offloading types are limited to local computing and full offloading.

By summarizing and analyzing existing work, we find that these algorithms describe the optimization goal as a function of a single state when performing offloading decisions while other networks' states and tasks' states make idealized assumptions, which reduces the applicability of the algorithms. This kind of distributed offloading decision is difficult to realize the rational use of the vehicle network edge network resources and reduces network performance. Based on the above analysis, this paper proposes a hybrid neural network algorithm that adopts the central decision method of MEC server centralized management and uses the advantages of deep learning technology. Starting from the comprehensive multi-dimensional state information, this paper takes minimizing time delay and energy consumption as the optimization goal, training and maintaining an offloading decision model suitable for the current network state, thereby realizing the global optimal offloading decision.

2 System Model

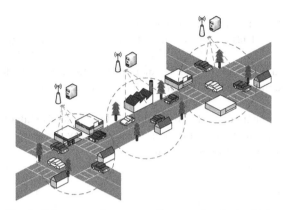

Fig. 1. Edge network structure diagram of typical vehicles.

The edge network structure diagram of typical vehicles adopted in this paper is shown in Fig. 1. The whole edge network system is composed of several MEC servers and vehicles. Each MEC server provides edge computing services for multiple vehicles within its coverage. The work in this paper only addresses problem of offloading decisions between a single MEC server and multiple vehicles within its coverage.

The notations and definitions involved in this paper are shown in Table 1.

Table 1. Notations and definitions

Notation	Definition	Notation	Definition
Λ	Weight of time delay	V_{local}	Speed of local computing
M	Weight of energy consumption	V_{wait}	Speed of waiting computing
a_k	Task split switch	R_{up}	Task upward transmission speed
b_k	Task parallel switch	R_{down}	Task downward transmission speed
M_i	Task data generated by vehicle	V_{MEC}	Computing speed on the MEC server
ρ_i	Mission launching power	P_{local}	Local computing power
h_i	Channel gain	P_{MEC}	MEC server computing power
Σ	Noise power	P_{up}	Task upward transmission Power
m_i	Offloading decision	P_{down}	Task downward transmission power
C_k	Queuing switch	Ω	Data transmission bandwidth
H	Processing rate on the MEC	P_{wait}	Task waiting power on MEC server
M_f	Idle computing ability on the MEC server		

2.1 Time Delay and Energy Consumption Model

There are three decision results when the vehicle performs task offloading.

- Local computing: Tasks are all calculated on the vehicle.
- Full offloading: All tasks are offloaded to the mobile edge computing server for calculation.
- Partial offloading: Part of the task is calculated on the local vehicle and the rest is offloaded to the mobile edge computing server for calculation.

In local computing mode, tasks are calculated directly in vehicular ECU (Electronic Control Unit). At this point, the time delay T_1 and energy consumption E_1 of the task calculation are shown in formulas (1) and (2).

$$T_1 = \frac{M_i}{V_{local}} \tag{1}$$

$$E_1 = T_1 \cdot P_{local} \tag{2}$$

All tasks of the vehicles are offloaded to the edge server for calculation. When the edge server completes the calculation, the calculation results are transmitted back to the vehicles from the edge server. According to Shannon's formula, its transmission rate can be obtained by

$$R_b = \omega log_2\left(1 + \frac{\rho_i \cdot h_i}{\sigma^2}\right) \tag{3}$$

In full offloading mode, the computation time delay T_2 of the tasks consists of four parts: upward transmission time delay t_1, queue time delay t_2, processing time delay t_3 and downward transmission time delay t_4.

$$T_2 = t_1 + C_k \cdot t_2 + t_3 + t_4 \tag{4}$$

where, $t_1 = M_i/R_{up}$, $t_2 = (M_i - M_f)/V_{wait}$, $t_3 = M_i/V_{MEC}$ and $t_4 = M_i \cdot \eta/R_{down}$. If M_i is greater than M_f, then C_K is 1 and vice versa.

The energy consumption E_2 of the tasks consists of four parts: upward transmission energy consumption e_1, queuing cache energy consumption e_2, processing energy consumption e_3 and downward transmission energy consumption e_4.

$$E_2 = e_1 + C_k \cdot e_2 + e_3 + e_4 \tag{5}$$

where, $e_1 = M_i \cdot P_{up}/R_{up}$, $e_2 = (M_i - M_f) \cdot P_{wait}/V_{wait}$, $e_3 = M_i \cdot P_{MEC}/V_{MEC}$ and $t_4 = M_i \cdot \eta \cdot P_{down}/R_{down}$.

If tasks can be divided, the partial offloading decision can be used. In partial offloading mode, the partial offloading delay includes two parts: the local delay t_5 and the offloading delay (t_6, t_7, t_8, t_9), as shown in formula (6).

$$T_3 = a_k \cdot [b_k \cdot T_{31} + (1 - b_k) \cdot T_{32}] \tag{6}$$

$$T_{31} = \max\{t_5, t_6 + C_k \cdot t_7 + t_8 + t_9\} \tag{7}$$

$$T_{32} = t_5 + t_6 + C_k \cdot t_7 + t_8 + t_9 \tag{8}$$

where, T_{31} is the time delay when the task is calculated in parallel and T_{32} is the time delay when the task is not calculated in parallel. where, $t_5 = M_i \cdot (1 - x)/V_{local}$, $t_6 = M_i \cdot x/R_{up}$, $t_7 = (M_i \cdot x - M_f)/V_{wait}$, $t_8 = M_i \cdot x/V_{MEC}$ and $t_9 = M_i \cdot \eta \cdot x/R_{down}$. x is the offloading ratio, a_k is equal to 1 if the task can be divided and 0 otherwise, b_k is equal to 1 if the task can parallel and 0 otherwise.

Energy consumption E_3 of partial offloading at this time is shown in formula (9).

$$E_3 = e_5 + e_6 + C_k \cdot e_7 + e_8 + e_9 \tag{9}$$

where, $e_5 = M_i \cdot (1 - x) \cdot P_{local}/V_{local}$, $e_6 = M_i \cdot x \cdot P_{up}/R_{up}$, $e_7 = (M_i \cdot x - M_f) \cdot P_{wait}/V_{wait}$, $e_8 = M_i \cdot x \cdot P_{MEC}/V_{MEC}$ and $e_9 = M_i \cdot \eta \cdot x \cdot P_{down}/R_{down}$.

2.2 System Overhead Model

System overhead is a representation of system performance under different offloading decision and is defined as the weighted sum of task computation time delay and energy consumption. The algorithm proposed in this paper aims to find the offloading decision that minimizes the system overhead and the specific description is shown in formula (10).

$$Q = \sum_{i=1}^{3} \lambda \cdot (m_i * T_i) + \mu \cdot (m_i * E_i) \tag{10}$$

S.T

$$0 \le \lambda, \mu \le 1 \tag{11}$$

$$\lambda + \mu = 1 \tag{12}$$

$$\sum_{i=1}^{3} m_i = 1 \tag{13}$$

$$m_i \in \{0, 1\} \tag{14}$$

Constraint (11) and (12) are time delay and energy consumption weights. Weight of energy consumption is inversely proportional to the remaining electric quantity of the vehicle. Constraint (13) and (14) indicate that the offloading decision can only be made in one of the ways: local computing, full offloading and partial offloading.

3   Offloading Decision Algorithm

Since solving the minimum system overhead problem needs to consider multi-dimensional dynamic state information, an offloading decision algorithm (multi-dimensional information deep learning, MDL) based on deep learning is proposed. The algorithm learns the existing state information through the neural network, trains the offloading decision model and then makes rapid decisions based on the current state information.

3.1   Convolutional Deep Neural Networks

Based on CNN [14] and DNN [15], this paper designs a convolutional deep neural networks (CDNN) to learn multi-dimensional state information and then train the required offloading decision model. The structure of CDNN is shown in Fig. 2.

Firstly, in the data preprocessing stage, the calculated labels of the generated data are divided into training set and test set according to 10:1, and sixteen-dimensions are divided into nine-dimensions and seven-dimensions. Then, the nine-dimensional information that is closely related is used to CNN extract features and use DNN model to reduce complexity for other seven-dimensional information parameters. What's more, the two models are processed in parallel and merged into a fully connected layer. Finally, the model is constructed to complete the offloading classification.

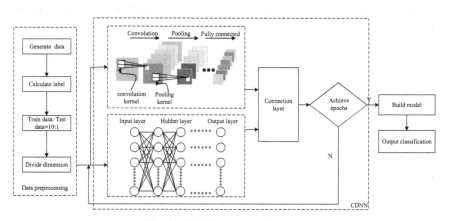

Fig. 2. CDNN structure diagram.

3.2   Offloading Decision Based on CDNN Model

The offloading decision model is the core of the algorithm in this paper. The model is learned and trained on known data through CDNN. The whole training process includes two stages: the forward propagation stage and the back propagation stage [16]. We separate task data and labels according to a ratio of 10:1 where the first 10 pieces of data and labels are used to train the model and the next piece of data and labels are used to test the model. The specific training process is shown in Algorithm 1.

Algorithm1.	Training CDNN algorithm
Input:	Data sets (16 parameters) and labels (1 result)
Output:	Weight matrix ω^l, offset matrix b^l
1:	Initialize all hidden layer parameters ω^i and b^i
2:	**For** n=1,2,…,N do
3:	Forward propagation stage
	Calculate weight product sum of each neuron and activation value in each hidden layer, Calculate the forward propagation loss function
4:	Back propagation stage
	Gradient descent method to update weights
5:	**If** loss function is minimized
6:	Turn to step 11
7:	**Else**
8:	Turn to step 3
9:	**End if**
10:	**End for**
11:	Return ω^l and b^l

It is worth noting that the data set input here will be updated with the appearance of new state information and the offloading decision model of CDNN will also be continuously learned and updated so as to reach the optimal state model.

3.3 MDL Algorithm

Using the offloading decision model obtained by training and taking the current state information as input, the offloading decision that meets the minimum system overhead can be quickly obtained.

Algorithm 2.	MDL algorithm
Input:	Sixteen-dimensional status information
Output:	Offloading decision
1:	Initialization
2:	Input current state information into the model calculate the probability of three offloading methods
3:	Choose the offloading method with the biggest probability
4:	Output the results of the offloading decision

4   Simulation and Result Analysis

4.1   Simulation Parameter Setting

The performance parameters of the mixed neural network (CDNN) model used in the simulation are shown in Table 2.

Table 2. Related parameters of CDNN.

Name	Parameter	Name	Parameter
Activation	Relu, softmax	Neurons	3–512
Learning rate	0.2	Epoch	30
Optimizer	Adadelta	Batch_size	16

4.2   Multidimensional Status Information Data Set

Due to the lack of a real state data set of vehicular edge network, matlab is used to generate a state information data set. Let the state information P $< \lambda, \mu, a_k, b_k, M_i, \eta, M_f, V_{local}, V_{wait}, R_{up}, R_{down}, V_{MEC}, P_{local}, P_{MEC}, P_{up}, P_{down} >$ be a collection of sixteen-dimensional information. According to the structure and characteristics of the typical vehicle edge network, the status information should meet the following five basic conditions.

- In general, the edge server CPU processes tasks faster than the local vehicle ECU processes tasks.
- The sum of the weight of the delay and the weight of the energy consumption is "1".
- Tasks are divided into two states: divisible and indivisible.
- Divisible tasks are divided into two cases: parallel processing and non-parallel processing.
- In edge computing, tasks are divided into cache waiting and direct computing.

As a result, a total of 17,600 groups of data are generated and each set of data is a collection of sixteen-dimensional state information. In addition, according to the system overhead formula, the task's offloading decision and offloading ratio are calculated. In this paper, the offloading decision is used as a label to be used as a result training and testing model. Table 3 exhibits three samples of data.

4.3   Results and Analysis

In this paper, the accuracy rate and the loss function are used to evaluate the offloading decision model.

Figure 3(a) shows the accuracy of DNN, CNN and CDNN at different epochs. Where, Train represents the training set and Val represents the test set. It can be seen that the

Table 3. Data set information

Parameters	1	2	3	Parameters	1	2	3
λ	0.52	0.19	0.92	V_{local} GB/s	0.956	0.568	0.459
μ	0.48	0.81	0.08	V_{MEC} GB/s	7.238	5.846	7.276
a_k	1	1	1	V_{wait} GB/s	7.997	6.988	8.920
b_k	0	1	1	R_{up} GB/s	0.564	1.413	1.499
η	0.84	0.80	0.71	R_{down} GB/s	0.883	1.445	0.899
P_{local} w	1.072	0.992	0.856	M_i GB	7.724	6.129	6.128
P_{MEC} w	2	2	2	M_f GB	9.440	91.150	76.69
P_{up} w	0.189	0.129	0.182	Label	1	2	3
P_{down} w	0.110	0.143	0.110				

accuracy of the three neural networks increases with the number of epochs and eventually stabilizes. Obviously, CDNN has the highest accuracy rate, followed by DNN and CNN has the lowest accuracy rate.

Figure 3(b) shows the loss function values of DNN, CNN and CDNN at different epochs. Where, Train represents the training set and Val represents the test set. It can be seen that the loss function values of the three neural networks all decrease as the number of epochs increase and eventually stabilize. Obviously, CDNN has the smallest loss function value, followed by DNN and CNN has the largest loss function value.

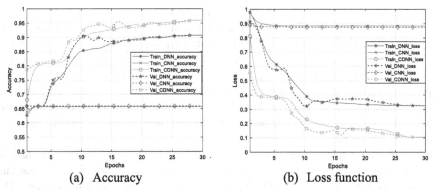

(a) Accuracy   (b) Loss function

Fig. 3. (a). The accuracy of CNN, DNN and CDNN at different epochs. (b). Loss functions of CNN, DNN and CDNN at different epochs.

Table 4 shows the accuracy and loss function when the three models reach their optimal performance.

As shown in Fig. 4(a) this is the value of the system overhead for the first 30 groups of local computing, full offloading, partial offloading and using MDL algorithm in the test set. Theoretically, the value of system overhead is greater than 0, and there is 0 in

Table 4. Comparison of the three models

Name	Accuracy	Loss function
DNN	0.9075000286102295	0.3253644686724874
CNN	0.6518750190734863	0.883384276330471
CDNN	0.9599999785423279	0.10568311154143885

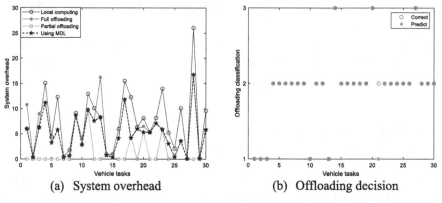

(a) System overhead   (b) Offloading decision

Fig. 4. (a). System overhead of different tasks. (b). Offloading decision result.

the figure because some tasks are inseparable and the value of Q is 0. Consequently, partial offloading is not applicable to all tasks. By comparison, it is found that the use of the MDL algorithm is the smallest overhead of each task.

Corresponding to the value of the system overhead in Fig. 4(a), this paper verifies the accuracy of the MDL algorithm in the offloading decision result in Fig. 4(b). Each task in the two pictures has a one-to-one correspondence. The minimum value in Fig. 4(b). is used to verify the type of offloading in Fig. 4(b). where, type 1 means local computing, type 2 means full offloading and type 3 means partial offloading. Correct represents the correct offloading decision in the tag and Predict represents the prediction result of the MDL algorithm. From this, it can be seen that the decision result of the MDL algorithm has a high correct rate.

Due to the segmentation of tasks, partial offloading is not common to every task so the values calculated by the task in local computing, full offloading and MDL algorithm are listed, as shown in Fig. 5. The MDL algorithm has the lowest system overhead. The reason is that each task overhead of the MDL algorithm is the optimal solution which is the lowest among the three offloading results. In contrast, other offloading methods are not the optimal solution for every task.

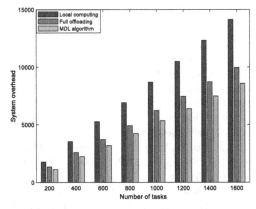

Fig. 5. Test set system overhead.

5 Conclusion

Task offloading calculation is one of the core technologies of the vehicular edge network. An efficient and intelligent offloading decision can effectively improve the service capability of the network. Different from the traditional distributed offloading decision, this paper proposes a centralized offloading decision (MDL) algorithm based on multi-dimensional information deep learning. The algorithm proposes a convolutional deep neural network (CDNN) to learn from the existing multi-dimensional information data and train to obtain an offloading decision model. Using this model, the offloading decision can be quickly implemented based on the current multi-dimensional status information. Simulation experiments show that the CDNN network proposed in this paper has high accuracy and small loss function value. The proposed MDL algorithm has the advantages of the low system overhead, short execution time and is more suitable for the edge network environment of the car network. However, considering that the static model of a single MEC server constructed by deep learning has certain limitations, the focus of the next step will be to try to start with multiple MEC servers, linking with reinforcement learning to build a dynamic model and achieve a more effective offloading decision.

Acknowledgement. This work was supported in part by the NSFC under Grant 61501102.

References

1. Ning, Z., et al.: Deep learning in edge of vehicles: exploring trirelationship for data transmission. IEEE Trans. Ind. Inf. **15**(10), 5737–5746 (2019)
2. Ning, Z., et al.: When deep reinforcement learning meets 5G-enabled vehicular networks: A distributed offloading framework for traffic big data. IEEE Trans. Ind. Inf. **16**(2), 1352–1361 (2019)
3. Shi, W., Cao, J., Zhang, Q., Li, Y., Xu, L.: Edge computing: vision and challenges. IEEE Internet Things J. **3**(5), 637–646 (2016)

4. De Souza, A.B., et al.: Computation offloading for vehicular environments: a survey. IEEE Access **8**, 198214–198243 (2020)
5. Xie, R., Lian, X., Jia, Q., Huang, T., Liu, Y.: Survey on computation offloading in mobile edge computing. J. Commun. **39**, 138 (2018)
6. Liu, J., Mao, Y., Zhang, J., Letaief, K.B.: Delay-optimal computation task scheduling for mobile-edge computing systems. In: 2016 IEEE International Symposium on Information Theory (ISIT), pp. 1451–1455 (2016)
7. Zhang, K., et al.: Energy-efficient offloading for mobile edge computing in 5G heterogeneous networks. IEEE Access **4**, 5896–5907 (2016)
8. Xu, Z., Zhang, Y., Qiao, X., Cao, H., Yang, L.: Energy-efficient offloading and resource allocation for multi-access edge computing. In: 2019 IEEE International Conference on Consumer Electronics-Taiwan (ICCE-TW), pp. 1–2 (2019)
9. Bi, S., Zhang, Y.J.: Computation rate maximization for wireless powered mobile-edge computing with binary computation offloading. IEEE Trans. Wirel. Commun. **17**(6), 4177–4190 (2018)
10. Chen, C., Zhang, Y., Wang, Z., Wan, S., Pei, Q.: Distributed computation offloading method based on deep reinforcement learning in ICV. Appl. Soft Comput. **103**, 107108 (2021)
11. Yuqing, M., Yi, X., Wanzhen, Z., Tonglai, L., Zheng, H.: Improved particle swarm algorithm for task offloading in vehicular networks. Appl. Res. Comput. **38**(07), 2050–2055 (2021)
12. Long, J., Luo, Y., Zhu, X., Luo, E., Huang, M.: Computation offloading through mobile vehicles in IoT-edge-cloud network. EURASIP J. Wirel. Commun. Netw. **2020**(1), 1–21 (2020). https://doi.org/10.1186/s13638-020-01848-5
13. Huang, L., Feng, X., Feng, A., Huang, Y., Qian, L.P.: Distributed deep learning-based offloading for mobile edge computing networks. Mob. Netw. Appl. 1–8 (2018)
14. Wu, J.: Introduction to convolutional neural networks. Natl. Key Lab Novel Softw. Technol. Nanjing Univ. China **5**, 495 (2017)
15. Kroiss, M.: Introduction to deep neural networks. In: Predicting the Lineage Choice of Hematopoietic Stem Cells, pp. 9–29 (2016)
16. Han, W., Zhang, X.W., Zhang, W., Cong-Ming, W.U., Yan-Jun, W.U.: Classical network models and training methods in deep learning. J. Mil. Commun. Technol. (2016)

Adversarial Examples Are Closely Relevant to Neural Network Models - A Preliminary Experiment Explore

Zheng Zhou, Ju Liu$^{(\boxtimes)}$, and Yanyang Han

Shandong University, Qingdao 266237, China
{zhouzheng,hanyanyang}@mail.sdu.edu.cn, juliu@sdu.edu.cn

Abstract. Neural networks are fragile because adversarial examples can readily assault them. As a result of the current scenario, academics from various countries have paid close attention to adversarial examples: many research outcomes, e.g., adversarial and defensive approaches and algorithms. However, numerous people are still baffled about how adversarial examples affect neural networks. We present hypotheses and devise extensive experiments to acquire more information about adversarial examples to verify this notion. By experiments, we investigate the neural network's sensitivity to adversarial examples in diverse aspects, e.g., model architectures, activation functions, and loss functions. The consequence of the experiment shows that adversarial examples are closely related to them. Peculiarly, sensitivity's property could help us distinguish the adversarial examples from the data set. This work will inspire the research of adversarial examples detection.

Keywords: Adversarial examples · The neural network · Attack & defense · The activation functions · The loss functions · Security of AI

1 Introduction

The adversarial examples were first proposed by Szegedy, Zaremba, and Sutskever in a paper titled "Intriguing properties of neural networks" [21] in 2014. Then, in recent years, research on adversarial attack and defense has made significant advances. The Fast Gradient Sign Method (FGSM) presented by Goodfellow, Shlens and Szegedy [6] is the most representative. To generate adversarial examples, this method makes use of the image's high-dimensional feature properties. FGSM's suggestion broadens the concept of adversarial example research and is credited with founding the discipline of adversarial white's box attack. Madry et al. presented Project Gradient Descent (PGD) [14], an innovative framework that combined the advantages of attack and defense algorithms

This work is supported by the National Science Foundation of China (Grant No. 6207 1275) and Shandong Province Key Innovation Project (Grant No. 2020CXGC010903 and Grant No. 2021SFGC0701).

Y. Tan et al. (Eds.): ICSI 2022, LNCS 13345, pp. 155–166, 2022.
https://doi.org/10.1007/978-3-031-09726-3_14

to improve the robustness of neural networks. There are still numerous varieties of adversarial attack and defense researches, including such C&W [1], deepfool [15], and JSMA [16]. There are studies on adversarial examples not just in the discipline of image, but also in the field of speech recognition. Yakura et al. [25] and Qin et al. [17], for example, have done extensive research on this. Regardless of the fact how we have investigated adversarial examples, we still do not grasp the nature of adversarial attacks. Ilyas, Andrew et al. recently concluded that the adversarial examples are not bugs, but rather features of images [8]. This viewpoint piqued our interest, therefore in order to further explore the interpretability of the adversarial example, we supplemented following four investigates as contributions to consummate the interpretability of the adversarial example:

- From the aspect of the model architecture: We change the neural network architecture with a simple structure to the DenseNet [7] neural network architecture, and then we find that the model more with more complex architecture occurs more stable performance.
- From the activation function aspect, we employ eight diverse activation functions of neural networks without any other changes, respectively, to be assaulted by adversarial examples. We discover that the neural networks model will be more sensitive in changing the diverse activation functions by calculating the standard deviation to measure fluctuation.
- From the loss function aspect, we utilize the Cross-Entropy and Focal as loss functions of the neural network, respectively, to be attacked by adversarial examples. Ultimately, the Focal loss function performs more sensitive than the Cross-Entropy loss function.
- We ulteriorly explore the variance fluctuation of attack success rates according to those mentioned above. The attack success rates will dramatically alter when we utilize diverse components, e.g., model architecture, activation function, and loss function. Therefore, we can investigate the strategy of adversarial examples detection based on this characteristic.

The rest of the paper is structured as follows. The strategy of adversarial attack is discussed in Sect. 2. Section 3 is dedicated to discussing the adversarial defense and adversarial examples detection. Then we present the method and experiment in Sect. 4 and Sect. 5. Ultimately, we conclude the main contributions with the future perspectives of this research direction in Sect. 6.

2   Adversarial Attacks

2.1   Identifying Adversarial Examples

Szegedy et al. [21] discovered a fascinating characteristic of neural networks. We can add a little disruption to the picture's surface that might not be readily perceived by human vision because of the high dimensional features of image. To validate the above hypothesis, we use the formula to construct a mathematical model.

$$Minimize\ \|\delta\|_2\ s.t. C(x+\delta) = I; x+\delta \in [0,1]^m\ , \tag{1}$$

where δ denotes a perturbation, x denotes the source image, C denotes a classifier, and I denotes the attacked target's class.

$$Minimize\ c|\delta| + J(x + \delta)\ s.t.x + \delta \in [0,1]^m\ ,\tag{2}$$

where the c denotes the $min\ C(x + \delta) = I$, and $c > 0$. We use the cross-entropy loss function J in the formula since the value of $\|\delta\|_2$ is difficult to optimize [21] and the J denotes the cross-entropy loss function [20].

That is the essential notion behind creating adversarial examples. The investigation of adversarial examples has ushered in rapid growth based on this notion.

2.2   Attack Algorithm Based on Gradient

Goodfellow et al. [6] proposed an attack algorithm named Fast gradient sign method (FGSM), which would be the originator of all gradient-based attack algorithms. We need to increase disturbance in the opposite direction of the gradient. Then we can generate an image with a disturbance that will interfere with the neural network. The FGSM algorithm is defined as follows [6]:

$$x' = x + \epsilon * sign(\nabla_x J(\theta, x, y))\ ,\tag{3}$$

where $\epsilon * sign(\nabla_x J(\theta, x, y))$ is a perturbation, ϵ is a parameter of step, the $sign$ is the SIGN function, x is the source image, J is the cross-entropy loss function, and θ is the class of the attacked target.

2.3   Attack Algorithm Based on Generative Adversarial Network

Xiao et al. [23] presented a novel attack algorithm with the current generative adversarial network (GAN) short for AdvGAN (Fig. 1). On the one hand, we put the raw image into the Generator of GAN. Then we add the perturbation to the generator's output and put the composite image into the discriminator of GAN. To keep the adversarial examples and authentic images as similar as possible, we minimize GAN loss (Eq. 4). On the other hand, we optimize the loss of the classifier to confuse the neural network to the point of misclassification (Eq. 5). To constrain the magnitude of the perturbation, we employ the Eq. 6 to constraint it. Ultimately, we combine all of the loss functions and minimize this loss function to provide adversarial examples (Eq. 7). The Fig. 1 demonstrates the entire process. The following formulas are derived from [23].

$$\mathcal{L}_{GAN} = \mathbb{E}_x log \mathcal{D}(x) + \mathbb{E}_x log(1 - \mathcal{D}(x + \mathcal{G}(x))),\tag{4}$$

$$\mathcal{L}_{adv} = \mathbb{E}_x \ell_\mathcal{C}(x + \mathcal{G}(x), t),\tag{5}$$

$$\mathcal{L}_{hinge} = \mathbb{E}_x \max(0, \|\mathcal{G}(x)\|_2 - c),\tag{6}$$

$$\mathcal{L} = \mathcal{L}_{adv} + \alpha \mathcal{L}_{GAN} + \beta \mathcal{L}_{hinge},\tag{7}$$

where the $\mathcal{L}_{GAN}$ represents the loss of generative adversarial network; the $\mathcal{L}_{adv}$ represents the loss of generating adversarial examples; the $\mathcal{L}_{hinge}$ represents the loss to limit the scale of the disturbance; $\mathcal{D}(x)$ and $\mathcal{G}(x)$ represent the discrimination function and generation function; the c denotes minimize scale of perturbation; the t denotes the target of adversary.

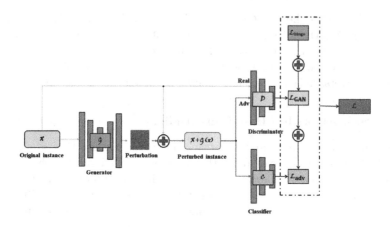

Fig. 1. AdvGAN process and architecture

3 Adversarial Defense

3.1 Transformation-Based Defenses

This method defends the adversarial attacks by transforming the input object. Kurakin et al. [10] presented the JPEG compression method to defend the adversarial attacks. Subsequently, Yang et al. [26] randomly deleted pixels of the input object and reconstructed the deleted pixels by using matrix estimation to restrain the effect of the adversarial examples. Raff et al. [18] utilized the random transformation of the input image to suppress the attack effect, which was generated by a potential adversary with the efficient computational strategy.

3.2 Manifold-Based Defenses

Ma et al. [13] investigated the relationship between the manifold of input data sets and the property of adversarial attacks. They proposed that the input samples closer to the boundary of the manifold were more vulnerable to being attacked easily. Wu et al. [27] presented a novel method to defend the adversarial attacks. They used the peculiarity of attacking gradient to defend the adversarial attacks.

3.3 Adversarial Examples Detection

Zhang et al. [28] visualized the attacking process and demonstrated how to take advantage of neuron sensitivity to detect the adversarial examples. Wang et al. [22] mutated the architecture of the model to recognize the adversarial examples. One of the fascinating research in this field was presented by Liu et al. [12]. They applied the perceptual hash to the field of adversarial examples detecting.

4   Method

4.1 Motivations

We propose the following three questions to be clarified or tested by the following designed hypotheses and experiment.

- The model architecture: Whether the adversarial examples are aggressive depends solely on the (non) robust feature of the original image from data set [8]?
- The activation function: Are the adversarial examples' aggression and the neural network model's traits, such as the hyperparameters, the activation functions, or the model's architecture, entirely separate?
- The loss function: Ultimately, we give this query with certain intuitions, in terms of prior studies, whether the aggression of adversarial examples correlates to the loss function?

4.2 Proposed Hypotheses

We believe that the aggressivity of the adversarial examples is connected to the various components from the model architecture, the activation function, and the loss function, based on our past work and relevant literature [1,6,14–16,21,23].

4.3 Design of Experiment

We intend to design an experiment that is separated into three phases to examine the reliability of the aforementioned hypotheses:

1. In terms of the model architecture, we use two diverse model architectures to test this hypothesis and investigate the attacking influence on the neural network model when we change diverse model architecture.
2. In terms of the activation function, we take the same data set and loss function in the experiment and establish a second neural network model that is different from the past one by modifying the model's structure, such as the hyperparameters, activation function, and so forth.
3. In terms of the loss function, we manage the data set in the same way as we control the neural network model, and we utilize the Cross-Entropy loss and the Focal loss as loss functions to investigate the relationship between adversarial examples and loss functions.

Combining our previous analysis and literature review above, we cannot generate the adversarial examples without the construction of model, the activation function and the loss function, so what roles do these three components play in adversarial examples? We carry out the experimental verification of the aspects of these three components. We create a four-layer neural network as experimental subject as the Fig. 2 shown, and the Fig. 3 express the entire processing of the experiment.

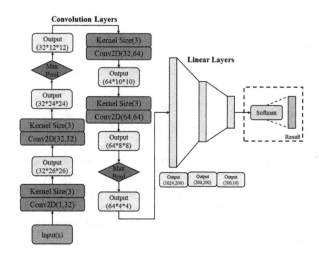

Fig. 2. The architecture of neural network

4.4   Introduction for Loss Functions

Cross-Entropy Loss Function. The Cross-Entropy loss function is commonly employed in machine learning, particularly for target classification and target detection. Shannon [20] introduced the all of the theories of Cross-Entropy in the paper "A mathematical theory of communication". The Cross-Entropy formula has been expressed as follows.
Binary classification [20]:

$$L = \frac{1}{N} \sum_i L_i = \frac{1}{N} \sum_i -[y_i \cdot \log(p_i) + (1 - y_i) \cdot \log(1 - p_i)] \ . \tag{8}$$

Multi-class classification [20]:

$$L = \frac{1}{N} \sum_i L_i = -\frac{1}{N} \sum_i \sum_{c=1}^{M} y_{ic} \log(p_{ic}) \ . \tag{9}$$

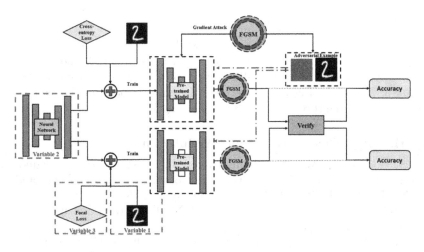

Fig. 3. The processing of experiment

Focal Loss Function. To balance the influence of positive samples and negative samples on the model, Lin et al. present the Focal loss function to replace the Cross-Entropy loss function [11].

$$FL(p_i) = -\alpha_t(1 - p_t)^\gamma \log(p_t) \ . \tag{10}$$

The α_t can balance the importance of positive samples and negative samples, and by adjusting the γ, we can balance the influence of the complex and straightforward examples. The p_t denotes the probability density.

5 Experiment

5.1 Environment

The environment of experiment is shown in the Table 1.

Table 1. The experiment environment

GPU	Tesla P100
CPU	12 * Xeon Gold 6271
Python	Python 3.7
Deep learning Frame	Pytorch1.8.1
Dataset	Mnist [24]/CIFAR-10 [9]

5.2   Process of Experiment

First, we create a four-layer neural network with ReLU [5], Sigmoid [19], LeakyReLU [2], Tanh [2], ELU [2], ReLU6 [2], Logsigmoid [3] and Softplus [4] as activation functions. The overall structure of the neural network can be seen in the Fig. 2, and the Fig. 3 demonstrates the entire process of the experiment. The details of these three controlled trials are as follows:

Experiment 1. Change the Model Architecture

- First, we use the Model we built before and DenseNet [7] with sigmoid as activation function, utilize the Cross-Entropy as the loss function, and use MNIST [24] and CIFAR10 [9] to test this hypothesis.
- We take advantage of the Model to fit MNIST [24] data and use DenseNet [7] to fit CIFAR10 [9].
- We train the diverse epoch of the neural network model with these two data sets separately.
- Ultimately, we attack both models with adversarial examples and gain the result of the attack as the Table 2 shown.

Table 2. The consequence of EXP.1

	Model+MNIST			DenseNet+CIFAR10		
	Sigmoid40	Sigmoid80	Sigmoid120	**Sigmoid40**	Sigmoid80	Sigmoid120
Ori.Tra.Acc.	0.996000	0.997617	0.998550	0.620000	0.830000	0.890000
Ori.Tes.Acc.	0.988800	0.990000	0.990100	0.590000	0.690000	0.690000
Clea.Acc.(TestData)	**0.994752**	**0.995459**	**0.996064**	**0.580600**	**0.688900**	**0.687700**
Att.Acc.(TestData)	**0.294278**	**0.287516**	**0.285296**	**0.542600**	**0.678800**	**0.679300**
	Sigmoid40	**Sigmoid80**	Sigmoid120	Sigmoid40	**Sigmoid80**	Sigmoid120
Clea.Acc.(TestData)	**0.996261**	**0.996767**	**0.996261**	**0.580500**	**0.688900**	**0.687700**
Att.Acc.(TestData)	**0.209255**	**0.196120**	**0.188239**	**0.569300**	**0.623500**	**0.625400**
	Sigmoid40	Sigmoid80	**Sigmoid120**	Sigmoid40	Sigmoid80	**Sigmoid120**
Clea.Acc.(TestData)	**0.996053**	**0.997268**	**0.996559**	**0.580500**	**0.688900**	**0.687700**
Att.Acc.(TestData)	**0.223133**	**0.207448**	**0.208561**	**0.573000**	**0.648300**	**0.615800**

[1] The Model+MNIST and DenseNet+CIFAR10 are the applied neural network architectures and datasets.

[2] Sigmoid40, Sigmoid80 and Sigmoid120 denote the number of training epochs with Sigmoid as activation function. **The objects in bold are the attacked models.**

[3] The Ori.Tra.Acc. and Ori.Tes.Acc. are the abbreviation for Original Training dataset Accuracy and Original Testing dataset Accuracy.

[4] The Clea.Acc(TestData) and Att.Acc.(TestData) are short for Clean samples recognition Accuracy and Attacked samples recognition Accuracy.

Experiment 1. Analysis. As the Table 2 illustrated, we investigate the sensitivity of diverse architecture of the neural network. When we use more complex architecture as a training object, better performance and more stability will occur, such as DenseNet is a more complex architecture than the Model we built before.

Experiment 2. Change the Activation Function

- We use the Mnist and Cross-Entropy loss function but change the activation function of the neural network.
- We utilize eight functions that is mentioned above as activation functions of this neural network model and train these models respectively.
- Then we use the FGSM algorithm to attack these models to obtain the adversarial examples. Afterwards, we let the adversarial examples attack the trained models separately. The result of experiment shows as Table 3.

Table 3. The consequence of EXP.2

Model+MNIST	**Sigmoid**	Relu	leaky_relu	tanh	ELU	Relu6	LogSigmoid	Softplus	Standard Deviation
Ori.Tra.Acc.	0.999800	0.999950	0.999967	0.999967	0.999950	0.999917	0.999883	0.999950	0.000053
Ori.Tes.Acc.	0.991000	0.993700	0.994700	0.992100	0.994300	0.993000	0.990300	0.992800	0.001443
Clea.Acc.(TestData)	0.991700	0.994500	0.994600	0.991700	0.993300	0.993900	0.990800	0.993800	**0.001349**
Att.Acc.(TestData)	0.862700	0.974700	0.976100	0.921000	0.983100	0.969600	0.963500	0.968900	**0.038283**
	Sigmoid	**Relu**	leaky_relu	tanh	ELU	Relu6	LogSigmoid	Softplus	Standard Deviation
Clea.Acc.(TestData)	0.992300	0.993500	0.995000	0.992000	0.994100	0.993700	0.989400	0.993200	0.001595
Att.Acc.(TestData)	0.976700	0.912500	0.970100	0.968400	0.979200	0.973600	0.981200	0.981300	0.021414
	Sigmoid	Relu	**leaky_relu**	tanh	ELU	Relu6	LogSigmoid	Softplus	Standard Deviation
Clea.Acc.(TestData)	0.992000	0.994200	0.994900	0.992400	0.994500	0.993900	0.989900	0.992900	**0.001542**
Att.Acc.(TestData)	0.974800	0.971400	0.911600	0.962700	0.977400	0.970700	0.979600	0.980100	**0.021250**
	Sigmoid	Relu	leaky_relu	**tanh**	ELU	Relu6	LogSigmoid	Softplus	Standard Deviation
Clea.Acc.(TestData)	0.991400	0.993900	0.994100	0.991900	0.993700	0.994100	0.990200	0.993400	**0.001382**
Att.Acc.(TestData)	0.974200	0.981400	0.976300	0.901600	0.983800	0.979000	0.979000	0.984400	**0.026042**
	Sigmoid	Relu	leaky_relu	tanh	**ELU**	Relu6	LogSigmoid	Softplus	Standard Deviation
Clea.Acc.(TestData)	0.990000	0.993300	0.994200	0.992400	0.993700	0.994000	0.990300	0.993000	**0.001519**
Att.Acc.(TestData)	0.982300	0.980300	0.980100	0.975500	0.937900	0.977900	0.983300	0.984700	**0.014437**
	Sigmoid	Relu	leaky_relu	tanh	ELU	**Relu6**	LogSigmoid	Softplus	Standard Deviation
Clea.Acc.(TestData)	0.992200	0.994000	0.994400	0.990800	0.994000	0.993200	0.990600	0.992300	**0.001367**
Att.Acc.(TestData)	0.976200	0.976800	0.974500	0.963700	0.982500	0.943100	0.979700	0.980900	**0.012245**
	Sigmoid	Relu	leaky_relu	tanh	ELU	Relu6	**LogSigmoid**	Softplus	Standard Deviation
Clea.Acc.(TestData)	0.991300	0.993600	0.994500	0.992200	0.994100	0.993600	0.989300	0.992500	**0.001605**
Att.Acc.(TestData)	0.967700	0.982700	0.983800	0.956800	0.984800	0.977900	0.908200	0.978300	**0.024114**
	Sigmoid	Relu	leaky_relu	tanh	ELU	Relu6	LogSigmoid	**Softplus**	Standard Deviation
Clea.Acc.(TestData)	0.991400	0.994200	0.994900	0.991600	0.993700	0.993900	0.989500	0.993100	**0.001687**
Att.Acc.(TestData)	0.962000	0.976500	0.974000	0.961200	0.982400	0.972600	0.971300	0.935800	**0.013502**

[1] The Model+MNIST denotes the applied neural network architecture and dataset.
[2] The Sigmoid, Relu, leakyrelu, ELU, Relu6, Logsigmoid and Softplus are the experimental activation function that we used and Standard Deviation reflects the fluctuation. **The objects in bold are the attacked models.**
[3] The Ori.Tra.Acc. and Ori.Tes.Acc. are the abbreviation for Original Training dataset Accuracy and Original Testing dataset Accuracy.
[4] The Clea.Acc(TestData) and Att.Acc.(TestData) are short for Clean samples recognition Accuracy and Attacked samples recognition Accuracy.

Experiment 2. Analysis. As the Table 3 shows, we implement training with eight diverse functions as activation functions and peculiarly compute the standard deviation to visualize the attack accuracy rates fluctuation. The attack success accuracy rate fluctuation of attacked example is more intensive than the clean sample's.

Experiment 3. Change the Loss Function

- To begin, we add diverse loss functions to the neural networks and train them with the Mnist data set. We employ Cross-Entropy loss function and Focal loss function for loss function as Fig. 3 describes it.
- We get two pre-trained classification models with different loss functions once we finish training.
- Furthermore, we utilize the FGSM [6] to attack the model that employs the Cross-Entropy as the loss function and generate the adversarial examples.
- Then we put the generating adversarial examples into the model that employs Focal as loss function and verify the accuracy of both models.
- Ultimately, we get the evaluation result of the two models. (Table 4)

Table 4. The consequence of EXP.3

Model+MNIST	**CrossEntropy**	Focal	Standard Deviation
Ori.Tra.Acc.	0.9903	0.9871	0.0016
Ori.Tes.Acc.	0.9805	0.9771	0.0017
Clea.Acc.(TestData)	0.9804	0.9776	**0.0014**
Att.Acc.(TestData)	0.8128	0.9000	**0.0436**
	CrossEntropy	**Focal**	Standard Deviation
Clea.Acc.(TestData)	0.9809	0.9760	**0.0025**
Att.Acc.(TestData)	0.9511	0.6547	**0.1482**

1 The Model+MNIST denotes the applied neural network architecture and dataset.

2 The CrossEntropy and Focal are the experimental loss function that we used and Standard Deviation reflects the fluctuation. **The objects in bold are the attacked models.**

3 The Ori.Tra.Acc. and Ori.Tes.Acc. are the abbreviation for Original Training dataset Accuracy and Original Testing dataset Accuracy.

4 The Clea.Acc(TestData) and Att.Acc.(TestData) are short for Clean samples recognition Accuracy and Attacked samples recognition Accuracy.

Experiment 3. Analysis. When we use different loss functions, Cross-Entropy and Focal, to train the classification model, they converge almost at the same time after several iterations, and the prediction accuracy is very similar. However, from the Table 4, we can conclude that diverse loss functions will affect the aggressivity of the adversarial examples, and the model with Focal loss will remain more sensitive than the model with Cross-Entropy.

6   Conclusion

This paper advances the research on the interpretability of adversarial examples and the principle of aggression. According to the consequence of experiments, we investigate the neural network's sensitivity to adversarial examples in diverse conditions. This property of sensitivity could be helpful to detect the adversarial examples. In addition, we introduce an innovative and necessary direction for research of adversarial examples to examine the internal mechanism of adversarial examples. We need to focus more on strengthening neural networks to reduce the impact of adversarial examples, such as creating more sophisticated and appropriate loss functions or hiding the loss function's information when the neural networks are infiltrated.

We intend to think more about the interpretation of the adversarial examples, and a further study is made on the relationship between the three variables discussed in this paper and the adversarial examples.

References

1. Carlini, N., Wagner, D.A.: Towards evaluating the robustness of neural networks. In: 2017 IEEE Symposium on Security and Privacy (SP), pp. 39–57 (2017)
2. Clevert, D.A., Unterthiner, T., Hochreiter, S.: Fast and accurate deep network learning by exponential linear units (elus). arXiv: Learning (2016)
3. Contributors, T.: Logsigmoid (2019). https://pytorch.org/docs/stable/generated/torch.nn.LogSigmoid.html. Accessed on 23 Mar 2022
4. Contributors, T.: Softplus (2019). https://pytorch.org/docs/stable/generated/torch.nn.Softplus.html. Accessed on 23 Mar 2022
5. Glorot, X., Bordes, A., Bengio, Y.: Deep sparse rectifier neural networks. In: 2011 Proceedings of the 14th International Conference on Artificial Intelligence and Statisitics (AISTATS), vol. 15, pp. 315–323 (2011)
6. Goodfellow, I.J., Shlens, J., Szegedy, C.: Explaining and harnessing adversarial examples. In: International Conference on Learning Representations (ICLR) abs/1412.6572 (2015)
7. Huang, G., Liu, Z., Van Der Maaten, L., Weinberger, K.Q.: Densely connected convolutional networks. In: 2017 IEEE Conference on Computer Vision and Pattern Recognition (CVPR), pp. 2261–2269 (2017). https://doi.org/10.1109/CVPR.2017.243
8. Ilyas, A., Santurkar, S., Tsipras, D., Engstrom, L., Tran, B., Madry, A.: Adversarial examples are not bugs, they are features. In: Advances in Neural Information Processing Systems abs/1905.02175, pp. 125–136 (2019)
9. Krizhevsky, A.: Learning multiple layers of features from tiny images. Technical report (2009)
10. Kurakin, A., Goodfellow, I.J., Bengio, S.: Adversarial examples in the physical world. Toulon, France (2017)
11. Lin, T.Y., Goyal, P., Girshick, R.B., He, K., Dollár, P.: Focal loss for dense object detection. IEEE Trans. Pattern Anal. Mach. Intell. **42**, 318–327 (2020)
12. Liu, C., et al.: Defend against adversarial samples by using perceptual hash. Comput. Mater. Continua **62**(3), 1365–1386 (2020)
13. Ma, X., et al.: Characterizing adversarial subspaces using local intrinsic dimensionality. In: International Conference on Learning Representations (2018). https://openreview.net/forum?id=B1gJ1L2aW
14. Madry, A., Makelov, A., Schmidt, L., Tsipras, D., Vladu, A.: Towards deep learning models resistant to adversarial attacks. ArXiv abs/1706.06083 (2018)

15. Moosavi-Dezfooli, S.M., Fawzi, A., Frossard, P.: Deepfool: a simple and accurate method to fool deep neural networks. In: 2016 IEEE Conference on Computer Vision and Pattern Recognition (CVPR), pp. 2574–2582 (2016)

16. Papernot, N., Mcdaniel, P., Jha, S., Fredrikson, M., Celik, Z.B., Swami, A.: The limitations of deep learning in adversarial settings. In: 2016 IEEE European Symposium on Security and Privacy (EuroS&P), pp. 372–387 (2016)

17. Qin, Y., Carlini, N., Goodfellow, I.J., Cottrell, G., Raffel, C.: Imperceptible, robust, and targeted adversarial examples for automatic speech recognition, vol. 97, p. 5231–5240, California, USA, 09–15 June 2019

18. Raff, E., Sylvester, J., Forsyth, S., McLean, M.: Barrage of random transforms for adversarially robust defense. In: 2019 IEEE/CVF Conference on Computer Vision and Pattern Recognition (CVPR), pp. 6521–6530 (2019). https://doi.org/10.1109/CVPR.2019.00669

19. Pearl, R., Reed, L.J.: On the rate of growth of the population of the united states since 1790 and its mathematical representation. Proc. Nat. Acad. Sci. **6**, 275–288 (1920)

20. Shannon, C.E.: A mathematical theory of communication. Bell Syst. Tech. J. **27**, 379–423 (1948)

21. Szegedy, C., et al.: Intriguing properties of neural networks. In: International Conference on Learning Representations (ICLR) abs/1312.6199 (2014)

22. Wang, J., Dong, G., Sun, J., Wang, X., Zhang, P.: Adversarial sample detection for deep neural network through model mutation testing. In: 2019 IEEE/ACM 41st International Conference on Software Engineering (ICSE), pp. 1245–1256 (2019)

23. Xiao, C., Li, B., Zhu, J.Y., He, W., Liu, M., Song, D.X.: Generating adversarial examples with adversarial networks. ArXiv abs/1801.02610 (2018)

24. LeCun, Y., Bottou, L., Bengio, Y., Haffner, P.: Gradient-based learning applied to document recognition. Proc. IEEE **86**(11), 2278–2324 (1998)

25. Yakura, H., Sakuma, J.: Robust audio adversarial example for a physical attack. In: International Joint Conference on Artificial Intelligence (IJCAI), vol. abs/1810.11793, Macao, China (2019)

26. Yang, Y., Zhang, G., Katabi, D., Xu, Z.: Me-net: towards effective adversarial robustness with matrix estimation, vol. 97, pp. 7025–7034. Long Beach, California, USA (2019)

27. Wu, Y., Arora, S.S., Wu, Y., Yang, H.: Beating attackers at their own games: Adversarial example detection using adversarial gradient directions. In: 35th AAAI Conference on Artificial Intelligence/33rd Conference on Innovative Applications of Artificial Intelligence/11th Symposium on Educational Advances in Artificial Intelligence, pp. 2969–2977 (2021)

28. Zhang, C., et al.: Interpreting and improving adversarial robustness of deep neural networks with neuron sensitivity. IEEE Trans. Image Process. **30**, 1291–1304 (2021). https://doi.org/10.1109/TIP.2020.3042083

Towards Robust Graph Convolution Network via Stochastic Activation

Zhengfei Yu, Hao Yang, Lin Wang, Lijian Sun, and Yun Zhou$^{(\boxtimes)}$

Science and Technology on Information Systems Engineering Laboratory,
National University of Defense Technology, Changsha, China
{yuzhengfei19,yanghao,wanglin_,ljsun,zhouyun}@nudt.edu.cn

Abstract. Graph Neural Networks (*GNNs* for short), a generalization of neural networks to graph-structured data, performance good in closed setting with perfect data for variety tasks, including node classification, link prediction and graph classification. However, GNNs are vulnerable to adversarial attacks, i.e., a small perturbation to the graph structure and node features in wild setting can lead to non-trivial performance degradation. Non-robustness is one of the main obstacle to applying GNNs in the wild. In this work, we focus on one of the most popular GNNs, Graph Convolutional Networks (*GCN* for short), and propose Stochastic Activation GCN (*SA-GCN* for short) to improve the robustness of GCN models. More specifically, we propose building a roust model by directly introducing a regularization term to the objective function and maximizing the feature distribution variance. Extensive experiments show that this simple design makes *SA-GCN* achieving significantly improved robustness against adversarial attacks. Moreover, our approach generalizes well and can be equipped with various models. Conducted empirical experiments demonstrate the effectiveness of *SA-GCN*.

Keywords: Graph neural networks · Adversarial attacks · Stochastic neural networks

1 Introduction

Graphs are the fundamental element of non-Euclidean data representation in real world. Many complex relationships between entities, including molecular networks [26], dating networks [16], physical networks [1] and even implicit networks in images [10], can be represented by graphs. Graph data analysis has become one of the core tasks in the deep learning community.

Recent years, Graph Neural Networks (*GNNs* for short), with the ability to effectively leverage the inherent structure, has attracted great attentions on graph-structured data analyzing task. Currently, GNNs have produced state-of-the-art performance on a variety of challenging tasks including node classification [6], link prediction [30], subgraph classification [25] and graph classification [12]. It can be said that GNNs have become one of the most essential tools for graph-structural data analyzing.

Supported by Huxiang Youth Talent Support Program (No. 2021RC3076), and Training Program for Excellent Young Innovators of Changsha (No. KQ2009009).

Despite successful use in in a wide range of applications, GNNs, known to be vulnerable to a variety of adversarial attacks, remain difficult to develop in wild setting. Recent studies [3,33] have shown that small imperceptible perturbations to graph data can rapidly degrade the classification performance of GNN. This limits the application in risk and safety-critical scenarios. For example, the popular Graph Convolutional Networks (*GCN* for short), which rely on aggregating message passes from a node's neighborhood, are not immune to structure attacks [3], wherein an attacker adds limited edges between nodes.

Though there exists a vast literature on adversarial attacks and defense on graph-structural data [23,32], we focus on designing defense mechanisms for graph attacks, which is a more critical research. While most existing GNN models assume that the given graph perfectlly depicts the ground-truth of the relationship between nodes, we proposed that such assumptions are bound to yield sub-optimal results as real-world graphs might be malicious perturbated. For example, Zheng et al. [29] proposed that the graph we obtained are almost with many missing edges or disconnected edges. In this case, GNNs employing deterministic propagation mechanisms are generally not robust to graph attacks.

In this work, we investigate GCN stochastic activation and consistency regularization strategies for improving the robustness of GCN models. Specifically, we present the Stochastic Activation GCN (*SA-GCN* for short), a simple yet robustness graph learning framework. We focus on semi-supervised node classification task. Through introducing a regularization term to the objective function and maximize the variance of the feature distribution, *SA-GCN* obtained more uncertainty while preserved abundant features. Extensive experiments also show empirically that *SA-GCN* improved robustness against adversarial attacks on semi-supervised node classification task on GCN benchmark datasets and mitigate the issue of non-robustness, which are commonly faced by existing GCNs, to a great extend.

The key contributions of this work are summarized as follows:

- A novel architecture for graph representation learning, *SA-GCN*, that is robust to attacks by design;
- An uncertainty learning strategy for achieving robustness to structural and features perturbations;
- Across a series of adversarial attacks, *SA-GCN* outperforms existing methods in most cases.

The paper is structured as follows. Section 2 is dedicated to the introduction of related work. It is followed by our finding in Sect. 3 on analysing the methodology of *SA-GCN*, which is the backbone of our metholody. The experiment results and conclusion are the last two section of this paper.

2 Related Work

2.1 Adversarial Attack and Defense on Graph Neural Networks

Graph Neural Networks have achieved excellent results in various graph-related tasks [6,12,25,30]. However, recent studies have shown that GNNs can be esaily fooled by small perturbation on the input graph data [3,18,21,33]. Dai et al. [3] attacks on both node and graph classification tasks by modifiying the graph structure with reinforcement learning. Zügner et al. [33] proposes a greedy approximation scheme to attack attributed graphs. Furthermore, researchers [18,21] present more realistic graph injection attacks. Interested readers please refer to the review [9].

While previous work has shown that GNNs are vulnerable to attacks, we focus on the more critical question, how can we improve the robustness of GNNs against adversarial attacks. One line to solve the problem is to design empirical defense mechanisms [22,24]. Another line suggests developing certified robustness guarantees [2,34]. Our work falls into the previous line.

2.2 Stochastic Neural Networks for Graph

Recent years, Stochastic Neural Networks (*SNNs* for short) have achieved great success in the deep learning community [7,8,14], inspiring several exploration of using SNNs on graph data [17,20,28,31]. Among them, Bayesian graph neural networks [28] uses a parametric random graph model to incorporate uncertainty information, and derive the graph stochastic neural network formulations theoretically. Graph Stochastic Neural Networks [20] employ variational inference to approximate the intractable joint posterior. Uncertainty-Matching Graph Neural Networks [17] utilize epistemic uncertainties to defend against poisoning attacks and achieves significant improvements. While the most relevant work is Robust GCN [31], which adopts Gaussian distributions in each layer and uses graph structure uncertainties to reduce the impacts of adversarial attacks, our method is indeed simpler and achieves similar performances.

3 Methodology

In this section, we present the proposed framework, *SA-GCN*, on semi-supervised node classification task. The idea is to design stochastic activation mechanism through adding Gaussian noise at each layer. With this simple mechanism, we can generate more uncertainty into the latent presentation of the nodes and induce an activation loss function to improve robustness under adversarial attacks.

3.1 Stochastic Mechanism

We consider a graph convolution network trained for semi-supervised node classification. Instead computing fixed hidden representation of feature vectors, we propose to use stochastic learning mechanism. As illustrated in Fig. 1, we describe the stochastic mechanism with the l-th layer and $(l+1)$-th layer. Given arbitrary entire graph, $\mathcal{G} = (\mathcal{V}, \mathcal{E})$ with features for all nodes $\mathbf{x}_v$, $\forall v \in \mathcal{V}$ are provided as input. In particular, we assume that we have obtained the hidden presentation $\mathbf{h}^l$ at layer l. For single layer $l + 1$, [11] directly aggregate the representations of nodes in its immediate neighborhood $\{\mathbf{h}_u^l, \forall u \in \mathcal{N}(v)\}$ into a single vector $\mathbf{h}_{\mathcal{N}(v)}^{l+1}$. In adversarial setting, due to the presence of attacks and noise in the graph structure and node features, we cannot use these representations directly. To alleviate the challenge, we propose to add a stochastic

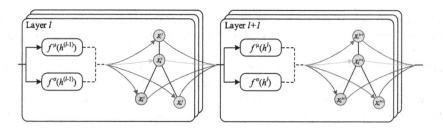

Fig. 1. Overall architecture of the proposed *SA-GCN* method for semi-supervised node classification.

learning module before the aggregation operation to output a series of univariate distributions. Then, we sample from the distribution with parameterized mean $f^\mu\left(h^l\right)$ and variance $f^\sigma\left(h^l\right)$ independently, and get a random hidden representation $\mathbf{h}_u^l$.

Finally, with the design described above, we choose to replace all the layer of the classical *GCN* and get a modified multi-layer message passing which obey the following layer-wise propagation rule:

$$H^{(l+1)} = \sigma\left(\tilde{D}^{-\frac{1}{2}}\tilde{A}\tilde{D}^{-\frac{1}{2}}Z^{(l)}W^{(l)}\right) \tag{1}$$

where $Z^{(l)} \sim \mathcal{N}\left(\mu^{(l)}, \sigma^{(l)}\right)$, $\mu^{(l)} = f^\mu\left(h^{(l-1)}\right)$ and $\sigma^{(l)} = f^\sigma\left(h^{(l-1)}\right)$. In this paper, we consider a three-layer *SA-GCN* for semi-supervised node classification task.

3.2   Loss Function

As shown in [11], the vanilla GCN use cross-entropy over all labeled nodes to calculate the classification loss:

$$Loss_C = -\sum_{i \in y_L}\sum_{j=1}^c Y_{ij}\ln Z_j(i) \tag{2}$$

where y_L denote the set of labeled nodes, and $Y_{ij} = \begin{cases} 1, & \text{if the label of node } i \text{ is } j \\ 0, & \text{otherwise} \end{cases}$.

Since stochastic learning module added in the framework, this design with Gaussian distribution derives a *max-entropy* regularization term. Thus, following [27], the loss function can be written as

$$Loss_{SA} = Loss_C - \lambda\sum_{i=1}^N \ln \sigma_i \tag{3}$$

where the first term $Loss_C$ corresponds to the cross-entropy loss defined in equation (2), the second term is used to generate randomness so that the model can obtain more robustness. Moreover, λ is the hyper-parameter used to control the ability of the proposed stochastic learning module, and the logarithmic operation $\ln \sigma_i$ is introduced to safeguard the numerical calculation. With this loss function, we can make robust classification predictions.

4   Experiment

In this section, we empirically evaluate the proposed *SA-GCN* method on semi-supervised node classification in adversarial setting.

4.1   Experimental Settings

Datasets. We conduct experiments on two widely used benchmark datasets, Cora [15] and Citeseer [4], where nodes represent documents and edges represent citation relationships, for classifying the research topic of papers. Each node is associated with a bag-of-words feature vector and a ground-truth label. Table 1 summarizes the statistics for the two benchmark datasets, including the number of nodes, edges, classes, the dimension of features and the datasets partitioning. Our preprocessing scripts for Cora and Citeseer are based on DeepRobust [13], a widely used pyTorch library for adversarial machine learning. By default, the loaded deeprobust.graph.data.Dataset will select the largest connect component of the graph. Follow the experimental setting of literature on semi-supervised graph mining, we sample 10% nodes for Cora and Citeseer to compose the training set. All the training set has no overlap with validation and test sets.

Table 1. Dataset Statistics. Following [12, 39, 40], we only consider the largest connected component (LCC).

	NLCC	ELCC	Classes	Features	Train/val./test
Cora	2,485	5,069	7	1,433	247/249/1988
Citeseer	2,110	3,668	6	3,703	210/211/1688

Baseline and Attack Used for Evaluation. In this paper, we compare the performance of GCN [11] and our proposed *SA-GCN* under NETTACK [33]. In particular, GCN is one of the most classic GNN models and defines the graph convolution in the spectral domain, which uses the first-order approximation to reduce the number of parameters when aggregating the neighbor information. For NETTACK, it is a well known targeted attack, which aim at degrading the overall performance of graph deep learning models.

More specifically, in this attack, new edges are randomly introduced between two previously unconnected nodes. Although simple, this attack is considered effective, especially with high noise ratios and sparse graphs. We implement NETTACK using the publicly available DeepRobust library.

Parameter Settings. For GCN and *SA-GCN*, we set the number of layers and hidden units as 3 and 32 in our experiments, respectively. We set the learning rate as 0.01 and weight decay as $5 * 10^{(-4)}$. The hyper-parameters λ for *SA-GCN* is set as 0.5 on all datasets. For a fair comparison, we follow literature [11] and set the dropout rate for GCN and *SA-GCN* as 0.8. Additionally, we initialize all weights with the widely used Xavier initialization [5].

4.2   Results Analysis

We evaluate the classification accuracy of different methods against NETTACK. In the experiments, we vary the number of perturbations per node in the original graph between 1 and 5 because more perturbations, which will lead to too low performance, is unbearable in practice.

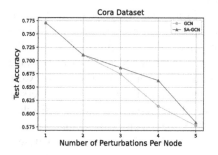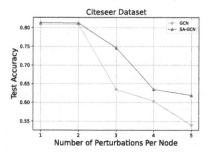

Fig. 2. Results of GCN and SA-GCN (ours) when adopting NETTACK as the attack method.

As shown in Fig. 2, We can see that SA-GCN outperforms GCN on all datasets most of the time. Experimental results show that, despite its simplicity, our proposed framework can significantly improve the robustness of GCNs.

5   Conclusion

In this paper, we presented *SA-GCN*, a simple and effective stochastic learning mechanism to enhance the robustness of GNN models. *SA-GCN* introduced the stochastic learning module with Gaussian distribution at GCN layers and derive an activation regularizer that encourages high activation variability via an entropy-maximization regularization term. Considerable experiments on GCN benchmark datasets have verified that *SA-GCN* can generally and consistently promote the performance of current popular GCNs in adversarial setting. Although we implement the stochastic learning mechanism using the graph convolution layers in [11], we can plug it in other message passing operators, such as GraphSAGE [6], GAT [19], etc. to further boost performance on variety datasets. We will carry out this work in the future.

References

1. Bapst, V., et al.: Unveiling the predictive power of static structure in glassy systems. Nature Physics (2020)
2. Bojchevski, A., Günnemann, S.: Certifiable robustness to graph perturbations. In: Proceedings of NeurIPS (2019)
3. Dai, H., et al.: Adversarial attack on graph structured data. In: Proceedings of ICML (2018)

4. Giles, C.L., Bollacker, K.D., Lawrence, S.: Citeseer: an automatic citation indexing system. In: Proceedings of the Third ACM Conference on Digital Libraries (1998)
5. Glorot, X., Bengio, Y.: Understanding the difficulty of training deep feedforward neural networks. In: Proceedings of AISTATS (2010)
6. Hamilton, W.L., Ying, Z., Leskovec, J.: Inductive representation learning on large graphs. In: Proceedings of NeurIPS (2017)
7. He, Z., Rakin, A.S., Fan, D.: Parametric noise injection: trainable randomness to improve deep neural network robustness against adversarial attack. In: Proceedings of CVPR (2019)
8. Jeddi, A., Shafiee, M.J., Karg, M., Scharfenberger, C., Wong, A.: Learn2perturb: an end-to-end feature perturbation learning to improve adversarial robustness. In: Proceedings of CVPR (2020)
9. Jin, W., et al.: Adversarial attacks and defenses on graphs. SIGKDD Explor. Newsl. **22**(2), 19–34 (2020)
10. Johnson, J., Gupta, A., Fei-Fei, L.: Image generation from scene graphs. In: Proceedings of CVPR (2018)
11. Kipf, T.N., Welling, M.: Semi-supervised classification with graph convolutional networks. In: Proceedings of ICLR (2017)
12. Lee, J.B., Rossi, R.A., Kong, X.: Graph classification using structural attention. In: Proceedings of KDD (2018)
13. Li, Y., Jin, W., Xu, H., Tang, J.: Deeprobust: a platform for adversarial attacks and defenses. In: Proceedings of AAAI (2021)
14. Liu, X., Cheng, M., Zhang, H., Hsieh, C.: Towards robust neural networks via random self-ensemble. In: Proceedings of ECCV (2018)
15. McCallum, A.K., Nigam, K., Rennie, J., Seymore, K.: Automating the construction of internet portals with machine learning. Inf. Retrieval **3**(2), 127–163 (2000). https://doi.org/10.1023/A:1009953814988
16. Pal, A., Eksombatchai, C., Zhou, Y., Zhao, B., Rosenberg, C., Leskovec, J.: Pinnersage: multi-modal user embedding framework for recommendations at pinterest. In: Proceedings of KDD (2020)
17. Shanthamallu, U.S., Thiagarajan, J.J., Spanias, A.: Uncertainty-matching graph neural networks to defend against poisoning attacks. In: Proceedings of AAAI (2021)
18. Sun, Y., Wang, S., Tang, X., Hsieh, T., Honavar, V.G.: Adversarial attacks on graph neural networks via node injections: a hierarchical reinforcement learning approach. In: Proceedings of WWW (2020)
19. Velickovic, P., Cucurull, G., Casanova, A., Romero, A., Liò, P., Bengio, Y.: Graph attention networks. In: Proceedings of ICLR (2018)
20. Wang, H., Zhou, C., Chen, X., Wu, J., Pan, S., Wang, J.: Graph stochastic neural networks for semi-supervised learning. In: Proceedings of NeurIPS (2020)
21. Wang, J., Luo, M., Suya, F., Li, J., Yang, Z., Zheng, Q.: Scalable attack on graph data by injecting vicious nodes. Data Min. Knowl. Discov. **34**(5), 1363–1389 (2020)
22. Wu, H., Wang, C., Tyshetskiy, Y., Docherty, A., Lu, K., Zhu, L.: Adversarial examples for graph data: deep insights into attack and defense. In: Proceedings of IJCAI (2019)
23. Xu, H., et al.: Adversarial attacks and defenses in images, graphs and text: a review. Int. J. Autom. Comput. **17**(2), 151–178 (2020). https://doi.org/10.1007/s11633-019-1211-x
24. Xu, K., et al.: Topology attack and defense for graph neural networks: an optimization perspective. In: Proceedings of IJCAI (2019)
25. Xu, M., Zhao, C., Rojas, D.S., Thabet, A.K., Ghanem, B.: G-TAD: sub-graph localization for temporal action detection. In: Proceedings of CVPR (2020)
26. You, J., Liu, B., Ying, Z., Pande, V.S., Leskovec, J.: Graph convolutional policy network for goal-directed molecular graph generation. In: Proceedings of NeurIPS (2018)

27. Yu, T., Yang, Y., Li, D., Hospedales, T.M., Xiang, T.: Simple and effective stochastic neural networks. In: Proceedings of AAAI (2021)
28. Zhang, Y., Pal, S., Coates, M., Üstebay, D.: Bayesian graph convolutional neural networks for semi-supervised classification. In: Proceedings of AAAI (2019)
29. Zheng, C., et al.: Robust graph representation learning via neural sparsification. In: Proceedings of ICML (2020)
30. Zhu, D., Cui, P., Wang, D., Zhu, W.: Deep variational network embedding in wasserstein space. In: Proceedings of KDD (2018)
31. Zhu, D., Zhang, Z., Cui, P., Zhu, W.: Robust graph convolutional networks against adversarial attacks. In: Proceedings of KDD (2019)
32. Zhu, Y., Xu, W., Zhang, J., Liu, Q., Wu, S., Wang, L.: Deep graph structure learning for robust representations: a survey. CoRR (2021)
33. Zügner, D., Akbarnejad, A., Günnemann, S.: Adversarial attacks on neural networks for graph data. In: Proceedings of KDD (2018)
34. Zügner, D., Günnemann, S.: Certifiable robustness of graph convolutional networks under structure perturbations. In: Proceedings of KDD (2020)

Machine Learning

Extreme Learning Machine Based on Adaptive Matrix Iteration

Yuxiang Li, Weidong Zou$^{(\boxtimes)}$, Can Wang, and Yuanqing Xia

School of Automation, Beijing Institute of Technology, Beijing 100081, China
zouweidong1985@163.com

Abstract. Under the continuous optimization and development of various algorithms in machine learning, the performance of the algorithm model on classification and regression prediction problems has become an important evaluation metric for the quality of algorithms. In order to solve the problems of low testing accuracy and unsatisfactory generalization performance of the models trained by the traditional extreme learning machine, this paper proposes an extreme learning machine algorithm based on adaptive convergence factor matrix iteration. This algorithm optimizes the calculation method of solving the hidden layer output weight matrix, while retaining the network structure model of the traditional extreme learning machine. This algorithm is implemented with a matrix iterative method that includes an adaptive convergence factor to compute the output weight matrix. As a result, it can adaptively select the optimal convergence factor according to the structure of the iterative equations, and thus use iterative method to solve linear equations efficiently and accurately upon ensuring the convergence of the equations. The experiment results show that the proposed algorithm has better performance in model training efficiency and testing accuracy, compared with the traditional extreme learning machine, the support vector machine, and other algorithms for data classification and regression prediction.

Keywords: Machine learning · Extreme learning machine · Adaptive convergence factor · Matrix iteration · Model optimization · Data classification

1 Introduction

The multi-layer neural network has the advantages of strong fitting ability and high training accuracy, but due to its complex network structure [1], the multi-layer neural network needs to set a large number of parameters during model training and needs to go through multiple iterations, making the training It is

This paper is partly supported by organization National Key R&D Program of China under Grant 2018YFB1003700 and National Natural Science Foundation of China under Grant 61906015.

© Springer Nature Switzerland AG 2022
Y. Tan et al. (Eds.): ICSI 2022, LNCS 13345, pp. 177–188, 2022.
https://doi.org/10.1007/978-3-031-09726-3_16

easy to fall into the local optimal solution, over-fitting the model and occupy a lot of computing resources and time in the learning process. In order to solve the above problems, Huang proposed extreme learning machine(ELM) [2], which is a single hidden layer feedforward neural network algorithm. Due to its simple structure and its characteristics that input weights and biases can be randomly generated during training and the output weights are determined by solving the minimum norm solution of linear equations. Compared with many traditional algorithms, the ELM algorithm can have faster training speed and better generalization ability on the basis of ensuring the learning accuracy [3]. It is used in image classification [4] and data prediction [5] and other fields.

In the development of ELM, the regularized extreme learning machine (RELM) solved the problem of model overfitting [6], and the ADMM-ELM proposed by Lai et al. [7] based on the convex optimization theory improved the convergence rate of the algorithm, Huang based on the kernel method proposed KELM [3] to promote the development of the algorithm. On the other hand, due to the large number of hidden layer nodes in the extreme learning machine, the randomly given weights and biases cannot obtain the best network structure. In order to solve this problem, researchers continue to optimize the network. Huang and Rong proposed the self-increasing [8] and pruning [9] algorithms respectively to optimize the network structure, and Ye et al. proposed the QRI-ELM algorithm based on the increment of QR decomposition [10] and so on. Although the above research has promoted the research process of ELM, it has not fundamentally solved the problem of insufficient ELM modeling accuracy.

Although ELM has the advantages of simple network structure and few setting parameters, due to the characteristics of its single-layer network structure and parameter randomness, the accuracy of the model trained by it has great instability. The Moore-Penrose generalized inverse method used by ELM to solve the output weight matrix also has certain defects. In many scenarios, the obtained model is often a nonlinear model [11], and the correlation matrix often has a large order. At this time, solving the model by generalized inverse is often inefficient and cannot obtain a more accurate solution. In this case, it is generally considered to use some common matrix iteration methods to solve the output weight matrix, such as Jacobi iteration algorithm, Gauss-Seidel iteration algorithm [12], successive over-relaxation iterative algorithm [13] and so on. Although these algorithms are more efficient in some cases [14], they are not suitable for all cases. For example, Gauss-Seidel iteration requires the coefficient matrix to be non-singular, and in the case where the dimension of the coefficient matrix is low, the solution obtained by iteration is much less accurate than that obtained by generalized inverse.

The shortcomings of the above optimization of the ELM algorithm are mainly manifested in two aspects: 1) These optimizations are all aimed at the ELM network structure [15], which has certain limitations. While the network structure is continuously optimized, the complexity of model training will increase. With the increase, the efficiency of data processing cannot meet the requirements of many scenarios. 2) Solving the output weight matrix is an extremely important step in the ELM algorithm, which has a great impact on the accuracy of the model and the time and space resources occupied. The method of using generalized inverse to solve the output weight matrix has some defects. How to solve the problem of insufficient model accuracy on the basis of retaining the fast modeling of the ELM algorithm should become the direction of algorithm research.

In response to the requirements of model operation efficiency and accuracy, this paper proposes an extreme learning machine algorithm based on adaptive convergence factor matrix iteration. The algorithm uses a matrix iteration method with adaptive convergence factors to calculate the output weight matrix, and uses the established model predictive analysis of the data. The experimental results show that the algorithm has higher test accuracy compared with extreme learning machine, regularized extreme learning machine and support vector machine. In experiments on large datasets, the resources and time consuming of our algorithm are less. In addition, the experiment also found that when training some datasets with specific characteristics, the accuracy of AC-ELM is even better than some multi-layer convolutional neural network structure algorithms.

2   Extreme Learning Machine

The extreme learning machine is a feedforward neural network (SLFN) composed of a single hidden layer. It consists of an input layer, a hidden layer and an output layer. Its input weights and biases are randomly generated, and the output weight is calculated from this. Unlike SLFN, ELM does not have output bias.

Given M training samples $(x_i, t_i), i = 1, 2...M$, $x_i = [x_{i1}, x_{i2}, \cdots, x_{im}]^T \in R^n$, $t_i = [t_{i1}, t_{i2}, \cdots, t_{il}]^T \in R^m$, where x_i and t_i represent the input feature vector of the i−th sample and the output feature vector corresponding to its label. Let $g(x)$ be the activation function, the network structure of ELM consists of an input layer with m nodes, a hidden layer with N nodes and an output layer with l nodes. Its mathematical model is

$$\sum_{i=1}^{N} \beta_i g\left(w_i x_j + b_i\right) = t_j, \ j = 1, 2 \cdots, M \tag{1}$$

where $w_i = [w_{i1}, w_{i2}, \cdots, w_{im}]^T$ represents the input weight vector connecting the ith hidden layer node and the input layer node, $\beta_i = [\beta_{i1}, \beta_{i2}, \cdots, \beta_{il}]$ is the output weight matrix connecting the i−th hidden layer node and the output layer node, and b_i is the bias of the i−th hidden layer node.

Let $\beta_i = [\beta_1, \beta_2, \cdots, \beta_N]^T$, $T_i = [t_1, t_2, \cdots, t_M]^T$ and

$$H = \begin{bmatrix} g\left(w_1 x_1 + b_1\right) & \cdots & g\left(w_N x_1 + b_N\right) \\ \cdots & \cdots & \cdots \\ g\left(w_1 x_M + b_1\right) & \cdots & g\left(w_N x_M + b_N\right) \end{bmatrix}_{M \times N}$$

then Eq. (1) can be simplified as

$$H\beta = T \tag{2}$$

where H is the output matrix of the hidden layer in the ELM network, and the least squares solution of the output weight can be easily obtained by solving the linear matrix Eq. (2),

$$\hat{\beta} = H^+ T \tag{3}$$

H^+ is the generalized inverse of H.

The specific process of the AC-ELM algorithm is as follows:

3   Adaptive Convergent ELM Algorithm

3.1   AC-ELM

Based on the matrix equation iteration method given in Ref. [16], this paper proposes a matrix iteration method including adaptive convergence factors, and combines this method with the ELM algorithm to obtain Obtain an extreme learning machine algorithm based on adaptive matrix iteration (AC-ELM). The algorithm uses the adaptive convergence factor to iterate the matrix equation, and iteratively solves the output weight matrix of the ELM. AC-ELM can be well applied to the situation that the output matrix of the hidden layer H is a high-order, non-square matrix. In many cases It can have higher accuracy and efficiency than solving matrix equations by generalized inverse and classical iterative methods of solving matrix equations, and has good performance in both classification and regression tests.

The implementation method of the core of the AC-ELM algorithm is as follows:

1) Obtain the matrix equation $H\beta = T$ according to the training samples.
2) Through the decomposition iterative method mentioned in references [17,18], this paper constructs the following adaptive iterative algorithm:

$$\beta(k) = \beta(k-1) + \mu N^{-1} H^T [T - H\beta(k-1)] \tag{4}$$

where μ is the convergence factor and N is the diagonal matrix of the matrix $H^T H$.

3) Set the number of loops k and randomly generate the initial solution $Best = \beta_0$, and use the following pseudocode to solve the output weight matrix β:

Algorithm 1: Search for the optimal solution of the output weight matrix

---
1 For $i = 0$ to k do
2       $Best \leftarrow f(Best)$

---

where $f(\cdot)$ is the iterative equation of formula (4).

Two points need to be added to the above algorithm flow:

For similar iterative algorithms, we find that users do not have a good strategy for determining the value of μ when solving practical problems. In this paper, a method of adaptively selecting the convergence factor is given by combining the sufficient and necessary conditions for the convergence factor to satisfy the convergence and the actual experimental effect.

$$\mu = \frac{2}{\lambda_{\max}\left[N^{-1}H^T H\right] + \left|\lambda_{\min}\left[N^{-1}H^T H\right]\right|} \tag{5}$$

2) The formula of formula (4) can be further simplified to obtain

$$\beta(k) = \left(I_m - \mu N^{-1}H^T H\right)\beta(k-1) + \mu N^{-1}H^T T \tag{6}$$

Here the notation $\tilde{\beta}(k) = \beta(k) - \beta(k-1)$ is introduced, and further simplification can be obtained as

$$\tilde{\beta}(k) = \left(I_m - \mu N^{-1}H^T H\right)\tilde{\beta}(1) \tag{7}$$

The above simplification is to facilitate the proof of the convergence of the algorithm in the next section.

3.2   Convergence of AC-ELM Algorithm

According to the two lemmas in the literature [18], we can know from the analysis of formula (7) that formula $\lim_{x \to \infty}[\beta(k) - \beta(k-1)] = 0$ holds if and only if $\rho\left[I_m - \mu N^{-1}H^T H\right] < 1$, and the m eigenvalues of matrix $N^{-1}H^T H$ can be expressed as

$$1 + \rho_1, 1 + \rho_2, \cdots, 1 + \rho_p, 1, 1 \cdots, 1, 1 - \rho_p, \cdots, 1 - \rho_2, 1 - \rho_1$$

where $\rho_1 \geq \rho_2 \geq \cdots \geq \rho_p$. Assuming that $\lambda_i, i = 1, 2, \cdots, m$ is the eigenvalue of matrix $N^{-1}H^T H$, the eigenvalue of the iterative matrix $I_m - \mu N^{-1}H^T H$ can be expressed as $1 - \mu\lambda_i$. Then the necessary and sufficient condition for the convergence of the iterative algorithm in Eq. (7) is if and only if the eigenvalues of the iterative matrix satisfy $-1 < 1 - \mu\lambda_i < 1$. From the above theorem, it is easy to verify that the adaptive parameter in Eq. (4) satisfies the convergence condition, so the adaptive convergence factor iteration algorithm proposed in this paper satisfies the convergence condition.

At this time, taking the limit on both sides of iterative Eq. (4) at the same time can obtain the unique solution.

4   Experiment

In this section, different standard datasets are used to conduct experimental analysis on the performance of the AC-ELM algorithm. The selected datasets are from the UCI database and the MedMNIST medical image analysis dataset [19]. In order to ensure the accuracy and authenticity of the experimental results, all the experimental indicators in this paper are repeated twenty times and the average is taken as the final result.

4.1   UCI Dataset Classification Experiment Test

In this section, eight sets of data sets for classification in the UCI database are selected for comparative experiments. The number of samples, training, testing, features, and categories of the dataset are shown in Table 1 below.

Table 1. Basic information on the eight datasets

Data set	Number of training	Number of tests	Number of features	Number of categories
Glass	150	64	9	7
Seeds	158	52	7	3
Breast cancer	455	114	30	2
Balance scale	500	125	5	3
Madelon	1600	1000	500	2
Spambase	3000	1600	58	2
Dry bean	12249	1362	16	7
Avila	10430	10437	10	12

In this paper, ELM, KELM [3], ADMM-ELM [20] and AC-ELM proposed based on convex programming problem are classified and compared with different numbers of nodes, and the test accuracy of the four algorithms is compared. In the experiment of each dataset, the interval of the number of hidden layer nodes is set to [10, 100], and ten different numbers of hidden layer nodes are uniformly selected in the interval. The activation function of the hidden layer of the four algorithms uses the sigmoid function, the iteration number of the AC-ELM algorithm is set to 200, and the KELM selection function is the RBF Gaussian kernel function. The cycle number K of ADMM-ELM is set to 1000, β is set to 0.001, and λ is set to 0.0001. The results of the comparative experiment are shown in Fig. 1. In this paper, four representative datasets of Breast Cancer, Spambase, Dry Bean and avila are selected for presentation.

From the analysis of the experimental results in Fig. 1, it can be seen that under the premise of the same number of hidden layer nodes, the accuracy of the model trained by the AC-ELM algorithm is generally better than that of the ELM. Compared with the KELM and ADMM-ELM algorithms, the model accuracy of AC-ELM is slightly lower than these two algorithms under individual nodes in a few datasets. At the same time, when the number of hidden nodes is small, the test accuracy of the model established by AC-ELM is often the best. In addition, with the increase of the number of nodes, the accuracy gap between the AC-ELM algorithm and the other two algorithms will gradually increase, which indicates that when the dimension of the correlation matrix is large, the accuracy of the model solved by the adaptive iterative equation is better than that by the generalized inverse method.

On this basis, the experiment added the BLS (width learning) [21] algorithm to conduct comparative experiments on the above eight datasets. BLS is also

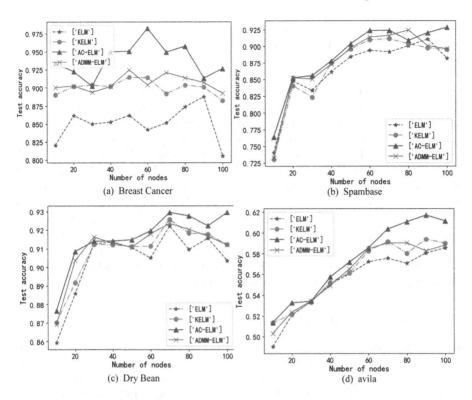

Fig. 1. Experimental results with different number of hidden layer nodes.

an algorithm that randomly generates weight bias, and its output weight is also obtained by generalized inversion. The experiments show the best test accuracy and training time of each algorithm on different datasets. Table 2 shows the experimental results of the five algorithms on the classification data set, and the experimental data with the highest accuracy is recorded for all algorithms.

Table 2. Comprehensive training results of datasets based on each algorithm

Data set	ELM			KELM			ADMM-ELM			AC-ELM			BLS	
Index	N	Acc (%)	T (s)	N	Acc (%)	T (s)	N	Acc (%)	T (s)	N	Acc (%)	T (s)	Acc (%)	T (s)
Glass	50	42.19	0.03	50	53.13	0.03	30	54.69	0.16	40	**68.75**	0.05	60.94	0.19
Seeds	50	96.15	0.07	50	96.46	0.06	50	96.46	0.23	50	**98.08**	0.05	90.38	0.34
Breast cancer	90	87.12	0.06	70	91.23	0.07	60	92.11	0.25	50	**98.25**	0.06	95.61	0.35
Balance scale	90	87.20	0.07	70	88.80	0.09	70	89.60	0.28	50	**91.20**	0.06	87.20	0.21
Madelon	90	53.40	0.29	70	**54.60**	0.30	80	53.60	0.98	70	54.10	0.23	52.70	0.92
Spambase	90	91.00	0.37	80	90.31	0.42	80	**92.44**	2.85	60	92.38	0.30	91.88	0.38
Dry bean	90	92.22	0.16	90	92.51	0.23	90	92.36	4.97	70	**92.95**	0.14	90.97	0.21
Avila	100	58.64	0.42	90	59.30	0.53	100	59.00	5.67	70	**61.76**	0.27	49.93	0.31

By analyzing the data in the table, it can be seen that, compared with ELM, the AC-ELM algorithm based on the adaptive convergence factor matrix iteration method has better accuracy on most data sets when the appropriate number of hidden layer nodes is set, and can have a faster model training speed when the data sample size is large. The KELM and ADMM-ELM algorithms are generally better than the traditional ELM algorithm in terms of accuracy when selecting suitable nodes, but they are still slightly lower than the AC-ELM algorithm, and due to iterative optimization, the time spent in training the model by the ADMM-ELM algorithm will be greatly increased. Compared with BLS, the convergence speed and accuracy of the AC-ELM algorithm are higher on all datasets used in the experiments. On the other hand, when training on datasets with large sample size such as Spambase, Dry Bean and avila, AC-ELM consumes less time than the other four algorithms, which is determined by the time complexity of the algorithm. In addition, from the point of view of the number of hidden layer nodes, the AC-ELM algorithm can train a model with higher accuracy with fewer nodes than ELM, KELM and ADMM-ELM, which also makes the training speed of the algorithm further improved.

4.2   MedMNIST Image Dataset Experiment

In order to more comprehensively verify the performance of the algorithm proposed in this paper in terms of classification, the MedMNIST dataset [?] is selected for the experiments in this section. In this experiment, four representative datasets, BreastMNIST, RetinaMNIST, DermaMNIST and PathMNIST, were selected for classification performance comparison with AC-ELM, ELM and the more advanced ResNet [18] and ResNet [50] respectively. Table 3 lists the basic features of the four datasets. This experiment combines the training accuracy, test accuracy and model training time to evaluate the performance.

Table 3. Basic information on the four datasets

Data set	Number of training	Number of tests	Number of features	Number of categories
Breast	546	156	784	2
Retina	1080	400	2352	5
Derma	7007	2005	2352	7
Path	89996	7180	2352	9

In the experiment, for the same data set, AC-ELM and ELM both use the sigmoid activation function and the same number of hidden layer nodes. On the BreastMNIST and RetinaMNIST datasets, the hidden layer nodes are set to 50, and the number of AC-ELM iterations is 500. On the DermaMNIST dataset, the hidden layer nodes are set to 80, and the number of AC-ELM iterations is selected 1000; On the set, the hidden layer nodes are set to 100, and the number of AC-ELM iterations is selected to be 3000. In all experiments in this section,

ResNet [18] and ResNet [50] use the SGD optimizer and the learning rate is set to 0.1.

The experimental results are shown in Table 4. It can be seen that compared with the traditional ELM algorithm, the AC-ELM algorithm has an improvement of more than 10% 20% in the model test accuracy. Compared with the neural network ResNet [18], the algorithm in this paper slightly improves the training accuracy, but greatly shortens the training time of the model. Compared with ResNet [50], the classification accuracy of the algorithm in this paper is significantly improved on small image datasets. Although AC-ELM performs generally on large-sample datasets, the training time of AC-ELM is greatly shortened compared to ResNet [50] in terms of convergence speed. Combined with the training accuracy, AC-ELM guarantees the generalization performance of the model under the premise of rapid convergence.

Table 4. Comprehensive results on the MedMNIST dataset

Data set	ELM			AC-ELM			ResNet [18]			ResNet [50]		
Index	Training	Testing	T (s)	Training	Testing	T (s)	Training	Testing	T (s)	Training	Testing	T (s)
Breast	78.02	74.13	**0.02**	82.98	**82.05**	0.06	81.50	77.56	32.34	78.57	74.80	136.56
Retina	56.67	46.21	**0.04**	56.11	**54.25**	0.08	53.98	52.75	72.34	55.33	53.00	215.26
Derma	69.05	67.33	4.05	73.00	**72.56**	**0.27**	72.73	71.02	285.45	72.00	71.87	1148.71
Path	40.43	52.34	21.36	68.01	77.89	**1.68**	68.98	77.51	4034.56	71.47	**82.67**	14109.04

Table 5. Basic information on the three datasets

Data set	Number of training	Predicted samples	Number of feature
Shanghai composite iindex	3450	60	5
Appliances energy prediction	10000	9735	28
SGEMM GPU kernel	141600	100000	14

4.3 Regression Prediction Experiment of AC-ELM

In the experiments in this section, regression predictions are performed on the Shanghai Composite Index opening, Appliances energy prediction and SGEMM GPU kernel performance datasets respectively. The number of training, prediction and features of the dataset are shown in Table 5.

In order to test the effect of the AC-ELM algorithm on data prediction compared with several mainstream ELM algorithms, the experiment sets different hidden layer nodes for each data set according to the scale of the data set to compare ELM, KELM, ADMM-ELM and AC-ELM. The number of iterations for the three data sets AC-ELM is set to 500, 800 and 1000 respectively, the kernel function of KELM is selected as the RBF Gaussian kernel function, and the parameter settings of ADMM-ELM are the same as those in Sect. 4.1. The time and error of several algorithms under different nodes are recorded respectively, and the experimental results are shown in Figs. 2, 3 and 4.

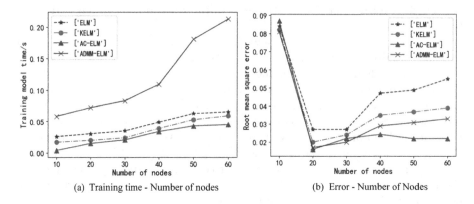

(a) Training time - Number of nodes   (b) Error - Number of Nodes

Fig. 2. Shanghai stock exchange opening index forecast curve.

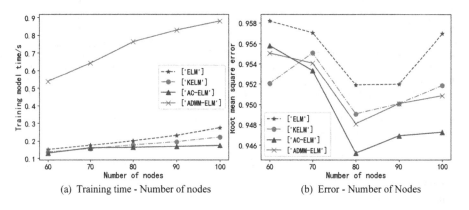

(a) Training time - Number of nodes   (b) Error - Number of Nodes

Fig. 3. Appliances energy prediction prediction curve.

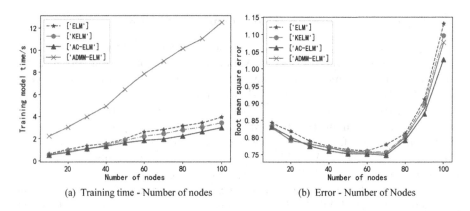

(a) Training time - Number of nodes   (b) Error - Number of Nodes

Fig. 4. SGEMM GPU kernel performance prediction curve.

Combining the results of the three figures (a), when the same number of nodes is set, the modeling efficiency of ADMM-ELM is significantly lower than that of the other three algorithms, while the modeling efficiency of AC-ELM algorithm is relatively high. As can be seen from Figure (b), compared with the other three algorithms, the root mean square error between the predicted value and the true value is basically smaller in the case of the same number of hidden layer nodes in the AC-ELM algorithm, and as the number of hidden layer nodes increases, the AC-ELM algorithm has more obvious advantages in terms of prediction accuracy. Combining graphs (a) and (b) for each data set, it can be found that the number of nodes with the smallest error is found in graph (b), and AC-ELM consumes less time under this number of nodes.

It is not difficult to see from the experiments in this section that the algorithm in this paper performs well in regression prediction, especially when predicting large data sets and large number of nodes, AC-ELM is better than the current mainstream ELM algorithm in terms of speed and accuracy.

5   Conclusion

Aiming at the problems that the traditional extreme learning machine has insufficient model accuracy in data classification and fitting prediction and takes too long to train a large data set model, this paper proposes an extreme learning machine algorithm based on adaptive convergence factor matrix iteration. The algorithm uses the model of the classic extreme learning machine, and on the basis of retaining the optimized network structure of the extreme learning machine, it solves the output weight matrix through a matrix iterative method including an adaptive convergence factor. The algorithm can obtain a model with higher accuracy by training with fewer iterations, which effectively solves the problems of low model learning accuracy and low model establishment efficiency in traditional algorithms. The algorithm in this paper performs well on both classification and regression datasets, which proves the effectiveness of the algorithm.

References

1. Rocco, I., Sivic, J.: Convolutional neural network architecture for geometric matching. In: IEEE Computer Society (2017)
2. Huang, G., Zhu, Q., Siew, C.K.: Extreme learning machine: a new learning scheme of feedforward neural networks. In: Proceeding of the IEEE International Joint Conference on Neural Networks. Piscataway, NJ. IEEE, pp. 513–529 (2004)
3. Huang, G.B., Zhou, H., Xiaojian, D., et al.: Extreme learning machine for regression and multiclass classification. IEEE Trans. Syst. Man Cybern. Part B **42**(2), 513–529 (2012)
4. Wang, S., Deng, C., Weisi, L., et al.: NMF-Based image quality assessment using extreme learning machine. IEEE Trans. Cybern. **47**(1), 232–243 (2016)
5. Hu, Y., Zhang, S.: Prediction of inlet NOx based on extreme learning machine of kernel mapping. Control Decis. **34**(1), 213–218 (2019)

6. Zhang, D., Zhang, L.: Evolutionary cost-sensitive extreme learning machine. IEEE Trans. Neural Netw. Learn. Syst. **28**(12), 3045–3060 (2017)

7. Xiaoping, L., Jiuwen, C., Xiaofeng, H., et al.: A maximally aplit and relaxed ADMM for regularized extreme learning machines. IEEE Trans. Neural Netw. Learn. Syst. **31**(6), 1899–1913 (2020)

8. Huang, G., Chen, L., Siew, C.K.: Universal approximation using incremental constructive feedforward networks with random hidden nodes. IEEE Trans. Neural Netw. **17**(4), 879–892 (2006)

9. Rong, H.J., Ong, Y.S., Tan, A.H., et al.: A fast pruned-extreme learning machine for classfication problem. Neurocomputing **72**, 359–366 (2008)

10. Ye, Y., Qin, Y.: QR factorization based incremental extreme learning machine with growth of hidden nodes. Pattern Recogn. Lett. **65**, 177–183 (2015)

11. Huijsmans, G., Loarte, A.: Non-linear MHD simulation of ELM energy deposition. J. Nucl. Mater. **438**, 57–63 (2013)

12. Tian, Z., Tian, M., Zhongyun, L., et al.: The Jacobi and Gauss-Seidel-type iteration methods for the matrix equation $AXB = C$. Appl. Math. Comput. **292**, 63–75 (2017)

13. Ford, W.: Basic iterative methods. Numerical Linear Algebra with Applications, pp. 469–490 (2015)

14. Ullah, M.Z., Soleymani, F., Al-Fhaid, A.S.: An efficient matrix iteration for computing weighted Moore-Penrose inverse. Appl. Math. Comput. **226**, 441–454 (2014)

15. Xu, R., Liang, X., Jinshan, Q., et al.: Advances and trends in extreme learning machine. Chin. J. Comput. **42**(7), 1640–1670 (2019)

16. Zhang, H.: Research on Matrix Equation Iteration Solution Methods. China University of Science and Technology Press, Hefei (2019)

17. Ding, F., Chen, T.: Iterative least-squares solutions of coupled Sylvester matrix equations. Syst. Control Lett. **54**(2), 95–107 (2005)

18. Ding, F., Liu, P.X., Ding, J.: Iterative solutions of the generalized Sylvester matrix equations by using the hierarchical identification principle. Appl. Math. Comput. **197**(1), 41–50 (2008)

19. Yang, J., Shi, R., Ni, B.: MedMNIST classification decathlon: a lightweight automl benchmark for medical image analysis. Applied Mathematics Computation. http://arxiv.org/abs/2021.14925(2020)

20. Liu, Z.: L1 regularization of algorithm based on algorithm algorithm algorithm algorithm research machine research. Beijing University of Chemical Technology (2018). https://doi.org/10.7666/d.Y3390087

21. Liu, Z., Chen, C.L., Feng, S., et al.: Stacked broad learning system: from incremental flatted structure to deep model. IEEE Trans. Syst. Man Cybern. Syst. **51**(1), 1–14 (2020)

An Incremental Extreme Learning Machine Prediction Method Based on Attenuated Regularization Term

Can Wang, Yuxiang Li, Weidong Zou$^{(\boxtimes)}$, and Yuanqing Xia

Beijing Institute of Technology, Beijing, China
7220190155@bit.edu.cn

Abstract. As a powerful tool for regression prediction, Incremental Extreme Learning Machine (I-ELM) has good nonlinear approximation ability, but the original model has the problem that the uneven output weights distribution affects the generalization ability of the model. This paper proposes an Incremental Extreme Learning Machine method based on Attenuated Regularization Term (ARI-ELM). The proposed ARI-ELM adds attenuation regularization term in the iterative process of output weights, reduces the output weights of the hidden node in the early stage of the iteration and ensuring that the new nodes after multiple iterations are not affected by the large regularization coefficient. Therefore, the overall output weights of the network reach a relatively small and evenly distributed state, which would reduce the complexity of the model. This paper also proves that the model still has convergence performance after adding the attenuated regularization term. Simulation results on the benchmark data set demonstrate that our proposed approach has better generalization performance than other incremental extreme learning machine variants. In addition, this paper applies the algorithm to specific weight prediction scene of intelligent manufacturing dynamic scheduling, and also gets good results.

Keywords: Incremental Extreme Learning Machine · Attenuated regularization term · Weight distribution · Generalization ability · Intelligent manufacturing · Dynamic scheduling

1 Introduction

As the earliest type of artificial neural network, feedforward neural network is widely used in many fields, such as image classification, target recognition and so on. Compared with recurrent neural network, the hidden layer of feedforward neural network has no horizontal connection, and its structure is simple

This paper was supported by National Key Research and Development Program of China under Grant 2018YFB1003700 and National Natural Science Foundation of China under Grant 61906015.

© Springer Nature Switzerland AG 2022
Y. Tan et al. (Eds.): ICSI 2022, LNCS 13345, pp. 189–200, 2022.
https://doi.org/10.1007/978-3-031-09726-3_17

[10, 15]. Feedforward neural network is generally composed of input layer, one or more hidden layers and output layer, and uses weighted average and activation function to transfer feature information to the next layer. Traditional neural network theory holds that all parameters in neural network need to be adjusted, and back-propagation algorithm is often used to train feedforward neural network. The feedforward neural network with many layers and updating the weight by back-propagation method can obtain better learning results, but it has the problems of long training time and many parameters. Subsequently, the researchers proved that the adjustment of Single Layer Feedforward Networks (SLFNs) parameters does not need to use the method of error back propagation, and SLFNs with random hidden layer nodes also have general approximation ability [4, 5, 7–9]. Huang et al. [6] proposed the limit learning machine algorithm, which is a SLFN, in which the input weight is obtained randomly, and the learning task can be completed efficiently by analytically solving the output weight. Extreme learning machine has a simple network structure and fewer network parameters. It does not need to update the iterative network weight through the back-propagation algorithm. The training speed is much faster than that of deep neural network. It overcomes the defects of traditional deep neural network, such as long training time, complex network structure and troublesome parameter adjustment. It shows great advantages and application prospects in machine learning [1, 3, 12, 14, 18].

Extreme Learning Machine (ELM) avoids the weight iterative updating process based on back propagation and simplifies the learning method, but there are also many problems. In particular, the number of ELM hidden layer nodes needs to be set by people themselves. How to select the optimal number of hidden nodes is still unknown. The number of hidden layer nodes needs to be determined through experience or a large number of experiments. Given a large number of training data, ELM has high computational complexity, and it is often necessary to calculate the generalized inverse of a large matrix. To solve the above problems, Huang et al. [4] proposed an incremental limit learning machine. Its core idea is to construct a SLFN that can add hidden nodes, randomly initialize the input matrix and deviation, add a single hidden layer node each time to fit the current residual of the network, calculate the output weight corresponding to the new node by using the least square method, and obtain the appropriate network structure through iteration. I-ELM solves the defects of fixed structure of traditional ELM network, difficult to obtain the optimal number of hidden layer nodes and high computational complexity. By adding hidden nodes, I-ELM can adaptively obtain the optimal fitting effect, and will not produce local optimal solution in theory, so it has good generalization ability. In addition, Chen et al. [5] added the convex optimization idea to I-ELM and proposed the Convex Incremental Extreme Learning Machine (CI-ELM) algorithm. After adding new hidden nodes, the convex optimization method is used to update and calculate the output weight of existing hidden nodes, so as to further improve the convergence speed of I-ELM algorithm. Feng et al. [2] proposed an EM-ELM algorithm based on minimum error. The algorithm allows to add hidden nodes one by one or group by group, and the output weight will be updated continuously when adding hidden nodes, which improves the learning speed of the algorithm. There are still some problems in I-ELM algorithm. In this paper, it is found that the output weight distribution of I-ELM network nodes is uneven. The weight of new nodes in the early stage of iteration is often large, while the weight of new nodes in the middle and late stage of iteration is small or even close to 0. The large weight distribution difference leads to the complexity and instability of the network [11,13], and the generalization ability needs to be further improved.

In order to solve the above problems, this paper adds the attenuatable regularization term to the weight solution process of I-ELM, and sets a large initial regularization coefficient to reduce the output weight of the new hidden nodes in the early stage of the iteration. With the increase of the number of iterations, the regularization coefficient decreases exponentially, so as to prevent the node weight from approaching 0 due to the excessive regularization coefficient in the middle and later stages of the iteration, so as to ensure that the node weight of the whole network is small and evenly distributed, It limits the complexity of the model and increases the generalization performance. The experimental results of this algorithm show that the performance of one or more groups of implicit regression algorithms is very good.

2   Proposed ARI-ELM

2.1   Model Framework

In order to solve the problem of uneven weight distribution of I-ELM and improve the generalization performance of the model, the attenuation regularization term is added to the original algorithm to reduce the network complexity. At the same time, new nodes can be added individually or in groups to improve the network learning efficiency.

Generally, L_2 penalty term is added after the loss function of standard limit learning machine to limit the output weight parameters of the model and obtain a smaller and more decentralized weight matrix [16]. The smaller the regularization coefficient, the closer the output parameter is to the real parameter.

The ARI-ELM topology of adding a single node is shown in Fig. 1(a), and the ARI-ELM topology of adding multiple nodes and multi-dimensional regression is shown in Fig. 1(b), where X is the input variable, y is the target output variable, a_n and b_n respectively represent the input weight vector and deviation of the nth hidden node, β_n represents the output weight obtained after adding the regularization term to the nth single hidden node, σ_n is the iteration coefficient, and $H_n(a_n, X, b_n)$ represents the activation value of the nth hidden node.

2.2   Convergence Analysis

Theorem 1. *Assuming that c_n and d_n are nonnegative sequences and B is constant and greater than zero. An incremental single hidden layer feedforward*

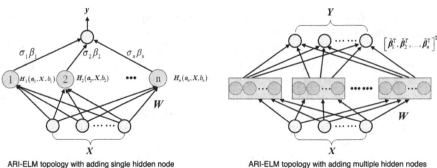

Fig. 1. Output weight distribution of I-ELM and ARI-ELM

neural network with arbitrary piecewise continuous activation function and Any non constant continuous objective function f are given. There are obtaining the hidden layer input matrix and offset parameter $(\hat{a}_n, \hat{b}_n)$ *randomly, and setting the hidden layer output matrix as* $G_n(\hat{a}_n \cdot x + \hat{b}_n)$. *If* $c_1 < B, c_2 < B, c_n < (1 + d_n^2) c_{n-1} - d_n^2 B$ *is satisfied, for any integer* $n, n \geq 2$, *when* $d_n^2 > \frac{1}{(n+1)(n^2-n-1)}$, *there must be* $c_n < \frac{B}{(n+1)n}$.

Proof. According to the mathematical induction, if $c_{n-1} < \frac{B}{(n+1)n}$, we have:

$$c_n - \frac{B}{(n+1)n} < (1 + d_n^2) \frac{B}{n(n-1)} - d_n^2 B - \frac{B}{(n+1)n}$$

$$= B \left(\frac{(n+1)(1 + d_n^2) - (n-1) - d_n^2(n^3 - n)}{(n-1)n(n+1)} \right) \qquad (1)$$

$$= \frac{B}{(n-1)n(n+1)} \left(1 - d_n^2(n^3 - 2n - 1) \right).$$

Thus, when $1 - d_n^2(n^3 - 2n - 1) < 0$, that is $d_n^2 > \frac{1}{(n+1)(n^2-n-1)}$, we can get $c_n < \frac{B}{(n+1)n}$, and the certificate is completed.

Theorem 2. *For a feedforward neural network with increasing hidden layer nodes, if an arbitrary non constant piecewise continuous function* $H : R \to R$ *is given as the hidden layer excitation function, a continuous and identically distributed random hidden layer output matrix* $\hat{H}_n(a_n \cdot x, b_n$ *is generated. Then, for the objective function f with any continuous non constant value, if the regularization coefficient* λ_n *and the output weight correction value* $\hat{\beta}_n$ *satisfy Eq. (2) and Eq. (3) there is* $\lim_{n \to \infty} \left\| f - \hat{\beta}_n \hat{H}_n (a_n \cdot x + b_n) \right\| = 0$:

$$\lambda_n = ce^{-n}. \qquad (2)$$

$$\hat{\beta}_n = \frac{\left\langle e_{n-1}, \hat{H}_n \left(a_n \cdot x + b_n\right)\right\rangle}{\left\|\hat{H}_n \left(a_n \cdot x + b_n\right)\right\|^2 + \lambda_n}. \tag{3}$$

Proof. First, we need to prove that $\|e_n\|$ is a monotonically decreasing function and has a lower bound of zero, that is $\lim_{n\to\infty} \|e_n\| = 0$.

Let $\hat{H}_n$ be the output matrix of a single hidden layer feedforward neural network with n hidden layer nodes. Let the network residual be $\Delta = \|e_{n-1}\|^2 - \|e_n\|^2$, where $e_{n-1} = f - \left[\hat{H}_1, \hat{H}_2, \ldots, \hat{H}_{n-1}\right] \cdot \left[\hat{\beta}_1, \hat{\beta}_2, \ldots, \hat{\beta}_{n-1}\right]^T$, then we can get $e_n = e_{n-1} - \hat{\beta}_n \hat{H}_n$, and then we can deduce:

$$\begin{aligned}
\Delta &= \|e_{n-1}\|^2 - \left\|e_{n-1} - \hat{\beta}_n \hat{H}_n\right\|^2 \\
&= 2\hat{\beta}_n \left\langle e_{n-1}, \hat{H}_n\right\rangle - \hat{\beta}_n^2 \left\|\hat{H}_n\right\|^2 = \hat{\beta}_n^2 \left\|\hat{H}_n\right\|^2 > 0.
\end{aligned} \tag{4}$$

Huang et al. have proved $\|e_n\|^2 - \left\|e_n - \hat{\beta}_{n+1} \hat{H}_{n+1}\right\|^2 \geq 0$, which means that $\|e_n\| > \|e_{n+1}\|$. Therefore, We can have $\|e_{n-1}\| > \|e_n\| > \|e_{n+1}\|$, and we can further draw the conclusion that $\|e_n\|$ is monotonically decreasing and bounded.

Thus, we have:

$$\begin{aligned}
\|e_n\|^2 &= \left\|e_{n-1} - \hat{\beta}_n \hat{H}_n\right\|^2 \\
&= \|e_{n-1}\|^2 + \hat{\beta}_n^2 \left\|\hat{H}_n\right\|^2 - 2\hat{\beta}_n \left\langle e_{n-1}, \hat{H}_n\right\rangle \\
&< \left(1 + \hat{\beta}_n^2\right) \|e_{n-1}\|^2 - \hat{\beta}_n^2 \left(2\frac{\left\langle e_{n-1}, \hat{H}_n\right\rangle}{\hat{\beta}_n} - \left\|\hat{H}_n\right\|^2\right) \\
&= \left(1 + \hat{\beta}_n^2\right) \|e_{n-1}\|^2 - \hat{\beta}_n^2 \left\|\hat{H}_n\right\|^2.
\end{aligned} \tag{5}$$

According to Theorem 1, we have:

$$\begin{aligned}
\|e_n\|^2 &< \left(1 + \hat{\beta}_n^2\right) \|e_{n-1}\|^2 - \hat{\beta}_n^2 \left\|\hat{H}_n\right\|^2 \\
&< \left(1 + \hat{\beta}_n^2\right) \|e_{n-1}\|^2 - \hat{\beta}_n^2 B.
\end{aligned} \tag{6}$$

Let $c_n = \|e_n\|^2, d_n^2 = \hat{\beta}_n^2$, when $n = 1$ we have:

$$\begin{aligned}
c_1 &= \|e_1\|^2 = \|f - f_1\|^2 = \left\|f - \hat{\beta}_1 \hat{H}_1\right\|^2 \\
&\leq \|f\|^2 + \hat{\beta}_1^2 \left\|\hat{H}_1\right\|^2 - 2\hat{\beta}_1 \left\langle f, \hat{H}_1\right\rangle.
\end{aligned} \tag{7}$$

According to the theorem, we have:

$$\hat{\beta}_1 = \frac{\left\langle e_0, \hat{H}_1 \right\rangle}{\left\| \hat{H}_1 \right\|^2 + \lambda_1}. \tag{8}$$

Bring the above-mentioned Eq. (8) into Eq. (7), we have:

$$
\begin{aligned}
\|e_1\|^2 &< \|f\|^2 + \frac{\left\langle f, \hat{H}_1 \right\rangle^2}{\left(\left\| \hat{H}_1 \right\|^2 + \lambda_1 \right)^2} \cdot \left\| \hat{H}_1 \right\|^2 - 2\frac{\left\langle f, \hat{H}_1 \right\rangle^2}{\left\| \hat{H}_1 \right\|^2 + \lambda_1} \\
&< \|f\|^2 + \frac{\left\langle f, \hat{H}_1 \right\rangle^2}{\left(\left\| \hat{H}_1 \right\|^2 + \lambda_1 \right)^2} \cdot \left(\left\| \hat{H}_1 \right\|^2 + \lambda_1 \right) - 2\frac{\left\langle f, \hat{H}_1 \right\rangle^2}{\left\| \hat{H}_1 \right\|^2 + \lambda_1} \\
&= \|f\|^2 + \frac{\left\langle f, \hat{H}_1 \right\rangle^2}{\left\| \hat{H}_1 \right\|^2 + \lambda_1} - 2\frac{\left\langle f, \hat{H}_1 \right\rangle^2}{\left\| \hat{H}_1 \right\|^2 + \lambda_1} \\
&= \|f\|^2 - \frac{\left\langle f, \hat{H}_1 \right\rangle^2}{\left\| \hat{H}_1 \right\|^2 + \lambda_1} < \|f\|^2 < B.
\end{aligned}
\tag{9}
$$

When $n = 2$, we can have:

$$
\begin{aligned}
c_2 = \|e_2\|^2 &= \|f - f_2\|^2 = \left\| f - f_1 - \hat{\beta}_2 \hat{H}_2 \right\|^2 \\
&= \left\| e_1 - \hat{\beta}_2 \hat{H}_2 \right\|^2 \le \|e_1\|^2 + \hat{\beta}_2^2 \left\| \hat{H}_2 \right\|^2 - 2\hat{\beta}_2 \left\langle e_1, \hat{H}_2 \right\rangle.
\end{aligned}
\tag{10}
$$

According to the theorem, we have:

$$\hat{\beta}_2 = \frac{\left\langle e_1, \hat{H}_2 \right\rangle}{\left\| \hat{H}_2 \right\|^2 + \lambda_2}. \tag{11}$$

Bring the above-mentioned Eq. (11) into Eq. (10), we have:

$$\|e_2\|^2 < \|e_1\|^2 + \frac{\left\langle e_1, \hat{H}_2 \right\rangle^2}{\left(\left\| \hat{H}_2 \right\|^2 + \lambda_2 \right)^2} \cdot \left\| \hat{H}_2 \right\|^2 - 2\frac{\left\langle e_1, \hat{H}_2 \right\rangle^2}{\left\| \hat{H}_2 \right\|^2 + \lambda_2}$$

$$< \|e_1\|^2 + \frac{\left\langle e_1, \hat{H}_2 \right\rangle^2}{\left\| \hat{H}_2 \right\|^2 + \lambda_2} - 2\frac{\left\langle e_1, \hat{H}_2 \right\rangle^2}{\left\| \hat{H}_2 \right\|^2 + \lambda_2} \tag{12}$$

$$= \|e_1\|^2 - \frac{\left\langle e_1, \hat{H}_2 \right\rangle^2}{\left\| \hat{H}_2 \right\|^2 + \lambda_2} < \|e_1\|^2 < B.$$

If the following inequality holds:

$$\hat{\beta}_n^2 > \frac{1}{(n+1)\left(n^2 - n - 1\right)}. \tag{13}$$

According to Eq. (9) and (12) and Theorem 1, we have:

$$\|e_n\| < \frac{\sqrt{B}}{\sqrt{(n+1)n}}. \tag{14}$$

where B is a constant, thus we have $\lim_{n \to \infty} \|e_n\| = \lim_{n \to \infty} \frac{\sqrt{B}}{\sqrt{(n+1)n}} = 0$.

2.3   Learning Steps

In ARI-ELM, when the network adds hidden nodes one by one and performs one-dimensional regression, $q = 1$, h and T are vectors. H and T are replaced by H and e respectively, and the corresponding output weight calculation formula is:

$$\beta_L = \frac{e \cdot h^{\mathrm{T}}}{\left(h \cdot h^{\mathrm{T}} + \lambda_L \right)}. \tag{15}$$

If the number of hidden layer nodes increased in each iteration is Q or when multi-dimensional regression training is carried out, the corresponding output weight can be:

$$\beta_L' = \left(H_L^{\mathrm{T}} H_L + \lambda_L I \right)^{-1} H_L^{\mathrm{T}} E, \tag{16}$$

where β_L' denotes the weight matrix corresponding to q new nodes in the Lth iteration, H_L represents the matrix after nonlinear activation of q new nodes, and E represents the residual matrix of the current network.

Table 1. Regression dataset

Regression datasets	Features	Training data	Test data
Short-term traffic flow prediction	4	276	92
Jialing river water quality evaluation	6	350	50
Shanghai stock index forecast	5	3663	916
Autompg	9	314	78
Airfoil Self-Noise	5	1202	301
Superconduct	81	17010	4253
PowerPlant	4	7654	1914
ConcreteCS	8	824	206
OnlineNewsPopularity	58	31715	7929
Residential Building	103	297	75
Winequality-white	11	3918	980
Winequality-red	11	1279	320

3   Experiment

3.1   Weight Distribution

In order to verify the generalization ability and convergence speed of the algorithm, we selects 12 regression data sets for comparative experiments. The first three data sets are from the book *43 case analysis of MATLAB neural network*, and the last nine data sets are from UCI database. The characteristic number, training number and test number of the data set are shown in Table 1.

The evaluation indexes of regression problems involved in this paper include RMSE, MAPE and R^2 [17]. we compares the output weight distribution of I-

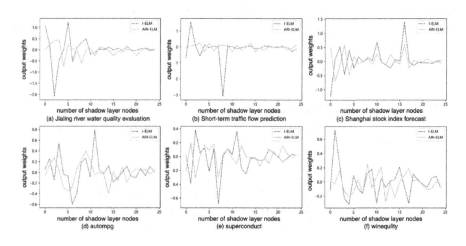

Fig. 2. Output weight distribution of I-ELM and ARI-ELM

ELM and ARI-ELM algorithms in benchmark data set prediction. Figure 2 shows
the weight comparison diagram of Short-term traffic flow prediction, Jialing river
water quality evaluation, Shanghai stock index forecast, Autompg, Superconduct
and Winequality datasets, in which there are 25 incremental nodes and one
hidden node is added each time. Compared with I-ELM algorithm, the output
weight distribution calculated by ARI-ELM algorithm is more uniform. The
weight is relatively small, and there is no obvious extreme weight. Figure 3 shows
the relative error of RMSE of I-ELM and ARI-ELM in training data and test
data. The maximum value of hidden nodes in this experiment is [1000:50:2000].
From the analysis, it can be seen that the relative error of ARI-ELM test data
RMSE and training data RMSE is smaller. The generalization performance of
ARI-ELM algorithm is better than I-ELM. Under the same accuracy, Table 2
compares the convergence time required by I-ELM and ARI-ELM under different
activation functions. It can be seen from Table 2 that on most datasets, the
convergence speed of ARI-ELM is faster than that of I-ELM.

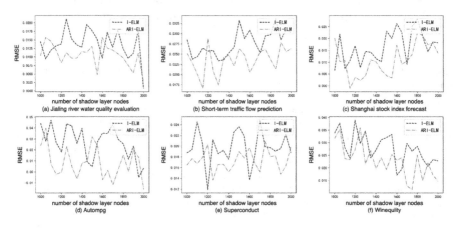

Fig. 3. RMSE relative error of test and training of I-ELM and ARI-ELM

3.2 Comparison of Experimental Results

In terms of model performance, we compared the generalization ability of I-ELM,
CI-ELM, EM-ELM and ARI-ELM algorithms on 12 regression datasets. Two
different activation functions, sigmoid function and RBF function, are applied.
The maximum number of hidden layer nodes of each algorithm is set to 300.
Finally, take the average value of RMSE of 200 experimental results. In Table 3,
the best results of the experiment are marked in bold, and the similar results
are underlined. Under the same maximum number of hidden nodes, except on
the winequality white dataset, the root mean square error of em-elm algorithm
is smaller than that of ARI-ELM algorithm. The root mean square error of
ARI-ELM algorithm in other datasets can reach the minimum or close to the
minimum. So, ARI-ELM has good generalization performance and stability.

Table 2. Comparison of training time between I-ELM and ARI-ELM under different activation functions

Regression datasets	Sigmoid		RBF	
	I-ELM	ARI-ELM	I-ELM	ARI-ELM
Short-term traffic flow prediction	0.0140	**0.0109**	0.5987	0.5578
Jialing river water quality evaluation	0.0879	0.0085	0.6188	0.5340
Shanghai stock index forecast	0.0279	0.0243	7.5548	7.2151
Autompg	0.0169	0.0159	0.8667	0.8256
Airfoil Self-Noise	**0.0360**	0.0410	5.2530	2.9303
Superconduct	3.6440	3.3637	31.3877	30.4426
PowerPlant	0.2673	**0.2038**	15.2263	14.7853
ConcreteCS	<u>0.0110</u>	<u>0.0110</u>	1.7274	1.0261
OnlineNewsPopularity	5.7203	**4.9370**	65.5311	64.3698
Residential Building	0.1046	**0.0159**	0.8846	0.7536
Winequality-white	0.0580	**0.0439**	6.7537	6.2951
Winequality-red	0.0139	0.0118	2.1281	2.0809

Table 3. Comparison of root mean square error of algorithms under different activation functions

Regression datasets	Sigmoid				RBF			
	I-ELM	CI-ELM	EM-ELM	ARI-ELM	I-ELM	CI-ELM	EM-ELM	ARI-ELM
Short-term traffic flow prediction	0.3134	0.3107	0.3124	**0.3011**	0.3212	0.3129	0.3102	0.3059
Jialing river water quality evaluation	0.0732	0.0714	0.0691	**0.0412**	0.1168	0.1222	0.0948	0.0742
Shanghai stock index forecast	0.1925	<u>0.1850</u>	0.1903	<u>0.1834</u>	0.3927	0.3801	0.3917	0.3923
Autompg	0.4242	0.4032	0.3846	**0.3198**	0.4683	0.4652	0.4598	0.4572
Airfoil Self-Noise	0.6833	<u>0.6798</u>	0.6814	<u>0.6772</u>	0.6125	0.6079	0.6201	0.6062
Superconduct	0.6099	0.6120	<u>0.6012</u>	<u>0.6061</u>	0.8706	0.8710	0.8413	0.8368
PowerPlant	0.3226	0.3124	<u>0.3189</u>	<u>0.3125</u>	0.3756	0.3645	0.3690	0.3621
ConcreteCS	0.6383	0.6140	0.5879	**0.5077**	0.7012	0.7121	0.6911	0.7004
OnlineNewsPopularity	0.8010	0.7918	0.7613	**0.6490**	0.8824	0.8637	0.8606	0.8579
Residential Building	0.5481	0.5180	0.4832	**0.4095**	0.8280	0.8081	0.7957	0.7909
Winequality-white	0.8464	0.8291	**0.8045**	0.8136	0.8996	0.9012	0.8736	0.8962
Winequality-red	0.8282	0.8187	0.8231	**0.7968**	0.9024	0.8945	0.8997	0.8874

4   Conclusion

This paper analyzes the principle and existing problems of incremental limit learning machine. Aiming at the uneven distribution of i-elm output weights, an incremental limit learning machine based on attenuation regularization term is proposed, which obtains small and evenly distributed output weights and improves the generalization ability of the model; At the same time, it allows a single or a group of hidden nodes to be added, which speeds up the efficiency of model training. Compared with the traditional algorithm, the algorithm proposed in this paper shows better generalization performance on the benchmark data set and production scheduling data set, and can effectively meet the real-time and diversity of production scheduling. However, there is no in-depth study on the choice of increasing the number of hidden nodes in this paper. How to select the optimal hidden nodes to increase the number and how to carry out more efficient regression on large-scale samples still need further research and experiments.

References

1. Cao, J.W., Lin, Z.P.: Extreme learning machines on high dimensional and large data applications: a survey. Math. Probl. Eng. **2015**(PT.12), 103796.1–103796.13 (2015)
2. Feng, G.R., Huang, G.B., Lin, Q.P., Gay, R.: Error minimized extreme learning machine with growth of hidden nodes and incremental learning. IEEE Trans. Neural Netw. **20**(8), 1352–1357 (2009)
3. Geng, Z., Dong, J., Chen, J., Han, Y.: A new self-organizing extreme learning machine soft sensor model and its applications in complicated chemical processes. Eng. Appl. Artif. Intell. **62**, 38–50 (2017)
4. Huang, G.B., Chen, L., Siew, C.K.: Universal approximation using incremental constructive feedforward networks with random hidden nodes. IEEE Trans. Neural Netw. **17**(4), 879–892 (2006)
5. Huang, G.B., Chen, L., Siew, C.K.: Convex incremental extreme learning machine. Neurocomputing **70**(16–18), 3056–3062 (2007)
6. Huang, G.B., Zhu, Q.Y., Siew, C.K.: Extreme learning machine: a new learning scheme of feedforward neural networks. In: International Joint Conference on Neural Networks, pp. 985–990. IEEE (2005)
7. Huang, G.B., Zhu, Q.Y., Siew, C.K.: Extreme learning machine: theory and applications. Neurocomputing **70**(1/3), 489–501 (2006)
8. Huang, G.B., Zhu, Q.Y., Siew, C.K.: Real-time learning capability of neural networks. IEEE Trans. Neural Netw. **17**(4), 863 (2006)
9. Liang, N.Y., Huang, G.B., Saratchandran, P., Sundararajan, N.: A fast and accurate online sequential learning algorithm for feedforward networks. IEEE Trans. Neural Netw. **17**, 1411–23 (2006)
10. Lukosevicius, M., Jaeger, H.: Reservoir computing approaches to recurrent neural network training. Comput. Sci. Rev. **3**(3), 127–149 (2009)
11. Tang, X.L., Han, M.: Partial Lanczos extreme learning machine for single-output regression problems. Neurocomputing **72**(13–15), 3066–3076 (2009)
12. Tang, Y.G., Li, Z.H., Guan, X.P.: Identification of nonlinear system using extreme learning machine based Hammerstein model. Commun. Nonlinear Sci. Numer. Simul. **19**(9), 3171–3183 (2014)
13. Tian, Z.D., Li, S.J., Wang, Y.H., Wang, X.D.: Network traffic prediction method based on improved ABC algorithm optimized EM-ELM. J. China Univ. Posts Telecommun. **25**(03), 37–48 (2018)
14. Wang, D., Wang, P., Ji, Y.: An oscillation bound of the generalization performance of extreme learning machine and corresponding analysis. Neurocomputing **151**, 883–890 (2015)
15. Williams, R.J., Zipser, D.: A learning algorithm for continually running fully recurrent neural networks. Neural Comput. **1**(2), 270–280 (1998)

16. Zhang, L., Zhang, D.: Evolutionary cost-sensitive extreme learning machine. IEEE Trans. Neural Netw. Learn. Syst. **28**(12), 3045–3060 (2017)
17. Zhongda, T., Shujiang, L., Yanhong, W., Yi, S.: A prediction method based on wavelet transform and multiple models fusion for chaotic time series. Chaos, Solitons Fractals **98**, 158–172 (2017)
18. Zhu, W., Huang, W., Lin, Z., Yang, Y., Huang, S., Zhou, J.: Data and feature mixed ensemble based extreme learning machine for medical object detection and segmentation. Multimed. Tools Appl. **75**(5), 2815–2837 (2015). https://doi.org/10.1007/s11042-015-2582-9

Kernel Discriminative Classifiers in Risk Prediction of Coronary Heart Disease

Hanli Qiao[1,2], Huazhou Chen[1,2], Jingyi Lyu[3], and Quanxi Feng[1,2(✉)]

[1] College of Science, Guilin University of Technology, Guilin 541004, China
[2] Center for Data Analysis and Algorithm Technology, Guilin 541004, China
fqx9904@163.com
[3] Department of Mathematical Sciences, Stevens Institute of Technology,
Hoboken, NJ 07030, USA

Abstract. Coronary heart disease is the most common type of organ lesion caused by atherosclerosis. It is a common disease that endangers the health of middle-elderly people. Predicting the future risk of coronary heart disease in advance is beneficial for efficient prevention. We provide an algorithm entitled by kernel discriminative analysis to complete the purpose in this paper. It is a kernel expansion of discriminative PCA, which enables the extraction of data features effectively. Results on the Framingham CHD dataset reveal its prediction ability.

Keywords: Kernel discriminative analysis · Coronary heart disease · Risk prediction · Feature extraction

1 Introduction

According to the findings in [1,2], Coronary heart disease (CHD) is the third leading cause of death globally and is associated with around 17.8 million deaths each year. Even the death probability declines widespread from age 70–90, mainly due to mortality reduction of cardiovascular diseases, they are still one of the most notable causes of disease burden in 2019 [3]. Precise future risk prediction of CHD is efficient for active prevention. Among diverse CHD causes, a complication is non-negligible. Such as in Type 2 diabetes mellitus (T2DM), a common chronic disease caused by insulin secretion disorder, CHD is the most common and severe complication [4]. The authors proposed an online predictive model to determine the risk probability of T2DM patients developing CHD in [4]. It's beneficial to precision diabetes mellitus care in providing early warning personalized guidance of CHD risk for T2DM patients and clinicians.

Predicting CHD risk is valuable for clinicians, healthcare planners, and researchers. It is a critical information source for individual patients. Study [5] provides a useful tool for clinical assessment. During the 10-year prediction for CHD-related issues like a heart attack or stroke, the scientists integrate multiple elements by the pooled cohort equation (PCE). The factors include age,

Y. Tan et al. (Eds.): ICSI 2022, LNCS 13345, pp. 201–208, 2022.
https://doi.org/10.1007/978-3-031-09726-3_18

sex, race, cholesterol, blood pressure, medication use, diabetes, and smoking history. It worked well for most people but overestimated the atherosclerotic cardiovascular disease (ASCVD) risk for moderate to severe obesity individuals. It concludes that as a risk-estimation tool, PCE enables guide prevention and treatment strategies in adults regardless of obesity status. Similarly, the work done in [6] is to assess the cardiovascular risk among seafarers and to compare lifestyle factors between Kiribati and European crew members. Results show a higher risk of cardiovascular diseases for Kiribati crew members due to alimentary habits.

Differing from the work concerning risk factors, machine learning, and AI-related techniques are attractive solutions to predict CHD from another view. They enable addressing the challenges of high complexity and correlations in CHD data for conventional techniques. In research [7], the scientists apply three supervised learners of Naïve Bayes, Support Vector Machine, and Decision tree, to improve the prediction rate for CHD data. Similarly, the authors exploit six machine learning algorithms to provide an open-source solution to detect coronary artery disease (CAD) [8]. Among various machine learners, random forest, K nearest neighbors (KNN), decision tree, artificial neural networks (ANN), and support vector machines (SVM) are popular choices like the work done in [9–12].

To effectively utilize machine learning-related models to predict CHD, we explore a useful feature extraction technique to improve the prediction performances. Providing an efficient feature extraction technique could fully mine latent information. We thereby aim to design such a model to extract intrinsic features. More specifically, getting nonlinear discriminative information. In detail, we transform original data into high-dimensional kernel space. The proceed implementation is using discriminative PCA proposed by [13] in the kernel space. We abbreviate this approach as KDPCA. The relevant information is described in the forthcoming section.

2   Kernel Discriminative Analysis

The essential tackle of KDPCA is firstly to find a nonlinear transformation for original data. Suppose that there are M training samples totally with c classes and l_i elements with dimensions D for each class, $M = \sum_{i=1}^{c} l_i$. Using ω_{ij} represents the jth observation of ith class. $\varphi : \mathbb{X} \to \mathbb{F}$ is an implicit transformation. The original data represented as

$$\Omega = [\omega_{11}, \ldots, \omega_{1l_1}, \cdots, \omega_{c1}, \ldots, \omega_{cl_c}]^{D \times M} . \tag{1}$$

KDPCA firstly map Ω into feature space $\mathbb{F}$ as the following way:

$$\Omega_\varphi = [\varphi(\omega_{11}), \ldots, \varphi(\omega_{1l_1}), \cdots, \varphi(\omega_{c1}), \ldots, \varphi(\omega_{cl_c})]^{F \times M} . \tag{2}$$

its mean vector by column is $\bar{\Omega}_\varphi = \frac{1}{M}\Omega_\varphi \mathbb{1}_{M \times 1}$. We use of the "kernel trick", i.e., $k(x_1, x_2) = \varphi^t(x_1)\varphi(x_2)$ to overcome the implicit limitation. The kernel matrix is

$$\mathbf{K}_1 = \Omega_\varphi^\mathbf{T} \Omega_\varphi = \begin{pmatrix} \varphi^t(\omega_{11})\varphi(\omega_{11}) & \cdots & \varphi^t(\omega_{11})\varphi(\omega_{cl}) \\ \vdots & \ddots & \vdots \\ \varphi^t(\omega_{cl})\varphi(\omega_{11}) & \cdots & \varphi^t(\omega_{cl})\varphi(\omega_{cl}) \end{pmatrix}^{M \times M} . \tag{3}$$

To facilitate the discriminative ability, we calculate the between- and within-class matrices $\mathbf{K}_b$, $\mathbf{K}_w$ of $\mathbf{K}_1$. Applying the idea of DLDA to obtain the discriminative matrix $W_\mathbf{K}^{M \times m}$ of $\mathbf{K}_1$, the satisfying processes ordered as the follow steps:

1. Calculating the eigenvectors e_b corresponding to their eigen values λ_b to construct spaces $\mathbf{E}_b = [e_1, \ldots, e_M]$ and $\Lambda_b = diag(\lambda_1, \ldots, \lambda_M)$
2. Discarding the eigenvectors corresponding to the zero eigenvalues to obtain the subspaces $\hat{\mathbf{E}}_b = [e_1, \ldots, e_m]$ and $\hat{\Lambda}_b = diag(\lambda_1, \ldots, \lambda_m)$ and then $\mathbf{B} = \hat{\mathbf{E}}_b \hat{\Lambda}_b^{-1/2}$
3. Computing the eigenvectors e_w of $\mathbf{B}^T \mathbf{K}_w \mathbf{B}$ and the corresponding eigenvalues λ_w to construct its eigen-subspaces $\mathbf{E}_w$, Λ_w and then $W_\mathbf{k} = \mathbf{B}\hat{\mathbf{E}}_w \hat{\Lambda}_w^{-1/2}$

Now the discriminative matrix W_{Ω_φ} can be derived from the following relationship:

$$W_{\Omega_\varphi} = \Omega_\varphi W_\mathbf{K}. \tag{4}$$

Note that φ is implicit, which means that we cannot compute W_{Ω_φ} directly. We skip the go-ahead to the next step, which is calculating the covariance $\mathbf{C}_W$ of centred W_{Ω_φ} and then diagonalise $\mathbf{C}_W$.

$$\mathbf{C}_W = \frac{1}{m}(W_{\Omega_\varphi} - \overline{W}_{\Omega_\varphi})(W_{\Omega_\varphi} - \overline{W}_{\Omega_\varphi})^\mathbf{T}$$
$$= \frac{1}{m}(\Omega_\varphi W_{\mathbf{K}_1} - \overline{\Omega_\varphi W}_{\mathbf{K}_1})(\Omega_\varphi W_{\mathbf{K}_1} - \overline{\Omega_\varphi W}_{\mathbf{K}_1})^\mathbf{T} . \tag{5}$$

However, $\mathbf{C}_W$ is still unknown because φ is implicit. We convert to calculate $\frac{1}{m}(\Omega_\varphi W_{\mathbf{K}_1} - \overline{\Omega_\varphi W}_{\mathbf{K}_1})^\mathbf{T}(\Omega_\varphi W_{\mathbf{K}_1} - \overline{\Omega_\varphi W}_{\mathbf{K}_1})$ instead of $\mathbf{C}_W$. Thanks to the fact that the eigenvectors of $AA^\mathbf{T}$ can be obtained by the eigenvectors of $A^\mathbf{T}A$ through

$$A^\mathbf{T}Au_k = \lambda_k u_k \rightarrow AA^\mathbf{T}(Au_k) = \lambda_k(Au_k) . \tag{6}$$

Therefore,

$$\widetilde{\mathbf{C}}_W = \frac{1}{m}(W_{\mathbf{K}_1} - \bar{W}_{\mathbf{K}_1})^\mathbf{T}\mathbf{K}_1(W_{\mathbf{K}_1} - \bar{W}_{\mathbf{K}_1}) . \tag{7}$$

Suppose $\tilde{U}_p$ is the eigenvectors of $\widetilde{\mathbf{C}}_W$ corresponding to the top p largest eigenvalues. Their eigenvectors of $\mathbf{C}_W$ are $U_p = (\Omega_\varphi W_{\mathbf{K}_1} - \Omega_\varphi \bar{W}_{\mathbf{K}_1})\tilde{U}_p$. Normalization is the proceed operation to get the coordinates of the projected samples in the following way:

$$U_p = (\Omega_\varphi W_{\mathbf{K}_1} - \Omega_\varphi \bar{W}_{\mathbf{K}_1})\tilde{U}_p' . \tag{8}$$

where $\tilde{U}'_p = \left[\frac{\tilde{u}_1}{\triangle_1} \cdots \frac{\tilde{u}_p}{\triangle_p}\right]$ and $\triangle_i = \sqrt{\tilde{u}_i^{\mathbf{T}}(W_{\mathbf{K}_1}^{\mathbf{T}} - \bar{W}_{\mathbf{K}_1}^{\mathbf{T}})\mathbf{K}_1(W_{\mathbf{K}_1} - \bar{W}_{\mathbf{K}_1})\tilde{u}_i}$, the project coordinates of training centred samples thereby are

$$\mathbf{Y}_\varphi = \tilde{U}'^{\mathbf{T}}_p(W_{\mathbf{K}_1}^{\mathbf{T}} - \bar{W}_{\mathbf{K}_1}^{\mathbf{T}})(\mathbf{K}_1 - \overline{\mathbf{K}}_1) . \tag{9}$$

Suppose Π is the testing dataset with N elements, t_i is the number of samples of each class, then

$$\Pi = [\pi_{11}, \ldots, \pi_{1t_1}, \cdots, \pi_{ci}, \ldots, \pi_{ct_c}]^{D \times N} . \tag{10}$$

therefore, the transformed Π in $\mathbb{F}$ is

$$\Pi_\varphi = [\varphi(\pi_{11}), \ldots, \varphi(\pi_{1t}), \cdots, \varphi(\pi_{ci}), \ldots, \varphi(\pi_{ct})]^{F \times N}. \tag{11}$$

analogy to $\bar{\Omega}_\varphi$, the mean of Π_φ can be obtained by

$$\Pi_\varphi = \frac{1}{N}\Pi_\varphi \mathbb{1}_{N \times 1}. \tag{12}$$

therefore, the project coordinates of centred test samples are

$$\mathbf{Y}'_\varphi = \tilde{U}'^{\mathbf{T}}_p(W_{\mathbf{K}_1}^{\mathbf{T}} - \bar{W}_{\mathbf{K}_1}^{\mathbf{T}})(\mathbf{K}_2 - \overline{\mathbf{K}}_2) . \tag{13}$$

where $\mathbf{K}_2 = \Omega_\varphi^{\mathbf{T}}\Pi_\varphi$.

Algorithm 1. Kernel Discriminative PCA for feature extraction

---

Input: Sample datasets as training Ω and testing Π
Output: The projected $\mathbf{Y}_\varphi$, $\mathbf{Y}'_\varphi$
1: Choosing various kernerls to generate the corresponding kernel matrix $\mathbf{K}_1$ and computing $\mathbf{K}_2$
2: Calculating $\mathbf{K}_b$ and $\mathbf{K}_w$ of $\mathbf{K}_1$
3: Obtaining the optimal discriminant matrix $W_{\mathbf{K}_1}$ by DLDA algorithm as described steps
4: Computing $\tilde{\mathbf{C}}_W$ as formula (7) and $\tilde{U}_p$
5: Normalizing $\tilde{U}_p$ to get $\tilde{U}'_p$
6: Projecting Ω_φ and Π_φ onto normalized U_p as formula (9) and (13) to output $\mathbf{Y}_\varphi$ and $\mathbf{Y}'_\varphi$
7: **return** $\mathbf{Y}_\varphi$, $\mathbf{Y}'_\varphi$

The processes of KDPCA are listed in Algorithm 1. The common kernels are Gaussian, Matérn, Wendland, and thin-plate spline (TPS). We choose in this paper are the specific Matérn with $v = \frac{1}{2}$ and TPS to analyze.

3   Experimental Results

We analyze the prediction results of multiple diseases and the ten-year risk for CHD to verify the proposed method. Experiments on the CHD dataset named

Framingham and the multiple disease dataset. The former is publicly available on the Kaggle website, and it is from an ongoing cardiovascular study on Framingham residents. It includes 4240 samples with 15 attributes and corresponding labels. After data cleaning, there are 3658 samples left. We delete the irrelevant attribute, i.e., education. The experimental purpose is to predict the 10-years risk of whether a patient will have CHD. To accomplish this task, we randomly choose 3000 samples for training and the remaining 658 records for testing.

There are 4962 records totally in the multiple disease dataset. Each record includes 132 attributes, and the disease category is 42. The experimental goal of this dataset is to classify which disease a patient takes. We train a classifier by randomly choosing 3000 samples. The testing performances use the remaining 1962 ones. To analyze the effectiveness of KDPCA as a feature extraction technique, we apply the KNN and the decision tree to evaluation.

Table 1. The average prediction accuracy of run five times on the Framingham CHD dataset.

KDPCA_KNN	KDPCA_Decision tree	KNN	Decision tree
0.7781	**0.8049**	0.7736	0.7863

Table 1 displays the average prediction accuracy on the Framingham CHD dataset five times running. The RBF kernel is TPS. It's clear that with the help of KDPCA to extract features, both KNN and the decision tree perform better than those without KDPCA. However, there only are 557 among 3658 who will take CHD disease in 10-years on the Framingham dataset. It shows such an imbalanced prediction task that accuracy is not enough. To overcome this limitation, we analytically compare the confusion matrix in Fig. 1. It displays the risk prediction ability on the Framingham CHD dataset when using KDPCA to extract features.

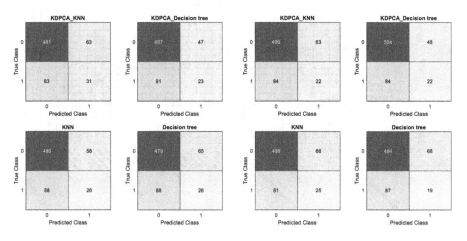

Fig. 1. Comparison of confusion matrix using KNN and Decision tree with- and without KDPCA on the Framingham CHD dataset.

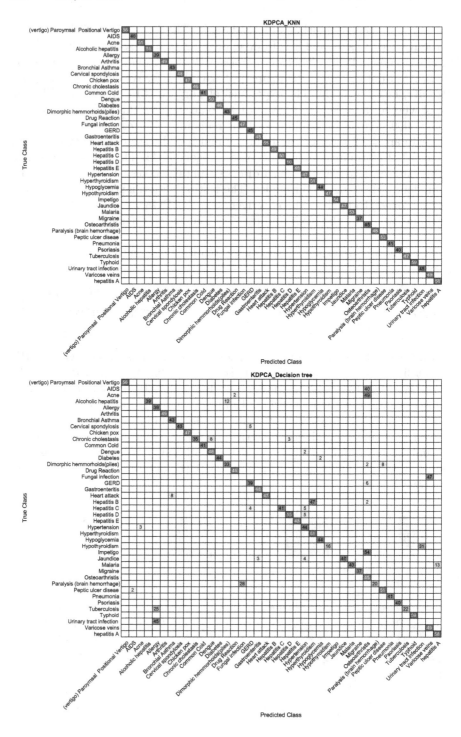

Fig. 2. Comparison of confusion matrix using KNN and Decision tree with KDPCA for multiple disease prediction.

Similar situations occur in the classification of multiple diseases, as Fig. 2 shows. In this case, the RBF kernel is Matérn when $v = \frac{1}{2}$. We can classify all diseases precisely when using KNN, whereas the decision tree performs worse a bit.

4 Conclusions

Kernel expansion is an effective technique to extract latent information in data features. In this paper, we map the original data into the kernel space. Then use the discriminative PCA to do feature extraction. We abbreviate this approach as KDPCA. Drawing support of various kernels like Matérn and TPS in KDPCA to bring better performances in multiple disease classification and the ten-year risk prediction for CHD disease. The evaluation results verify the effectiveness of the proposed method. However, the shape parameter in kernels influences the results. How to choose an optimized shape parameter for practical applications is a promising study.

Acknowledgements. This work is jointly supported by the Guangxi Science and Technology Base Foundation [AD18281039], the Natural Science Foundation of Guangxi [2020-GXNSFBA297011].

References

1. Brown, J.C., Gerhardt, T.E., Kwon, E.: Risk Factors For Coronary Artery Disease. StatPearls Publishing, Treasure Island (FL) (2022)
2. GBD 2017 Causes of Death Collaborators: Global, regional, and national age-sex-specific mortality for 282 causes of death in 195 countries and territories, 1980–2017: a systematic analysis for the Global Burden of Disease Study 2017. Lancet **392**(10159), 1736–1788 (2018)
3. GBD 2019 Ageing Collaborators: Global, regional, and national burden of diseases and injuries for adults 70 years and older: systematic analysis for the Global Burden of Disease 2019 Study. BMJ **376**, e068208 (2022). https://doi.org/10.1136/bmj-2021-068208
4. Fan, R., Zhang, N., Yang, L., Ke, J., Zhao, D., Cui, Q.: AI-based prediction for the risk of coronary heart disease among patients with type 2 diabetes mellitus. Sci. Rep. **10**, 14457 (2020)
5. Khera, R., Pandey, A., Ayers, C.R., et al.: Performance of the pooled cohort equations to estimate atherosclerotic cardiovascular disease risk by body mass index. JAMA Netw. Open **3**(10), e2023242 (2020)
6. von Katzler, R., Zyriax, B.C., Jagemann, B., et al.: Lifestyle behaviour and prevalence of cardiovascular risk factors - a pilot study comparing Kiribati and European seafarers. BMC Public Health **19**(1), 855 (2019)
7. Gonsalves, A.H., Thabtah, F., Mohammad, R.M.A., Singh, G.: Prediction of coronary heart disease using machine learning: an experimental analysis. In: Proceedings of the 2019 3rd International Conference on Deep Learning Technologies, pp. 51–56 (2019). https://doi.org/10.1145/3342999.3343015

8. Akella, A., Akella, S.: Machine learning algorithms for predicting coronary artery disease: efforts toward an open source solution. Future Sci. OA **7**(6), FSO698 (2021). https://doi.org/10.2144/fsoa-2020-0206

9. Du, Z., et al.: Accurate prediction of coronary heart disease for patients with hypertension from electronic health records with big data and machine-learning methods: model development and performance evaluation. JMIR Med. Inform. **8**(7), e17257 (2020)

10. Ayatollahi, H., Gholamhosseini, L., Salehi, M.: Predicting coronary artery disease: a comparison between two data mining algorithms. BMC Public Health **19**, 448 (2019). https://doi.org/10.1186/s12889-019-6721-5

11. Krishnani, D., Kumari, A., Dewangan, A., Singh, A., Naik, N.S.: Prediction of coronary heart disease using supervised machine learning algorithms. In: 2019 IEEE Region 10 Conference (TENCON 2019), pp. 367–372 (2019). https://doi.org/10.1109/TENCON.2019.8929434

12. Li, W., Chen, D., Le, J.: Coronary heart disease prediction based on combined reinforcement multitask progressive networks. In: 2020 IEEE International Conference on Bioinformatics and Biomedicine (BIBM 2020), pp. 311–318 (2020). https://doi.org/10.1109/BIBM49941.2020.9313275

13. Qiao, H., Blech, J.O., Chen, H.: A machine learning based intrusion detection approach for industrial networks. In: 2020 IEEE International Conference on Industrial Technology (ICIT 2020), pp. 265–270, (2020). https://doi.org/10.1109/ICIT45562.2020.9067253

Genetic Programming for Ensemble Learning in Face Recognition

Tian Zhang[1], Lianbo Ma[1,2(✉)], Qunfeng Liu[3], Nan Li[1], and Yang Liu[4]

[1] College of Software, Northeastern University, Shenyang, China
malb@swc.neu.edu.cn
[2] Key Laboratory of Smart Manufacturing in Energy Chemical Process, Ministry of Education,
East China University of Science and Technology, Shanghai, China
[3] School of Computer Science and Technology, Dongguan University of Technology,
Guangdong, China
[4] Shenyang Institute of Automation, Chinese Academy of Sciences, Shenyang, China

Abstract. Ensemble learning has recently been explored to achieve a better generalization ability than a single base learner through combining results of multiple base learners. Genetic programming (GP) can be used to design ensemble learning via different strategies. However, the challenge remains to automatically design an ensemble learning model due to complex search space. In this paper, we propose a new automated ensemble learning framework, based on GP for face recognition, called Evolving Genetic Programming Ensemble Learning (EGPEL). This method integrates feature extraction, base learner selection, and learner hyperparameter optimization, into several program trees. To this end, multiple program trees, a base learner set, and a hyperparameter set are developed in EGPEL. Meanwhile, an evolutionary approach to results integration is proposed. The performance of EGPEL is verified on face benchmark datasets of difficulty and compared with a large number of commonly used peer competitors, including state-of-the-art competitors. The results show that EGPEL performs better than most competitive ensemble learning methods.

Keywords: Ensemble learning · Genetic programming · Multiple program trees · Face recognition

1 Introduction

Ensemble learning, which is one of the most popular machine learning approaches, shows its strong learning and generalization capability while applied to various real-world prediction problems [1, 2]. It aims to improve the performance of a single learner by combining the prediction results of multiple base learners [2]. Many studies have been designed to achieve a batter ensemble, such as stacking, bagging and boosting methods [3]. However, existing ensemble methods for selection and combination of base learners are often manually determined [4]. This process needs a considerable amount of expert experience and trial and error, limiting the further development of the algorithm [5].

© Springer Nature Switzerland AG 2022
Y. Tan et al. (Eds.): ICSI 2022, LNCS 13345, pp. 209–218, 2022.
https://doi.org/10.1007/978-3-031-09726-3_19

Thus, designing an automated ensemble learning model can be extremely useful for community development.

During recent years, the application of evolutionary computation has gained a lot of progress [6–9]. Especially, automated ensemble learning using evolutionary computation has been designed [4]. For example, [14] proposed a multi-objective evolutionary optimization algorithm for ensemble learning to obtain a set of Pareto solutions with good diversity and accuracy. Genetic programming (GP) [15] as an automatically evolving technique can easily solve recognition problem due to flexible representation and good search ability, where each individual is often represented by a tree [16]. However, most of the existing approaches use the principle of ensemble learning to design multi-GP trees and then solve downstream tasks (i.e., image classification tasks).

This paper aims at designing a new evolving ensemble learning for face recognition. The new approach will provide end-to-end solutions for given problem via performing base learner selection, and learner hyperparameter optimization automatically and simultaneously. To achieve above points, a multiple program trees, a base learner set, and a hyperparameter set are designed in EGPEL.

The rest of this work is organized as follows. Sections 2 and 3 introduce the related work and framework of EGPEL. Section 4 experimental results are shown to prove the performance of the EGPEL on the AR face dataset and give some discussion and analysis on the relation between the EGPEL and relevant methods. Finally, Sect. 5 concludes the work.

2   Related Work

2.1   Ensemble Learning

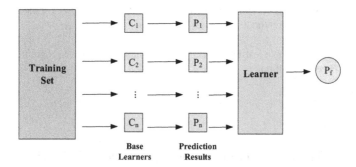

Fig. 1. Framework of stacking.

Ensemble learning has become a promising technique for model performance enhancement [1, 2]. It combines multiple base models (e.g., Multilayer Perceptron (MLP) and Decision Tree (DT)) to obtain a better supervised model. The idea of ensemble learning is that if one base learner gets an incorrect prediction, other weak classifiers can also correct the error back [17–20]. We can generally divide ensemble learning into four categories, i.e., Bagging, Boosting, Stacking, and Blending [21]. This paper focuses on the Stacking method. Readers who are interested in other methods can refer to [18, 22, 23].

For Stacking method, the entire dataset is divided into several sub-datasets, which are divided into a training set and a validation set. The base learner fits the data in the training set to generate the underlying model. The predictions of the learner on the validation set are used as input to the second layer [24–26]. In this way, the higher-level learner is able to further generalize and enhance the model, which is the reason why the Stacking approach always achieves good prediction performance on the test set [27–29]. Figure 1 illustrates the framework of Stacking.

Algorithm 1: Framework of GP

Input: m: population size; n: Maximum No. of Run;
Output: the best-so-far individual;
1: **for** i = 1 to n **do**
2:      Generate Initial Population Randomly
3:      **while not** (Terminate Condition for Run)
4:          Evaluate fitness of each individual;
5:          **for** j = 1 to m **do**;
6:              Op ← Select Genetic Operator;;
7:              **Case** Op is UNARY;
8:                  Select one individual based on fitness;
9:                  Perform Reproduction with Probability P_r;
10:                 Copy the offspring into new Population;
11:             **Case** Op is BINARY;
12:                 Select two individuals based on fitness;
13:                 Perform Crossover with Probability P_c;
14:                 $j = j + 1$;
15:                 Insert two offspring into new Population;
16:         **end**
17:     **end**
11: **end**

2.2 Genetic Programming

Evolutionary computation and swarm intelligence algorithms have attracted a lot of attention in various application scenarios [30]. There are many emerging directions to be addressed by evolutionary computing as well [31]. Genetic Programming (GP) is a type of Evolutionary Algorithms (EA), which inherits the basic idea of Genetic Algorithms (GA), i.e., reproducing offspring from parents according to the fitness value [32]. Different from GA (e.g., fixed-length encoding), the most common is tree-based representations, including three types of nodes i.e., root node, internal node, and leaf node, where the root node and internal node include some functions from predefined function set and the leaf node often contains variables and constants [33]. The Algorithm 1 shows framework of GP.

In recent years, many works [34–36] have been undertaken on GP and ensemble learning. In [37], a novel GP-based approach is proposed to generate an ensemble with low computational complexity from a single run. Karakatič et al. [38] designed a GPAB approach for classification problem, where AdaBoost and GP-based classifier are combined together to improve model performance.

3 Proposed Approach

3.1 Overall Algorithm

In Algorithm 2, we describe the overall algorithm of EGPEL for face recognition. The inputs of the EGPEL approach are image dataset. The algorithm consists of three main parts, namely, the feature extraction module (see Sect. 3.2), the base learner selection and optimization module (see Sect. 3.3), and the result integration module (see Sect. 3.4).

Algorithm 2: Framework of EGPEL

Input: N: population size; D: face data; T: Number of iterations
Output: the best-so-far program tree;
1:   $t = 0$;
2:   **for** $t < T$ **do**:
3:     *Features* ← Reduce the original image feature dimension by feature extraction module;
4:     $i = 0$;
5:     **for** $i < T$ **do**:
6:       *trees* ←Initialize trees through a set of learners and hyperparameters;
7:       *result* ←Calculate the fitness value for each tree model on test data;
8:       Results of integrating base learners using evolutionary computation;
9:       Generation of offspring using genetic operators;
10:      $i = i + 1$;
11:     **end**
12:     $t = t + 1$;
13:   **end**
14:   **Return** the best-so-far program tree;

3.2 Feature Extraction Module

In order to extract effective features from the images as well as to reduce the input dimensionality of the base learner, we construct feature extraction module, including two convolutional operations, one pooling operation and one flatten operation, as shown in Fig. 2.

The convolution operation is used to extract features; the pooling operation is used to reduce dimensionality, remove redundant information, compress features, and simplify network complexity; flatten operation converts the features from matrix format to vector format.

3.3 Base Learners Selection and Optimization Module

To batter select and optimize base learners from learner and hyperparameter set, we design the base learner as a binary tree, where the root node indicates the base learner

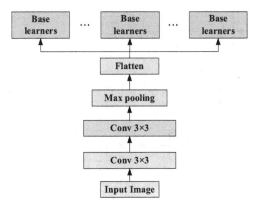

Fig. 2. Feature extraction module.

type and the terminal nodes represent the hyperparameter to be optimized, as shown in Fig. 3(a). Each individual can contain multiple learners (see Fig. 3(b)) with different lengths.

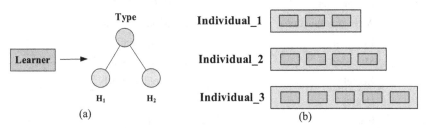

Fig. 3. Learner encoding. (a): An example of a tree-based encoding learner, where H_1 and H_2 denote hyperparameters. (b): Individuals of different lengths

Crossover: it is that two randomly selected crossover points on two chromosomes are crossed without changing the hyperparameters of the internal learner.

Mutation: it is mainly involved in adding, deleting and modifying information about base learners in chromosomes.

3.4 Results Integration Module

By validating on the dataset we can obtain the predicted values for each base learner. Unlike traditional methods for integrating results (e.g., average and voting method), we use an evolutionary approach to optimize the weights of each result and thus improve the accuracy of the final prediction, as shown in Fig. 4.

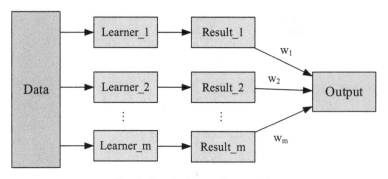

Fig. 4. Results integration module

Table 1. The hyperparametric collections

Type	Name	Value
KNN	K	[6, 7, 8]
	p	[Euclidean, Minkowski]
ANN	Learning rate	[0.01, 0.05, 0.07]
	Activation function	[Sigmoid, Tanh, ReLu]
DT	Criterion	[Entropy, Gini]
	Max_depth	[5, 9, 10, None]
SVM	Kernel	[Sigmoid, Linear]
	Degree	[2, 4, 6]
LR	Penalty	[L1, L2]
	Optimizer	[NAG, AdaGrad]
RBF	Kernel	[Polynomial, Laplacian, Gaussian]
	Number of hidden layer neurons	[3, 5, 7]

4 Experiments and Results

4.1 Parameter Settings and Dataset

The parameter settings for EGPEL are based on the commonly applied settings for GP. In EGPEL, the population size is 100 and the number of iterations is 50. The rates for elitism, crossover and mutation are 0.05, 0.9 and 0.15, respectively. An individual can contain up to 7 base learners. Table 1 collates the hyperparameter values needed for each type of base learner, including K-Nearest Neighbor (KNN), Artificial Neural Network (ANN), Decision Tree (DT), Support vector machine (SVM), Linear regression (LR), Radial basis function (RBF).

Face recognition experiments have been run on the AR face dataset, which includes over 4,000 color images corresponding to 126 people's faces images (70 men and 56 women) with different facial expressions, illumination conditions, and occlusions (e.g., glasses and scarf). The proportion of training set, validation set, and test set is 6:2:2.

4.2 Result and Discussion

Table 2 reports the comparison between EGPEL and the advanced competitors on AR. Training, validation, test accuracy obtained by all approaches are listed advanced in Table 2. The models obtained by the algorithm achieved a superiorities in all three types of accuracy. Training accuracy is 2% higher than the best competitor (i.e., EGPEL-A); Verification accuracy and test accuracy higher than AdaBoost 1.7%, 1.6% respectively. Meanwhile, the results integration using our proposed method improves 2.2% and 2% on test set over the voting (i.e., EGPEL-V) and average (i.e., EGPEL-V), respectively.

Figure 5 shows the search result, which contains a total of 15 base learners for a total of 3 lengths of base learner sets. The root node of the base learner indicates the selected hyperparameter. We can find that ANN and RBF appear more often.

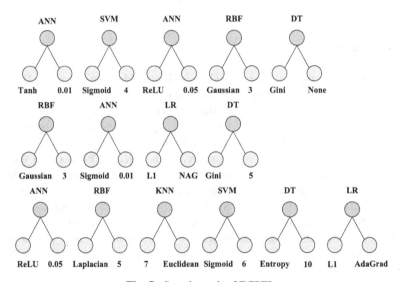

Fig. 5. Search result of EGPEL.

Table 2. Comparison with other baselines on AR face dataset

Method	Accuracy		
	Training set (%)	Validation set (%)	Test set (%)
SVM	89.3	89.7	89.1
RBF	90.2	90.1	90.4
Random Forest	91.8	91.6	92.0
AdaBoost	94.5	94.2	94.6
XGBoost	93.4	92.9	92.7
GBDT	92.9	92.4	92.2
LightGBM	92.8	92.6	92.0
EGPEL-V	93.4	93.1	93.3
EGPEL-A	94.8	94.2	94.5
EGPEL	96.8	96.4	96.6

5 Conclusion

The objective of this paper is to propose a GP-based ensemble learning method (EGPEL) that can automatically build an ensemble learning model which can improve the performance of a single weak learner on image datasets. The goal was achieved by designing base learners and hyperparameters sets for constructing ensemble learning configuration. A new results integration is proposed to better improve model performance.

Performances of the EGPEL are compared with advanced methods on the Face datasets. Results on the two image datasets show that EGPEL is not only more effective in almost all performance indicators but also more efficient than existing methods.

In the future, more base learners and hyperparameters will be optimized. To this end, some effective local search strategies may be needed for GP-based approach. A more efficient method should be used for image feature extraction. Finally, the construction of automated ensemble learning on large-scale datasets (i.e., ImageNet) will be a challenging direction [39].

Acknowledgments. This research was funded by the National Key R&D Program of China under Grant 2018YFB2003502, the Fundamental Research Funds for the Central Universities, and Guangdong Universities' Special Projects in Key Fields of Natural Science under Grant 2019KZDZX1005.

References

1. Dong, X., Yu, Z., Cao, W., et al.: A survey on ensemble learning. Front. Compt. Sci. **14**(2), 241–258 (2020). https://doi.org/10.1007/s11704-019-8208-z
2. Zhang, C., Ma, Y. (eds.): Ensemble Machine Learning: Methods and Applications. Springer, Heidelberg (2012). https://doi.org/10.1007/978-1-4419-9326-7

3. Gomes, H.M., Barddal, J.P., Enembreck, F., et al.: A survey on ensemble learning for data stream classification. ACM Comput. Surv. (CSUR) **50**(2), 1–36 (2017)
4. Bi, Y., Xue, B., Zhang, M.: An automated ensemble learning framework using genetic programming for image classification. In: Proceedings of the Genetic and Evolutionary Computation Conference, pp. 365–373 (2019)
5. Hutter, F., Kotthoff, L., Vanschoren, J.: Automated Machine Learning: Methods, Systems, Challenges. Springer, Heidelberg (2019). https://doi.org/10.1007/978-3-030-05318-5
6. Ma, L., Huang, M., Yang, S., et al.: An adaptive localized decision variable analysis approach to large-scale multiobjective and many-objective optimization. IEEE Trans. Cybern. (2021). https://doi.org/10.1109/TCYB.2020.3041212
7. Ma, L., Li, N., Guo, Y., et al.: Learning to optimize: reference vector reinforcement learning adaption to constrained many-objective optimization of industrial copper burdening system. IEEE Trans. Cybern. (2021). https://doi.org/10.1109/TCYB.2021.3086501
8. Ma, L., Wang, X., Huang, M., et al.: Two-level master–slave RFID networks planning via hybrid multiobjective artificial bee colony optimizer. IEEE Trans. Syst. Man Cybern. Syst. **49**(5), 861–880 (2019)
9. Ma, L., Wang, X., Huang, M., et al.: A novel evolutionary root system growth algorithm for solving multi-objective optimization problems. Appl. Soft Comput. **57**, 379–398 (2017)
10. Chen, H., Zhu, Y., Hu, K., et al.: Bacterial colony foraging algorithm: combining chemotaxis, cell-to-cell communication, and self-adaptive strategy. Inf. Sci. **273**, 73–100 (2014)
11. Su, W., Ma, L., Hu, K., et al.: A research on integrated application of RFID-based lean manufacturing. In: 2009 Chinese Control and Decision Conference, pp. 5781–5784. IEEE (2009)
12. Liu, Y., Ma, L., Yang, G.: A survey of artificial bee colony algorithm. In: 2017 IEEE 7th Annual International Conference on CYBER Technology in Automation, Control, and Intelligent Systems (CYBER), pp. 1510–1515. IEEE (2017)
13. Ma, L., Li, X., Gao, T., et al.: Indicator-based multi-objective bacterial foraging algorithm with adaptive searching mechanism. In: International Conference on Bio-Inspired Computing: Theories and Applications, pp. 271–277. Springer, Singapore (2016). https://doi.org/10.1007/978-981-10-3614-9_33
14. Chandra, A., Yao, X.: Ensemble learning using multi-objective evolutionary algorithms. J. Math. Model Algorithms **5**(4), 417–445 (2006). https://doi.org/10.1007/s10852-005-9020-3
15. Espejo, P.G., Ventura, S., Herrera, F.: A survey on the application of genetic programming to classification. IEEE Trans. Syst. Man Cybern. Part C (Appl. Rev.) **40**(2), 121–144 (2009)
16. Bi, Y., Xue, B., Zhang, M.: Dual-tree genetic programming for few-shot image classification. IEEE Trans. Evol. Comput. **PP**(99), 1 (2021)
17. Alam, K.M., Siddique, N., Adeli, H.: A dynamic ensemble learning algorithm for neural networks. Neural Comput. Appl. **32**(12), 8675–8690 (2020). https://doi.org/10.1007/s00521-019-04359-7
18. Wang, Y., Wang, D., Geng, N., et al.: Stacking-based ensemble learning of decision trees for interpretable prostate cancer detection. Appl. Soft Comput. **77**, 188–204 (2019)
19. Cui, S., Yin, Y., Wang, D., et al.: A stacking-based ensemble learning method for earthquake casualty prediction. Appl. Soft Comput. **101**, 107038 (2021)
20. Wang, Y., Liu, S., Li, S., et al.: Stacking-based ensemble learning of self-media data for marketing intention detection. Future Internet **11**(7), 155 (2019)
21. Zhou, Z.H.: Ensemble Learning. Machine Learning, pp. 181–210. Springer, Singapore (2021)
22. Awad, A., Bader-El-Den, M., McNicholas, J., et al.: Early hospital mortality prediction of intensive care unit patients using an ensemble learning approach. Int. J. Med. Inf. **108**, 185–195 (2017)

23. Mohammed, A.A., Yaqub, W., Aung, Z.: Probabilistic forecasting of solar power: an ensemble learning approach. In: International Conference on Intelligent Decision Technologies, pp. 449–458. Springer, Cham (2017). https://doi.org/10.1007/978-3-319-19857-6_38

24. Wang, S., Zhu, J., Yin, Y., et al.: Interpretable multi-modal stacking-based ensemble learning method for real estate appraisal. IEEE Trans. Multimedia **PP**(99), 1 (2021)

25. Aboneh, T., Rorissa, A., Srinivasagan, R.: Stacking-based ensemble learning method for multi-spectral image classification. Technologies **10**(1), 17 (2022)

26. El-Rashidy, N., Abuhmed, T., Alarabi, L., et al.: Sepsis prediction in intensive care unit based on genetic feature optimization and stacked deep ensemble learning. Neural Comput. Appl. **34**(5), 3603–3632 (2022). https://doi.org/10.1007/s00521-021-06631-1

27. Madhu, G., Bharadwaj, B.L., Boddeda, R., et al.: Deep stacked ensemble learning model for COVID-19 classification. Comput. Mater. Continua, 5467–5486 (2022)

28. Obasi, T.G., Shafiq, M.O.: CARD-B: a stacked ensemble learning technique for classification of encrypted network traffic. Comput. Commun. (2022). https://doi.org/10.1016/j.comcom.2022.02.006

29. Dong, Y., Zhang, H., Wang, C., et al.: Wind power forecasting based on stacking ensemble model, decomposition and intelligent optimization algorithm. Neurocomputing **462**, 169–184 (2021)

30. Ma, L., Cheng, S., Shi, Y.: Enhancing learning efficiency of brain storm optimization via orthogonal learning design. IEEE Trans. Syst. Man Cybern. Syst. **51**(11), 6723–6742 (2020)

31. Ma, L., Wang, X., Wang, X., et al.: TCDA: truthful combinatorial double auctions for mobile edge computing in industrial internet of things. IEEE Trans. Mobile Comput. (2021)

32. Santoso, L., Singh, B., Rajest, S., et al.: A genetic programming approach to binary classification problem. EAI Endorsed Trans. Energy Web **8**(31), e11 (2020)

33. Devarriya, D., Gulati, C., Mansharamani, V., et al.: Unbalanced breast cancer data classification using novel fitness functions in genetic programming. Expert Syst. Appl. **140**, 112866 (2020)

34. Tran, C.T., Zhang, M., Xue, B., et al.: Genetic programming with interval functions and ensemble learning for classification with incomplete data. In: Australasian Joint Conference on Artificial Intelligence, pp. 577–589. Springer, Cham (2018). https://doi.org/10.1007/978-3-030-03991-2_53

35. Park, J., Mei, Y., Nguyen, S., et al.: An investigation of ensemble combination schemes for genetic programming based hyper-heuristic approaches to dynamic job shop scheduling. Appl. Soft Comput. **63**, 72–86 (2018)

36. Chen, Q., Xue, B., Zhang, M.: Instance based transfer learning for genetic programming for symbolic regression. In: 2019 IEEE Congress on Evolutionary Computation (CEC), pp. 3006–3013. IEEE (2019)

37. Dick, G., Owen, C.A., Whigham, P.A.: Evolving bagging ensembles using a spatially-structured niching method. In: Proceedings of the Genetic and Evolutionary Computation Conference, pp. 418–425 (2018)

38. Karakatič, S., Podgorelec, V.: Building boosted classification tree ensemble with genetic programming. In: Proceedings of the Genetic and Evolutionary Computation Conference Companion, pp. 165–166 (2018)

39. Recht, B., Roelofs, R., Schmidt, L., et al.: Do imagenet classifiers generalize to imagenet? In: International Conference on Machine Learning. PMLR, pp. 5389–5400 (2019)

A Multi-constraint Deep Semi-supervised Learning Method for Ovarian Cancer Prognosis Prediction

Hua Chai[1], Longyi Guo[3], Minfan He[1], Zhongyue Zhang[2(✉)], and Yuedong Yang[2(✉)]

[1] School of Mathematics and Big Data, Foshan University, Foshan 528000, China
[2] School of Data and Computer Science, Sun Yat-Sen University, Guangzhou 510000, China
zhangzhy58@mail2.sysu.edu.cn, yangyd25@mail.sysu.edu.cn
[3] Department of Gynecology, Guangdong Provincial Hospital of Traditional Chinese Medical, Guangzhou, China

Abstract. Evaluating ovarian cancer prognosis is important for patients' follow-up treatment. However, the limited sample size tends to lead to overfitting of the supervised evaluation task. Considering to get more useful information from different perspectives, we proposed a semi-supervised deep neural network method called MCAP. MCAP introduced the heterogeneity information of the tumors through unsupervised clustering constraint, to help the model better distinguish the difference in the prognosis of ovarian cancer. Besides, the data recovering constraint is used to ensure learning a high-quality and low-dimensional representation of the genes in the network. For making a comprehensive analysis for ovarian cancer, we applied MCAP to seven gene expression datasets collected from TCGA and GEO databases. The results proved that the MCAP is superior to the other prognosis prediction methods in both 5-fold cross-validation and independent test.

Keywords: Bioinformatics · Survival analysis · Ovarian cancer

1 Introduction

Ovarian cancer is the deadliest gynecological tumor, which caused 207,252 deaths in 2020 based on the global cancer statistics [1]. Clinical studies found the same treatment used for ovarian cancer patients may lead to different treatment outcomes, which is the main reason for the high mortality of ovarian cancer. It is an ideal condition that patients should be administered different treatment regimens, based on the evaluated prognosis risks. Therefore, an accurate method is required for ovarian cancer prognosis prediction.

Currently, different machine learning methods have been used for cancer prognosis prediction. The traditional proportional hazard model proposed by Cox was the first used method in medical research for cancer survival analysis [2]. It was widely used to evaluate the impact of a few clinical features on the prognosis of patients. Based on the proportional hazard model, Wang *et al.* designed the random survival forests (RSF) for predicting cancer outcomes by utilizing the bootstrapping strategy [3]. With

Y. Tan et al. (Eds.): ICSI 2022, LNCS 13345, pp. 219–229, 2022.
https://doi.org/10.1007/978-3-031-09726-3_20

the fast development of gene sequencing technology, more and more high-dimensional genomic data were made available [4]. However, the high-dimensional omics data limits the prediction accuracy of these methods. Different solutions were designed to solve this challenge. One way is to use the reconstructed low-dimensional representation instead of the high-dimensional features. Jhajharia *et al.* used PCA to reconstructed the high-dimensional gene data, and the compressed features were used to predict breast cancer outcomes [5]. In another way, different penalized functions were added to the loss of the Cox proportional hazard model [6]. For example, Wang et al. used a group lasso-based Cox regression model to predict cancer prognosis and identified risk-related protein complex [7]. However, the high-dimensional nonlinear features limit the performance of these methods.

In recent years, deep neural networks were shown to have advantages in dealing with nonlinear features[8]. Katzman implemented DeepSurv by combining the proportional hazards function and deep neural network [9]. It was proved that DeepSurv achieved better cancer prognosis prediction performances by comparing with the traditional machine learning methods. In another way, Chaudhary applied the Cox model to predict liver cancer outcomes by using the Autoencoder processed low-dimensional features (AE-Cox) [10]. Nevertheless, the two-step framework affects the robustness of the model. Chai *et al.* designed TCAP, a multi-tasks deep learning-based method by combing the data reconstruction loss and risk prediction loss [11]. To improve the performance of the deep learning method with small size samples, Qiu proposed a meta-learning-based method for cancer outcomes prediction (MTLC) [12]. The results proved that MTLC can speed up the convergence of deep neural networks effectively. Though these studies have been carefully designed for predicting the prognosis of different cancers, the cancer heterogeneity and the small sample size of cancer data hinder the prognosis prediction performance. In addition, a deeply analyzing of ovarian cancer is lacking in these studies.

To improve the prognosis prediction accuracy of ovarian cancer, we made a comprehensive analysis by collecting seven ovarian cancer datasets from TCAG and GEO databases, and proposed a novel deep semi-supervised learning method with multi-constraint (MCAP). MCAP can get more useful information for cancer prognosis analysis from different perspectives. The unsupervised similarity constraint introduced the heterogeneity information of the tumors, to help the model better distinguish the differences between the ovarian cancer patients. Besides, the data recovering constraint is used to ensure learning a high-quality and low-dimensional representation of the features in the middle-hidden layer. The optimized representation was used to estimate the ovarian cancer prognosis in the proportional hazard module.

The experimental results show MCAP achieved a 2.64% higher C-index value than the state-of-the-art method, and 7.39% higher than the commonly used methods on average. The higher $|\log 10(P)|$ proved that by using the heterogeneity information between the tumors in different patients, MCAP can separate the high-risk patients from low-risk ones more significantly. The independent test showed it can accurately predict ovarian cancer prognosis (C-index > 0.6) and divided the patients into different risk subgroups significantly ($p < 0.05$). Based on the divided risk subgroups, we identified the top 10 differential expressed genes related to ovarian cancer prognosis: *CCL21, SERPINB7,*

LRRC15, OMD, ITGBL1, PRSS1, PKHD1, DRYSL5, GSTT1, and *HNF1B*, among which 6 genes have been proved by literature review.

2 Methods

2.1 Datasets

In this study, we collected seven ovarian cancer datasets from TCGA and GEO databases. We used the common 10326 gene features shared by all these ovarian cancer datasets and normalized the expression data by log transformation. The batch effect was removed by using the *"limma"* package [13]. Details about the used data are given in Table 1.

Table 1. Details about the used 7 ovarian cancer datasets

Dataset	Sample	Uncensored
TCGA-OV	296	177 (59.8%)
GSE17260	108	46 (42.6%)
GSE26193	107	76 (71.0%)
GSE26712	185	129 (69.7%)
GSE32062	260	121 (46.5%)
GSE53963	174	153 (87.9%)
GSE63885	75	66 (88.0%)
ALL	1205	768 (63.7%)

2.2 The Deep Neural Network in MCAP

As shown in Fig. 1a, after the ovarian cancer gene expression data X is input into the deep neural network, the extracted low-dimensional representation Z is compressed in the middle-hidden layer. In this layer, a new feature matrix is constructed by combining the compressed Z and the tumor heterogeneity label C of the ovarian cancer patients. The initial values of the tumor heterogeneity label C are clustered by the k-means, and are updated in every training epoch in the deep neural network by the unsupervised similarity constraint. In this neural network, multiple different constraints are optimized through a multi-task learning strategy [14]. It is worth noting that, the heterogeneity labels optimized by the similarity constraint will be seen as the extra features in the middle-hidden layer, and be updated with model training, for giving the cancer patients information from another perspective, to improve the accuracy of the ovarian cancer prognosis prediction.

Figure 1b shows that MCAP contains three constraints: The data recovering constraint can ensure the deep neural network gets the high-quality compressed representation in the middle-hidden layer. The similarity constraint is used to offer patients' tumor heterogeneity information about the ovarian cancer patients by the KL divergence-based loss function. Besides, the proportional hazard constraint can predict the cancer patients' prognosis.

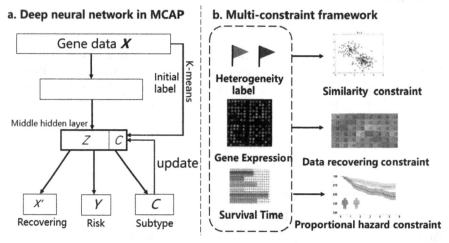

Fig. 1. The workflow of MCAP for ovarian cancer prognosis prediction. (a) The deep neural network in MCAP. (b) The multi-constraint framework in MCAP for cancer prognosis prediction

Supposing $X = (x_1, x_2, \ldots x_n)$ represents the gene expression of the ovarian cancer patients, Z is the compressed low dimensional features of X in the middle-hidden layer, for the data recovering constraint, it can be seen as the encoder-decoder part. Supposing E is the encoder function and D is the decoder function, the compressed Z is written as: $Z = E(X)$, and the recovered X' can be expressed: $X' = D(Z)$. The loss of the data recovering constraint is written as:

$$L_R = \sum_{i=1}^{n} \left(x_i - x_i'\right)^2. \tag{1}$$

The similarity constraint is used to extract the tumor heterogeneity information of ovarian cancer patients. The initial clustering labels were given by k-means ($k = 2$). In the middle-hidden layer, the reconstructed feature matrix F was formed by merging the compressed representation of the high-dimensional gene expression features and the clustered labels. In the deep neural network, the loss of the similarity constraint is defined as KL divergence between the two distributions P and Q by the following [15]:

$$L_c = KL(P||Q) = \sum_i \sum_j p_{ij} log \frac{p_{ij}}{q_{ij}}. \tag{2}$$

where q_{ij} is used to describe the similarity between the cluster center μ_j and the cluster point s_j, and the initial μ_j and s_j were obtained by k-means, p_{ij} is the target distribution derived from q_{ij}:

$$q_{ij} = \frac{\left(1 + \|s_j - \mu_j\|^2\right)^{-1}}{\sum_j \left(1 + \|s_j - \mu_j\|^2\right)^{-1}}; p_{ij} = \frac{q_{ij}^2 / \sum_i q_{ij}}{\sum_j \left(q_{ij}^2 / \sum_i q_{ij}\right)}. \tag{3}$$

After obtaining the tumor heterogeneity information, the proportional hazard constraint is used to predict ovarian cancer patient's prognosis. Supposing $S(t) = Pr(T > t)$ represents the survival probability that the ovarian cancer patient will survive before time t. The time interval T is the time elapsed between data collection and the patient's last contact. The risk function of the risk probability at time t can be expressed as:

$$\lambda(t) = \lim_{\delta \to 0} \frac{Pr(t \le T \langle t + \delta | T \ge t)}{\delta}. \tag{4}$$

By following Katzman's work [9], the loss function of the proportional hazard constraint is written as:

$$L_P = -\sum_{i=1} \left(h_\theta(x) - log \sum_{j \in \Re(T_i)} exp^{h_\theta(x_j)} \right). \tag{5}$$

where the neural network updates the $h(x)$ by the weight θ, and $\Re(T_i)$ is the risk set of patients still alive at time T_i.

As the network structure shown in Fig. 1, the objective of MCAP can be expressed as:

$$l_{MCAP} = \gamma L_R + \beta L_C + L_P. \tag{6}$$

The γ and β were used to adjust the importance of three constraints. They are seen as the hyper-parameters that can be chosen by the 5-fold cross-validation (CV). In this study, the values of γ and β were set 1 and 10, respectively.

2.3 Hyper-parameter Selection by CV

In this study, the average C-index (CI) values of the 5-fold CV were used for parameters selection and methods comparison. The parameter list in 5-fold CV was given in below: The number of nodes in hidden layer 1 was set 1000, the number of nodes in hidden layer 2 was set 500, and the dimension of Z in middle-hidden layer 3 was selected in [50, 20, 10]. The learning rate (LR) was set to [1e−6, 5e−7, 1e−7], and the max iteration in the neural network was set to 2000. The L1-norm regularization and L2-norm regularization coefficients were both set 1E−5. After a 5-fold CV performance comparison, we excluded the single ovarian cancer dataset from the combined datasets as the independent tests.

3 Results

3.1 5-fold CV for Method Comparison

In Table 2, we compared the 5-fold CV C-index values obtained by different methods in seven ovarian datasets. The TCGA_OV represents the ovarian cancer dataset collected from TCGA, and the OV_ALL is the dataset that includes all ovarian cancer patients in this study. MCAP was compared with six methods mentioned in the introduction section: The Cox proportional hazard model with elastic net (Cox-EN), Deep_surv, Cox proportional hazard model with Autoencoder (AE-Cox), MTLC, and TCAP. For the compared methods, Cox-EN achieved the lowest C-index value 0.544 on average. By Comparing with Cox-EN, the average C-index value achieved by Deepsurv is significantly improved (6.2%). MTLC performed better than other methods, but worse than our proposed method MCAP. MCAP achieved the highest C-index values between 0.601 (TCGA_OV) and 0.647 (GSE63885), with an average of 0.622. Compared with MTLC, it improved 2.64% C-index value on average.

For comparing the difference significance of the divided risk subgroups, in Table 3 we give the $|log10(P)|$ values obtained by different methods. The higher $|log10(P)|$ value indicates more significant differences in survival between different prognosis risk groups. As illustrated in Table 3, we got similar conclusions as in Table 2: MCAP achieved the best performance for separating the high-risk ovarian cancer patients from low-risk ones, MTLC and TCAP ranked 2nd and 3rd, respectively.

Table 2. The C-index values of different methods in ovarian cancer datasets

	Cox-EN	Deepsurv	AE-Cox	TCAP	MTLC	MCAP
TCGA _OV	0.547 (± 0.053)	0.564 (± 0.057)	0.572 (± 0.044)	0.575 (± 0.061)	0.585 (± 0.030)	0.601 (± 0.016)
GSE 17260	0.490 (± 0.077)	0.565 (± 0.091)	0.573 (± 0.013)	0.582 (± 0.104)	0.587 (± 0.088)	0.614 (± 0.119)
GSE 26193	0.538 (± 0.094)	0.582 (± 0.085)	0.599 (± 0.021)	0.584 (± 0.077)	0.612 (± 0.104)	0.624 (± 0.090)
GSE 26712	0.596 (± 0.083)	0.570 (± 0.037)	0.607 (± 0.022)	0.593 (± 0.035)	0.597 (± 0.041)	0.620 (± 0.058)
GSE 32062	0.566 (± 0.036)	0.619 (± 0.071)	0.598 (± 0.080)	0.631 (± 0.075)	0.639 (± 0.077)	0.621 (± 0.077)
GSE 53963	0.562 (± 0.029)	0.572 (± 0.042)	0.552 (± 0.054)	0.578 (± 0.058)	0.575 (± 0.034)	0.606 (± 0.043)
GSE 63885	0.504 (± 0.119)	0.584 (± 0.064)	0.572 (± 0.160)	0.602 (± 0.078)	0.620 (± 0.025)	0.647 (± 0.022)
OV _ALL	0.551 (± 0.030)	0.585 (± 0.017)	0.582 (± 0.013)	0.611 (± 0.016)	0.629 (± 0.006)	0.645 (± 0.020)
AVE	0.544	0.580	0.582	0.595	0.606	0.622

3.2 Parameter Sensitivity Study

In Fig. 2a, we show the C-index evaluated by 5-fold CV in MCAP for convergence analysis in OV_ALL dataset as an example. We show the C-index curves obtained with

Table 3. The |log10(P)| obtained by different methods in ovarian cancer datasets

	Cox-EN	Deepsurv	AE-Cox	TCAP	MTLC	MCAP
TCGA _OV	0.425 (± 0.265)	0.969 (± 0.857)	0.668 (± 0.792)	0.832 (± 0.769)	0.422 (± 0.547)	0.771 (± 0.429)
GSE 17260	0.196 (± 0.149)	0.226 (± 0.182)	0.581 (± 0.414)	0.212 (± 0.184)	0.347 (± 0.138)	0.783 (± 0.507)
GSE 26193	0.456 (± 0.178)	0.468 (± 0.273)	0.483 (± 0.283)	0.688 (± 0.493)	0.454 (± 0.496)	1.008 (± 0.496)
GSE 26712	0.586 (± 0.182)	0.349 (± 0.191)	0.600 (± 0.516)	0.536 (± 0.502)	0.922 (± 0.819)	1.364 (± 0.819)
GSE 32062	0.515 (± 0.494)	0.939 (± 0.963)	0.858 (± 0.704)	0.733 (± 0.798)	1.04 (± 0.882)	0.794 (± 0.895)
GSE 53963	0.462 (± 0.306)	1.063 (± 0.798)	0.677 (± 0.451)	0.876 (± 0.887)	0.694 (± 0.693)	0.990 (± 0.701)
GSE 63885	0.321 (± 0.218)	0.481 (± 0.314)	0.574 (± 0.283)	0.972 (± 0.841)	0.782 (± 0.729)	0.598 (± 0.479)
OV _ALL	1.259 (± 0.691)	2.884 (± 0.814)	2.323 (± 0.326)	3.570 (± 0.150)	4.756 (± 1.389)	4.947 (± 0.181)
AVE	0.527	0.922	0.846	1.052	1.180	1.407

different nodes number in the middle-hidden layer with LR = 1E−7. It indicated that the C-index increased sharply with an increase of epoch when the epoch is less than 400, and the values increased slowly until the epoch reached 1400. Then the curves flatten out or even decreased when the epoch reached 2000.

In Fig. 2b, we analyzed the effects of the parameters on the OV_ALL dataset. We show the C-index values with different LR and middle-hidden node sizes. It shows that the LR has a greater influence on the prediction performance of MCAP than the number of nodes. Compared with LR, the effect of middle-hidden node size is small. Hence, for achieving the best performance in ovarian cancer prognosis prediction, the parameters were selected by 5-fold CV.

3.3 Independent Test

In Table 4, we show the C-index values by excluding the single ovarian cancer dataset from the combined datasets as the independent tests. Deepsurv performed better than Cox-EN, which is similar to AE-Cox. MCAP achieved the C-index values ranging from 0.601 to 0.693 with the highest one in -GSE1726 and the lowest one in -GSE63885. By comparison, MCAP performed best with an average C-index value of 0.617, the MTLC and TCAP ranked 2nd and 3rd, respectively. The results show that MCAP achieved a 7.62% higher C-index value (0.617) than other methods in the independent test (C-index = 0.570 on average). The |log10(P)| values in Table 5 also supported our conclusion:

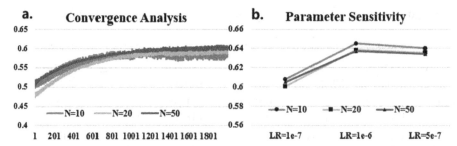

Fig. 2. The parameter sensitivity study in MCAP. (a) The convergence analysis in OV_all dataset. (b) The parameter sensitivity about the learning rate and middle-hidden node size.

MCAP achieved better prediction performances in independent tests by comparing with other methods.

Table 4. The C-index values in independent test obtained by different methods

	Cox-EN	Deepsurv	AE-Cox	TCAP	MTLC	MCAP
-TCGA_OV	0.573	0.562	0.570	0.570	0.591	0.619
-GSE17260	0.558	0.593	0.618	0.616	0.688	0.693
-GSE26193	0.553	0.608	0.536	0.577	0.616	0.604
-GSE26712	0.525	0.551	0.568	0.584	0.583	0.628
-GSE32062	0.586	0.578	0.591	0.581	0.572	0.604
-GSE53963	0.564	0.523	0.520	0.519	0.533	0.570
-GSE63885	0.482	0.531	0.553	0.563	0.615	0.601
AVE	0.549	0.564	0.565	0.573	0.600	0.617

Table 5. The $|\log10(P)|$ values in independent test obtained by different methods

	Cox-EN	Deepsurv	AE-Cox	TCAP	MTLC	MCAP
-TCGA_OV	1.846	1.389	1.634	0.642	3.20	1.929
-GSE17260	1.004	0.768	1.517	1.490	4.038	4.819
-GSE26193	0.002	0.181	0.097	0.597	2.057	1.801
-GSE26712	0.306	0.952	2.103	1.599	1.220	3.240
-GSE32062	1.140	1.880	1.984	2.668	1.532	3.542
-GSE53963	1.816	0.771	0.727	0.601	0.570	1.435
-GSE63885	0.125	0.280	1.373	0.477	1.685	1.835
AVE	0.891	0.889	1.348	1.153	2.043	2.657

3.4 Ovarian Cancer Prognosis Related Gene Identification

Based on the divided risk subgroups of ovarian cancer patients by MCAP, we performed differential expression analysis by using R package "*limma*" (Fig. 4). The differential expressed genes which I log2 fold changeI>0.5 and the p-values < 0.05 are seen as the cancer-realted genes. The top 5 genes with the highest log2 fold change values in up-regulated and down-regulated groups were seen as the important targets related to ovarian cancer prognosis. As given in Fig. 4, 10 ovarian cancer prognosis related genes (*CCL21, SERPINB7, LRRC15, OMD, ITGBL1, PRSS1, PKHD1, DRYSL5, GSTT1*, and *HNF1B*) were identified by MCAP. Among these genes, *CCL21, LRRC15, ITGBL1, PRSS1, GSTT1*, and *HNF1B* have been proved by literature review. For the remained genes, *SERPINB7, PKHD1*, and *DRYSL5* have been proved to be associated with other cancers, they may also be potential prognostic targets for ovarian cancer. This result illustrated that MCAP has advantages in potential ovarian cancer prognosis targets identification.

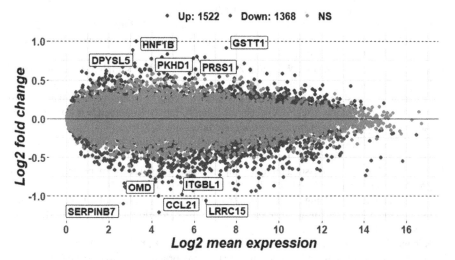

Fig. 4. The ovarian cancer prognosis related genes identification by MCAP.

4   Conclusion and Discussion

Ovarian cancer is the deadliest gynecological cancer. Accurately assessing the patients' prognosis can help clinicians choose appropriate treatment. In this research, we designed a deep semi-supervised learning method MCAP with three constraints: the data recovering constraint, the similarity constraint, and the proportional hazard constraint. The data recovering constraint can make the deep neural network learn a high-quality compressed representation of the high-dimensional gene expression. The similarity constraint is used to learn the heterogeneity information of the tumors, to help the deep neural network better distinguish the differences in the prognosis of the ovarian cancer patients from another perspective.

By comparing with the state-of-the-art methods in cancer prognosis prediction, the results demonstrated that MCAP achieved higher C-index values both in the 5-fold CV experiments and independent test, respectively. Though these results proved the accuracy and robustness of MCAP, there are still some interesting questions worth to be discussed. Firstly, the purity issue in ovarian cancer datasets may increase the difficulty of accurately predicting patients' outcomes. We will further investigate the effect of different ages and races on the prognosis of ovarian cancer. Secondly, in the future, we will collect more different omics data for ovarian cancer patients, and update our method and model by multi-omics integration.

References

1. Sung, H., Ferlay, J., Siegel, R.L., Laversanne, M., Soerjomataram, I., Jemal, A., et al.: Global cancer statistics 2020: GLOBOCAN estimates of incidence and mortality worldwide for 36 cancers in 185 countries, **71**(3), 209–49 (2017)
2. Cox, D.: Regression models and life-tables. J. Roy. Stat. Soc. B **34**(2), 187–202 (1972)
3. Wang, H., Zhou, L.: Random survival forest with space extensions for censored data. Artif. Intell. Med. **79**, 52–61 (2017)
4. Wang, Q., Zhou, Y., Zhang, W., Tang, Z., Chen, X.: Adaptive sampling using self-paced learning for imbalanced cancer data pre-diagnosis. Expert Syst. Appl. **152**, 113334 (2020)
5. Jhajharia, S., Varshney, H.K., Verma, S., Kumar, R. (eds.) A neural network based breast cancer prognosis model with PCA processed features. In: 2016 International Conference on Advances in Computing, Communications and Informatics (2016)
6. Simon, N., Friedman, J., Hastie, T., Tibshirani, R.: Regularization paths for Cox's proportional hazards model via coordinate descent. J. Stat. Softw. **39**(5), 1–13 (2011)
7. Wang, W., Liu, W.: PCLasso: a protein complex-based, group lasso-Cox model for accurate prognosis and risk protein complex discovery. Briefings in Bioinf. (2021)
8. Chai, H., Zhou, X., Zhang, Z., Rao, J., Zhao, H., Yang, Y.: Integrating multi-omics data through deep learning for accurate cancer prognosis prediction. Comput. Biol. Med. **134**, 104481 (2016)
9. Katzman, J.L., Shaham, U., Cloninger, A., Bates, J., Jiang, T., Kluger, Y.: DeepSurv: personalized treatment recommender system using a Cox proportional hazards deep neural network. BMC Med Res Methodol. **18**(1), 24 (2018)
10. Chaudhary, K., Poirion, O.B., Lu, L., Garmire, L.X.: Deep learning–based multi-omics integration robustly predicts survival in liver cancer. Clin. Cancer Res. **24**(6), 1248–1259 (2018)
11. Chai, H., Zhang, Z., Wang, Y., Yang, Y.: Predicting bladder cancer prognosis by integrating multi-omics data through a transfer learning-based Cox proportional hazards network. CCF Trans. High Perform. Comput. **3**(3), 311–319 (2021). https://doi.org/10.1007/s42514-021-00074-9
12. Qiu, Y.L., Zheng, H., Devos, A., Selby, H., Gevaert, O.: A meta-learning approach for genomic survival analysis. Nat. Commun. **11**(1), 1–11 (2020)
13. Ritchie, M.E., Phipson, B., Wu, D., Hu, Y., Law, C.W., Shi, W., et al.: limma powers differential expression analyses for RNA-sequencing and microarray studies. Nucleic Acids Res. **43**(7), e47 (2015)

14. Zhang, Y., Yang, Q.: A survey on multi-task learning. IEEE Trans. Knowl. Data Eng. (2021)
15. Guo, X., Gao, L., Liu, X., Yin, J. (eds.) Improved Deep Embedded Clustering with Local Structure Preservation. Ijcai (2017)

Modelling and Analysis of Cascade Digital Circuit System Based on DNA Strand Displacement

Hui Lv[1]([✉]), Tao Sun[1], and Qiang Zhang[1,2]([✉])

[1] Key Laboratory of Advanced Design and Intelligent Computing,
Ministry of Education, School of Software Engineering, Dalian University,
Dalian 116622, China
`lh8481@tom.com, zhangq@dlu.edu.cn`
[2] School of Computer Science and Technology, Dalian University of Technology,
Dalian 116024, China

Abstract. The cascade digital circuit system based on DNA strand displacement is investigated in this paper. In order to more accurately represent the reaction process, the loss of reaction substrate caused by base pair mismatch and "hairpin structure" in the process of DNA strand displacement are taken into account. Meanwhile, the time delay of DNA double-strand molecular breaks is also added to the constructed system model. The positivity of solutions, the stability and bifurcation at equilibrium point are investigated at length. It can be observed that when the time delay parameter pass some critical values, Hopf bifurcation may appear near the equilibrium point. By choosing different initial parameters, the equilibrium state reached by the system is different, and the stability of the system under equilibrium state will also be different.

Keywords: Cascade digital circuit · DNA strand displacement · Time delay · Hairpin structure · Hopf bifurcation

1 Introduction

In recent years, more and more researchers have paid attention to DNA molecular technology [5,17,27], in which the research on DNA digital circuit technology [20,24,30] is particularly prominent. As an excellent material for biochemical circuit engineering, DNA molecule is easy to chemically synthesize and provides

Supported by the National Key Technology R&D Program of China (No. 2018YFC0910500), the National Natural Science Foundation of China (Nos. 61425002, 61751203, 61772100, 61972266, 61802040), the Natural Science Foundation of Liaoning Province (Nos. 2020-KF-14-05, 2021-KF-11-03), High-level Talent Innovation Support Program of Dalian City (No. 2018RQ75), State Key Laboratory of Light Alloy Casting Technology for High-end Equipment (No. LACT-006), the Innovation and Entrepreneurship Team of Dalian University (No. XQN202008) and LiaoNing Revitalization Talents Program (No. XLYC2008017).

© Springer Nature Switzerland AG 2022
Y. Tan et al. (Eds.): ICSI 2022, LNCS 13345, pp. 230–241, 2022.
https://doi.org/10.1007/978-3-031-09726-3_21

convenience for practical operation in vitro. At the same time, the unique organizational structure and pairing principles of DNA molecules make it dynamic compilability [4,23]. This inspired the abundant theories based on DNA strand replacement [6,9,15,26] and promoted its considerable development and practice.

A large number of studies have shown that time delay [13,21] is an important research parameter of nonlinear system models. Considering the transfer process of input strand, time delay is added to the model of DNA strand displacement reaction based on toehold exchange, and dynamic behaviors such as local stability, Hopf bifurcation and chaos are discussed [14]. About the digital circuit model of cascade amplification involved in [19], because the separation of DNA double-strands and the combination with "invasion" single-strands take a certain amount of time to complete, and the multi-layer cascade circuit [16,25] includes two processes of thresholding and catalysis, the time delay of this process also cannot be ignored. In addition, the sequence of bases in the DNA molecular chain design process is very lengthy, and it is inevitable that base pair mismatches occur during the binding process [1,12]. Simultaneously, because the base sequence may have a certain repetitiveness, if small fragments of complementary base pairs appear at the head and tail of the DNA molecule, then during the reaction process, the DNA single-stranded molecule may bind itself head and tail to form a "hairpin structure" [28,29]. No matter what kind of situation occurs, the DNA strand cannot be combined with the "invading" strand. Therefore, the loss of reaction substrate cannot be ignored in the analysis of the system model. Adding the reaction substrate loss item to the constructed nonlinear system model makes constructed mathematical model more accurate and in line with reality.

The stability of system at the equilibrium point determines the direction of system model improvement. Bifurcation [7,8,11,18] always appears under the condition of system instability. In dynamic system [10], the phenomenon that the topological structure of each variable changes suddenly with change of control parameters is bifurcation, and the stability of system changes. It is the premise for bifurcation to occur. Bifurcation has multiple branches, among which Hopf bifurcation has more research value. Many new methods have been proposed for the study of Hopf bifurcation [2,3,10,22].

Motivated by the above discussions, a system model is established based on the cascade digital circuit. According to the characteristics of the DNA strand displacement reaction, the loss of the reaction substrate has influence on the circuit. Thus, the loss term and time delay parameter are added to the system model. Furthermore, the positivity of solutions, the performances of stability and bifurcation of the system are investigated. Finally, by setting input parameters, the state variable response diagrams of the systems are drawn to analyse the relationships and differences between them. The main contributions of this article are as follows: (1) The process of cascade digital circuit based on DNA strand displacement is modeled and presented in the form of differential equations. (2) According to the actual reaction mechanism, the loss item and the time delay parameter related to reaction substrate are added to the system model of DNA digital circuit. (3) The performance of stability and Hopf bifurcation are analysed under different initial situations, which helps to better understand the reaction law of cascaded digital circuits based on DNA strand displacement.

2   Model Establishment

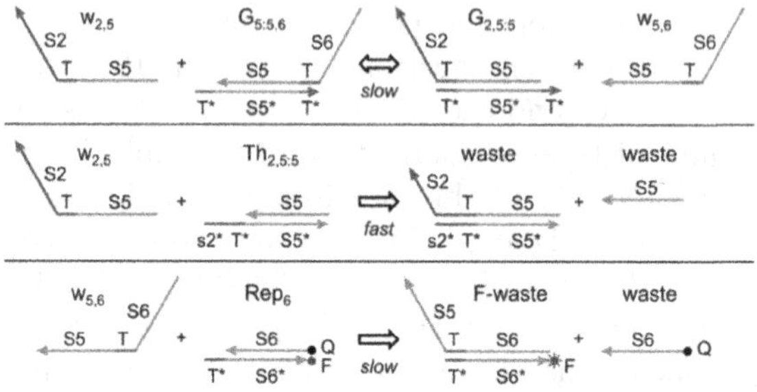

Fig. 1. Digital circuit model of DNA strand displacement cascade amplification [19]

The realization process of cascade-amplification digital circuit [19] is shown in Fig. 1. $W_{2,5}$, $G_{5,5,6}$, $Th_{2,5,5}$ and Rep_6 represent the initially added reactants and the fluorescent reporter product, respectively.

The reaction process can be simplified into the following form

$$\begin{cases} A + B \underset{k_{-1}}{\overset{k_1}{\rightleftharpoons}} C + D \\ A + E \overset{k_2}{\longrightarrow} F + G \\ D + H \overset{k_3}{\longrightarrow} I + J. \end{cases} \tag{1}$$

A nonlinear system model describing cascaded digital circuit can be built as

$$\begin{cases} \dot{A}(t) = -k_1 A(t)B(t) + k_{-1}C(t)D(t) - k_2 A(t)E(t) \\ \dot{B}(t) = -k_1 A(t)B(t) + k_{-1}C(t)D(t) \\ \dot{C}(t) = k_1 A(t)B(t) - k_{-1}C(t)D(t) \\ \dot{D}(t) = k_1 A(t)B(t) - k_{-1}C(t)D(t) - k_3 D(t)H(t) \\ \dot{E}(t) = -k_2 A(t)E(t) \\ \dot{F}(t) = k_2 A(t)E(t) \\ \dot{G}(t) = k_2 A(t)E(t) \\ \dot{H}(t) = -k_3 D(t)H(t) \\ \dot{I}(t) = k_3 D(t)H(t) \\ \dot{J}(t) = k_3 D(t)H(t). \end{cases} \tag{2}$$

where $A(t), B(t), E(t), H(t)$ are the concentrations of the reaction substrates A, B, E and H at time t; $D(t)$ represents the concentration of intermediate product D at time t; $C(t), F(t), G(T), J(t)$ respectively represent the concentration of the accompanying product at time t; $I(t)$ represents the concentration of the fluorescent reporter product I at time t; k_1, k_{-1}, k_2 and k_3 represent the reaction rate of each stage.

Let A_0, B_0, E_0 and H_0 be the reaction substrate the initial concentration of $W_{2,5}$, $G_{5,5,6}$, $Th_{2,5,5}$ and Rep_6. Because the reaction is carried out in a separate space, the consumption of reactants should be equal to the amount of products produced. According to this principle, the following relationship can be derived

$$\begin{cases} A_0 = A(t) + D(t) + I(t) + F(t) \\ B_0 = B(t) + C(t) \\ E_0 = E(t) + F(t) \\ H_0 = H(t) + I(t). \end{cases} \tag{3}$$

Substituting the equivalent relationship of Eq. (3) into the system Eq. (2) can get the three-dimensional system model as shown below

$$\begin{cases} \dot{A}(t) = - k_1 A(t)[B_0 - A_0 + E_0 + A(t) - E(t)] \\ \qquad + k_{-1}(A_0 - A(t) - E_0 + E(t))(A_0 - E_0 - A(t) + E(t) \\ \qquad - I(t)) - k_2 A(t)E(t) \\ \dot{E}(t) = - k_2 A(t)E(t) \\ \dot{I}(t) = k_3(A_0 - E_0 - A(t) + E(t) - I(t))(H_0 - E(t)). \end{cases} \tag{4}$$

It is shown that the reaction between substrates $W_{2,5}$ and $G_{5,5,6}$ of the DNA strand displacement reaction requires a transfer process [14]. What's more, due to its special structure, a single-stranded DNA molecule has multiple base pairs at the head and tail of the molecule. The single-stranded molecule is likely to combine head and tail to form a "hairpin structure". Unable to participate in subsequent reactions. Part of the reaction substrate will be consumed in this process. Therefore, in order to describe the system more accurately, the time delay parameter and the loss part are added to the model.

$$\begin{cases} \dot{A}(t) = - k_1 A(t - \tau)[B_0 - A_0 + E_0 + A(t - \tau) - E(t)] \\ \qquad + k_{-1}(A_0 - A(t - \tau) - E_0 + E(t))(A_0 - E_0 - A(t - \tau) + E(t) - I(t)) \\ \qquad - k_2 A(t - \tau)E(t) - K A(t - \tau)^2 \\ \dot{E}(t) = - k_2 A(t - \tau)E(t) - K E(t)^2 \\ \dot{I}(t) = k_3(A_0 - E_0 - A(t - \tau) + E(t) - I(t))(H_0 - E(t)) - K I(t)^2. \end{cases} \tag{5}$$

where $A(t - \tau)$ represents the concentration of the logic operation input chain A at time $t - \tau$, and τ is the time lag constant term.

3 Positivity of Solutions

The initial conditions of the state variables are $A_0 > 0$, $E_0 > 0$ and $H_0 = 0$. $H(t)$ is the concentration of product Rep_6 at time t, so $H(t) > 0$. $A(t)$ represents the initially added reactant, assume $t = t_1 > 0$, $A(t_1) = 0$ can be obtained. From Eq. (4) we can get

$$A'(t_1) = k_{-1}(A_0 - E_0 + E(t_1))(A_0 - E_0 + E(t_1) - I(t_1)) > 0.$$

which shows that $A(t) < 0$ for $t \in (t_1 - \delta, t_1)$. But $A_0 > 0$, δ is an arbitrarily small positive number. There is at least a $t_2 \in (0, t_1)$ ensures $A(t_2) = 0$. This contradicts to the hypothesis, so $A(t) > 0$. From system (4), $E(t) = E(0) \exp(\int_0^t (-k_2 A(\theta)) d\theta)$ can be obtained. It is obvious that $A(t) > 0$. As a result, all of the solutions of system (4) are positive.

4 System Stability and Hopf Bifurcation Analysis

In this section, we will focus on the stability and Hopf bifurcation performance of system (5) near its positive equilibrium point (A^*, E^*, I^*). Let $A^*(t) = A(t) - A^*$, $E^*(t) = E(t) - E^*$, $I^*(t) = I(t) - I^*$. Linearize the system equation.

$$\begin{cases} \dot{A}^*(t) = a_1 A^*(t - \tau) + a_2 E^*(t) + a_3 I^*(t) \\ \dot{E}^*(t) = b_1 A^*(t - \tau) + b_2 E^*(t) \\ \dot{I}^*(t) = c_1 A^*(t - \tau) + c_2 E^*(t) + c_3 I^*(t). \end{cases} \tag{6}$$

where

$$\begin{cases} a_1 = -k_1(B_0 - A_0 + E_0 - I^*) - k_{-1}(2A_0 - 2E_0 - 2A^* + 2E^* - I^*)E^*(t) - k_2 E^* - 2KA^* \\ a_2 = k_1 A^* - k_2 A^* + k_{-1}(2A_0 - 2E_0 - 2A^* + 2E^* - I^*) \\ a_3 = -k_{-1}(A_0 - E_0 - A^* + E^*) \\ b_1 = -k_2 E^* \\ b_2 = -k_2 A^* E^*(t) - 2KE^* \\ c_1 = -k_3 H_0 \\ c_2 = -k_3(A_0 - E_0 - H_0 - A^* + E^* - I^*) \\ c_3 = -k_3 H_0 - 2KI^*. \end{cases}$$

The characteristic equation of system (6) is

$$\begin{aligned} |\lambda e - J| &= (\lambda - b_2)[\lambda^2 - c_3\lambda - a_1\lambda e^{-\lambda\tau} + a_1 c_3 e^{-\lambda\tau} - c_1 a_3 e^{-\lambda\tau}] \\ &= \lambda^3 - (b_2 + c_3)\lambda^2 - b_2 c_3 \lambda + [(a_1 c_3 - a_3 c_1 - a_1\lambda)\lambda \\ &\quad - b_2(a_1 c_3 - a_3 c_1 - a_1\lambda)]e^{-\lambda\tau} = 0. \end{aligned} \tag{7}$$

Suppose there is a τ_0 that makes the characteristic equation of system have a pair of pure imaginary roots, denoted as $\lambda = i\omega$ ($\omega > 0$), we can get

$$\begin{cases} \sin \omega\tau = \dfrac{g_1\omega^5 + g_2\omega^3 + g_3\omega}{g_4\omega^6 + g_5\omega^4 + g_6\omega^2 + g_7} \\[2mm] \cos \omega\tau = \dfrac{g_8\omega^5 + g_9\omega^4 + g_{10}\omega^3 + g_{11}\omega^2 + g_{12}}{g_4\omega^6 + g_5\omega^4 + g_6\omega^2 + g_7}. \end{cases} \tag{8}$$

where

$$g_1 = a_1, \; g_2 = -a_1b_2c_3 + a_3b_2c_1 + a_1b_2, \; g_3 = -b_2c_3(a_1b_2c_3 - a_3b_2c_1)$$

$$g_4 = a_1^3, \; g_5 = a_{(}(a_1b_2c_3 - a_3b_2)^2 + (a_1b_2c_3 - a_3b_2c_1)a_1 + 2a_1^2((a_1b_2c_3 - a_3b_2c_1)$$

$$g_6 = a_1(a_1b_2c_3 - a_3b_2c_1) + (-a_1b_2c_3 + a_3b_2c_1)(-a_1c_3 - a_3c_1 + a_1b_2)^2$$

$$g_7 = (a_1b_2c_3 - a_3b_2c_1)(a_1b_2c_3 - a_3b_2c)^2, \; g_8 = (b_2 + c_3)(a_1b_2c_3 - a_3b_2c_1 - a_1b_2)$$

$$g_9 = (a_1c_3 - a_3c_1 + a_1b_2)(a_1b_2c_3 - a_3b_2c_1 - a_1b_2,$$

$$g_{10} = b_2c_3(b_2 + c_3)(-a_1b_2c_3 + a_3b_2c_1), g_{11} = b_2c_3(a_1c_3 - a_3c_1 + a_1b_2),$$

$$g_{12} = a_1c_3 - a_3c_1 + a_1b_2.$$

In the light of $\cos^2\theta + \sin^2\theta = 1$, it can be solved

$$h_1\omega^{12} + h_2\omega^{10} + h_3\omega^8 + h_4\omega^6 + h_5\omega^4 + h_6\omega^2 + h_7 = 0. \tag{9}$$

where

$$h_1 = g_4^2, h_2 = g_8^2 - 2g_2 + g_1^2, h_3 = 2(g_3 + g_1g_2 - g_8g_9)g_4^2$$

$$h_5 = 2(g_1g_4 + g_2g_2 - g_5g_7 - g_9g_{10}) - g_9^2, h_6 = 2(g_3g_4 - g_{10}g_{11}) - g_7^2$$

$$h_7 = -(g_7^2 + g_{10}^2).$$

Substitute $z = \omega^2$ into Eq. (9) to get

$$f(z) = h_1z^6 + h_2z^5 + h_3z^4 + h_4z^3 + h_5z^2 + h_6z + h_7 = 0. \tag{10}$$

In line with Intermediate Value Theorem, $\lim\limits_{x \to +\infty} f(z) = +\infty$ can be got, if $h_7 < 0$, then there at least one point $z_0 \in (0, +\infty)$ exists to make $f(z_0) = 0$ hold.

There is a series of $\tau_k^i > 0(k = 1, 2, ...6; i = 0, 1, 2...)$ such that characteristic Eq. (7) has a pair of pure imaginary roots $\pm i\omega$.

$$\tau_k^i = \frac{1}{\omega_0} a \cos \frac{g_8\omega^5 + g_9\omega^4 + g_{10}\omega^3 + g_{11}\omega^2 + g_{12}}{g_4\omega^6 + g_5\omega^4 + g_6\omega^2 + g_7} + \frac{i\pi}{\omega_k}. \tag{11}$$

where $\tau_0 = \min\{\tau_k^i\}$ and correspond to τ_0.

Consulting Hopf Bifurcation Existence Theorem, the following conclusions can be drawn.

Theorem 1. *When $\tau \in [0, \tau_0)$, the system (6) is locally asymptotically stable. When $\tau \geq \tau_0$, the system (6) loses its steady state at the equilibrium point and Hopf bifurcation occurs. The system switches between a stable equilibrium state and a limit cycle, where τ_0 is the critical parameter.*

5   Numerical Simulations

Because adding different concentrations of the reaction substrate to the same reaction process will reach different equilibrium states, so here are two different equilibrium states.

Case 1. Initial reaction substrate $Th_{2,5,5}$ is excessive, and $W_{2,5}$ reacts completely. Parameters are presented in Table 1 and Table 2.

Table 1. The initial concentration of substrate in Case 1.

Reaction substrate	$W_{2,5}$	$G_{5,5,6}$	$Th_{2,5,5}$	Rep_6
Concentration (nm)	4 * (1e−7)	1 * (1e−7)	1.2 * (1e−7)	1 * (1e−7)

Table 2. Approximate the reaction rate constant in Case 1.

Reaction rate	k_1	k_{-1}	k_2	k_3	K
nm/s	1 * (1e6)	2 * (1e5)	2 * (1e6)	1 * (1e6)	1 * (1e6)

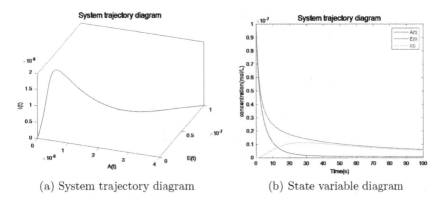

(a) System trajectory diagram   (b) State variable diagram

Fig. 2. System trajectory diagram and state variable diagram when $\tau = 1$.

After calculation, it can be seen that the critical value of bifurcation control parameter is $\tau_0 = 3.576$ in Case 1. Comparing Fig. 2 and Fig. 3, it can be known that changing control parameter τ, the state of system will change accordingly, but the system is still in stable state.

Case 2. Initial reactant $W_{2,5}$ is excessive and $Th_{2,5,5}$ reacts completely. Parameters are demonstrated in Table 3 and Table 4.

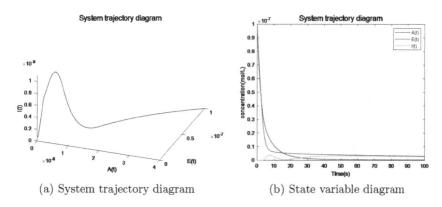

(a) System trajectory diagram   (b) State variable diagram

Fig. 3. System trajectory diagram and state variable diagram when $\tau = 3.576$.

Table 3. The initial concentration of substrate in Case 2.

Reaction substrate	$W_{2,5}$	$G_{5,5,6}$	$Th_{2,5,5}$	Rep_6
Concentration (nm)	4 * (1e−7)	1 * (1e−7)	1 * (1e−7)	1 * (1e−7)

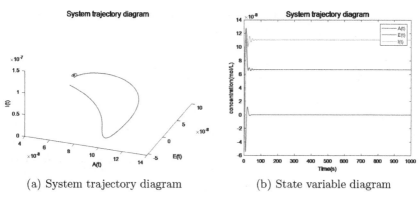

(a) System trajectory diagram   (b) State variable diagram

Fig. 4. System trajectory diagram and state variable diagram when $\tau = 4$.

Similarly, the critical value can be calculated as $\tau_0 = 5.524$ in Case 2. When $\tau = 4 < \tau_0$, the bifurcation control parameter is less than the critical value, and the system should be in stable state. From system trajectory diagram and state variable diagram in Fig. 4, it can be seen that system has a small period of fluctuation at the beginning of the reaction, and finally tends to be locally asymptotically stable near the equilibrium point $(A_0 - B_0 - E_0, 0, B_0)$.

When $\tau = 5.524 = \tau_0$, the control parameter reaches the critical value, It can be seen from Fig. 5 that the shock of system is obviously aggravated, and the state of system is switched between a stable equilibrium point and a limit cycle.

When $\tau = 6 > \tau_0$, the bifurcation control parameter exceeds the critical value. At this time, the system has lost stability near the equilibrium point and

Table 4. Approximate the reaction rate constant in Case 2.

Reaction rate	k_1	k_{-1}	k_2	k_3	K
nm/s	1 * (1e6)	2 * (1e5)	2 * (1e6)	1 * (1e6)	1 * (1e6)

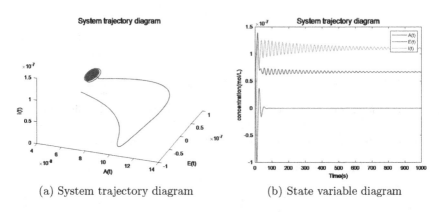

(a) System trajectory diagram   (b) State variable diagram

Fig. 5. System trajectory diagram and state variable diagram when $\tau = 5.524$.

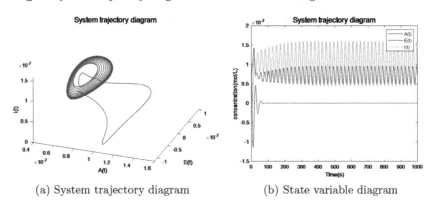

(a) System trajectory diagram   (b) State variable diagram

Fig. 6. System trajectory diagram and state variable diagram when $\tau = 6$.

Hopf bifurcation occurs. The simulation results in Fig. 6 can verify the correctness of conclusions.

Remark 1. Combining Case 1 and Case 2, for the improved model (5), the reaction will reach a different equilibrium state when the initial substrate concentration is different. When the initial reaction substrate $Th_{2,5,5}$ is excessive and $W_{2,5}$ is completely reacted, the system is always in a stable state during the reaction process. When the initial reactant $W_{2,5}$ is excessive and $Th_{2,5,5}$ is fully reacted, the equilibrium state reached by the system will change with the time delay parameter τ. When the value of parameter τ reaches the critical value τ_0, the system will change from a locally asymptotically stable state to a stable equilibrium point and limit cycle switching state, and Hopf bifurcation occurs. Therefore, choice of initial concentration of the reaction substrate is also important for design of reaction model, and the correct initial concentration configuration can avoid erroneous results. This result indicates that the choice of initial reaction substrate concentration for the same reaction process will have a great impact on the reaction process. For the DNA cascade amplification digital circuit model discussed in this article, selecting a sufficient amount of $Th_{2,5,5}$ will be more conducive to the observation and control of the reaction process.

6 Conclusions

In this article, our research focuses on the modelling and dynamics of cascaded digital circuit systems based on DNA strand displacement. Firstly, a nonlinear system model was established and improved, the loss and time delay parameters of the reaction substrate were added to the system model. The positivity of state variables of the nonlinear system is proved. After that, time delay is used as a control parameter to analyze the stability of system and the Hopf bifurcation near the equilibrium point. Given different initial concentrations and reaction rates, the system will reach different equilibrium states, and the stability of system under distinct equilibrium states will be different. By changing the bifurcation control parameters, the original stability of the system will be broken, and Hopf bifurcation will appear near the equilibrium point. In the future, feedback control for DNA strand replacement reaction with time delay will be the focus of our research.

References

1. Broadwater, J.D.W.B., Kim, H.D.: The effect of basepair mismatch on DNA strand displacement. Biophys. J. **110**, 1476–1484 (2016)
2. Cao, J., Guerrini, L., Cheng, Z.: Stability and Hopf bifurcation of controlled complex networks model with two delays. Appl. Math. Comput. **343**, 21–29 (2019)
3. Chaplain, M., Ptashnyk, M., Sturrock, M.: Hopf bifurcation in a gene regulatory network model: molecular movement causes oscillations. Math. Models Methods Appl. Sci. **25**, 1179–1215 (2015)
4. Chen, R.P., Blackstock, D., Sun, Q.: Wilfred: Dynamic protein assembly by programmable DNA strand displacement. Nat. Chem. **10**, 474–481 (2018)
5. Frykholm, K., Nyberg, L.K., Westerlund, F.: Exploring DNA-protein interactions on the single DNA molecule level using nanofluidic tools. Integr. Biol. **8**, 650–661 (2017)
6. Fu, H., Lv, H., Zhang, Q.: Using entropy-driven amplifier circuit response to build nonlinear model under the influence of lévy jump. BMC Bioinform. **22**, 437 (2022)
7. Huang, C., Zhang, H., Cao, J., Hu, H.: Stability and Hopf bifurcation of a delayed prey-predator model with disease in the predator. Int. J. Bifurcat. Chaos **29**, 1950091–1950091 (2019)
8. Jiang, Y., Korpas, L.M., Raney, J.R.: Bifurcation-based embodied logic and autonomous actuation. Nat. Commun. **10**, 1–10 (2019)

9. Johann, E., Oleg, L., Fuan, W., Remacle, F., Levine, R.D., Willner, I.: DNA computing circuits using libraries of DNAzyme subunits. Nat. Nanotechnol. **5**, 417–422 (2010)

10. Kou, B., Chai, Y., Yuan, Y., Yuan, Y.L.: Dynamical regulation of enzyme cascade amplification by a regenerated DNA nanotweezer for ultrasensitive electrochemical DNA detection. Anal. Chem. **90**, 10701–10706 (2018)

11. Kozłowska, E., Puszynski, K.: Application of bifurcation theory and siRNA-based control signal to restore the proper response of cancer cells to DNA damage. J. Theor. Biol. **408**, 213–221 (2016)

12. Liu, D., Keijzers, G., Rasmussen, L.J.: DNA mismatch repair and its many roles in eukaryotic cells. Mutat. Res./Rev. Mutat. Res. **773**, 174–187 (2017)

13. Liu, Y.H., Lv, H., Wang, B., Yang, D.Y., Zhang, Q.: Modelling and analysis of haemoglobin catalytic reaction kinetic system. Math. Comput. Model. Dyn. Syst. **26**, 306–321 (2020)

14. Liu, Y.H., et al.: Stability analysis and Hopf bifurczhonglong zheng and ation research for dna strand displacement with time delay. In: 2018 IEEE SmartWorld, Ubiquitous Intelligence & Computing, Advanced & Trusted Computing, Scalable Computing & Communications, Cloud & Big Data Computing, Internet of People and Smart City Innovation (SmartWorld/SCALCOM/UIC/ATC/CBDCom/IOP/SCI) 2018, pp. 228–233 (2018)

15. Lv, H., Li, H.W., Zhang, Q.: Analysis of periodic solution of DNA catalytic reaction model with random disturbance. IEEE Open J. Nanotechnol. **2**, 140–147 (2021)

16. Marasso, S.L., et al.: A multilevel lab on chip platform for DNA analysis. Biomed. Microdevices **13**, 19–27 (2011)

17. Matsuura, Y.: Coherent spin transport in a DNA molecule. Chem. Phys. **60**, 915–935 (2020)

18. Nathans, J.F., Cornwell, J.A., Afifi, M.M., Paul, D., Cappell, S.D.: Cell cycle inertia underlies a bifurcation in cell fates after DNA damage. Sci. Adv. **7**, eabe3882–eabe3882 (2021)

19. Qian, L., Winfree, E.: Scaling up digital circuit computation with DNA strand displacement cascades. Science **332**, 1196–1201 (2011)

20. Qian, L., Winfree, E., Bruck, J.: Neural network computation with DNA strand displacement cascades. Nature **475**, 368–372 (2011)

21. Seuret, A., Gouaisbaut, F.: Wirtinger-based integral inequality: application to time-delay systems. Automatica **49**, 2860–2866 (2013)

22. Tanouri, Z., Motlagh, O.R., Mohammadinezhad, H.M.: On the conditions of Hopf bifurcation for ATM protein and DNA damage signal model; cuts off the DNA healing process. J. Adv. Math. Comput. Sci. 395–400 (2015)

23. Thuronyi, B.W., Koblan, L.W., Levy, J.M., Yeh, W.H., Liu, D.R.: Continuous evolution of base editors with expanded target compatibility and improved activity. Nat. Biotechnol. **37**, 1070–1079 (2019)

24. Wang, F., Lv, H., Li, Q., Li, J., Fan, C.: Implementing digital computing with DNA-based switching circuits. Nat. Commun. **11**, 1–8 (2020)

25. Wang, S., Sun, J., Zhao, J., Lu, S., Yang, X.: Photo-induced electron transfer-based versatile platform with g-quadruplex/hemin complex as quencher for construction of DNA logic circuits. Anal. Chem. **90**, 3437–3442 (2018)

26. Yuan, Y., Lv, H., Zhang, Q.: DNA strand displacement reactions to accomplish a two-degree-of-freedom PID controller and its application in subtraction gate. IEEE Trans. Nanobiosci. **20**, 554–564 (2021)

27. Yuan, Y., Lv, H., Zhang, Q.: Molecular device design based on chemical reaction networks: state feedback controller, static pre-filter, addition gate control system and full-dimensional state observer. J. Math. Chem. **60**, 915–935 (2022)
28. Zhang, C., Shen, L., Liang, C., Dong, Y., Xu, J.: DNA sequential logic gate using two-ring DNA. ACS Appl. Mater. Interfaces **8**, 9370–9376 (2016)
29. Zhou, W., Li, D., Yuan, R., Xiang, Y.: Programmable DNA ring/hairpin-constrained structure enables ligation-free rolling circle amplification for imaging mRNAS in single cells. Anal. Chem. **91**, 3628–3635 (2019)
30. Zou, C., Wei, X., Zhang, Q., Liu, C., Zhou, C., Liu, Y.: Four-analog computation based on DNA strand displacement. ACS Omega **2**, 4143–4160 (2017)

Two-Replacements Policy Island Model on GPU

Faiza Amin[1] and Jinlong Li[2(✉)]

[1] University of Computer Science and Technology of China, Hefei, Anhui, China
faizach222@mail.ustc.edu.cn
[2] Department Computer Science, University of Computer Science and Technology of China, Hefei, Anhui, China
jlli@ustc.edu.cn

Abstract. The island model is one technique to tackle complex and critical difficulties of evolutionary algorithms. This paper will design a two-replacements policy and warp-based island mapping mechanism in TRPIM with ring topology on GPU Nvidia's CUDA programming. Each thread in the warp-based island executes the same instruction sequence in parallel to eliminate thread divergence. The two-replacement policy would replace the worse individuals with the better ones asynchronously and synchronously, reducing the waiting duration. We conduct experiments on the knapsack problem to verify the warp-based island mapping mechanism's effectiveness and two-replacement policy in TRPIM. And the results show that the proposed TRPIM improves the speedup time and solution quality on the GPU version compared to the CPU.

Keywords: Graphics processing units · Island model · CUDA programming · Knapsack problem · Warp

1 Introduction

Evolutionary algorithms (EAs) have been successfully applied in real-world optimization [14]. Island Models (IM) are one possibility considered for improving EAs' problem-solving capabilities [13]. An island model is a distributed EAs in which the population spreads onto multiple islands [7]. This model introduces a new evolutionary operator migration that allows individuals to migrate from one island to another. To accelerate Island Model optimization, we examine how Island Model implementations could be implemented on Graphics Processing Units (GPU) using CUDA programming.

The GPU has initially used for computer graphics, but it is now widely used for scientific computing, machine learning, etc. The GPUs are perfect for tasks requiring many threads to execute parallel, using the same instruction set on heterogeneous input data. GPU chips contain a set of streaming multiprocessors (SMs). Each SM is divided into N threads per block. The CUDA kernel uses a thread hierarchy to support hundreds of threads [8]. Through the CUDA

© Springer Nature Switzerland AG 2022
Y. Tan et al. (Eds.): ICSI 2022, LNCS 13345, pp. 242–253, 2022.
https://doi.org/10.1007/978-3-031-09726-3_22

architecture, blocks are executed in warps, which allows the execution of single instructions multiple times (SIMT). For more details on GPU/CUDA, architecture can review in [2]. The island model (IM) approach for EAs is practical and powerful in solving real-world optimization problems, and GPU-based implementation of IM has become attractive. The authors proposed a GPU-based genetic algorithm implementation [9], where every thread controls one individual and each block has one unique island. The sub-population is saved in shared memory, while the global memory is used for migration. Griewank's, Michalewicz's, and Rosenbrock's functions assessed speed and convergence. They reported the results that speed up 2600 times by CUDA software model with 100 iterations. To improve the performance of EAs by using GPUs, the authors proposed three different schemes [6]. The CPU is responsible for all evaluations on any given island in the first scheme, and the GPU is responsible for the parallel assessment. In the second scheme, all sub-populations of the island model (IM) are stored in global memory, and one block per island is parallelized. A third scheme distributes the IM through shared memory on GPU. A shared memory holds all sub-populations, while a global memory allows migration between islands. The approach solved the Weierstrass-Mandelbrot numerical optimization problem 2074 times faster than the CPU. However, for these results, the number of iterations is fixed at 100. Based on the island model and simulated annealing [5], the authors proposed a parallel Genetic algorithm on GPU. A benchmark study is conducted on travelling salesman problems (TSP). In a global memory, the sub-populations of each island are mapped by blocks. Recently, the authors proposed [15,16] island model genetic algorithm (IMGA) on GPU/CUDA for solving the unequal area facility layout problems (UA-FLP) and achieving a performance ratio as high as 84. This strategy executes the algorithm based on GPU with one block per island. The global memory stores all populations and is used for migration among blocks. And the shared memory is used for the selection phase data. In [4], the authors proposed a genetic Algorithm on GPU/CUDA platform. In this algorithm, multiple islands are mapped to the thread block on each SM and assigned numerous threads for each individual. The migration is performed using global memory. This method is used for the N-Queen problem and achieved a speedup of 45.5 times faster. Furthermore, in [3], the authors proposed GPU acceleration of the IMGA using the CUDA programming platform and achieved speedups as much as 18.9 times faster than their own work published in [4]. However, numerous research has attempted the algorithm speedup and the best solution. The focus is on the length of blocks to exchange data via global memory or shared memory. And use the different number of generations, crossover rate, mutation rate, and replacement according to the environment. The core will wait for the crossover and mutation operation to complete so the replacement for the best solution quality can do. It takes a lot of time to replace them. As a result, choosing a good individual and replacing its offspring affect the solution's speed, time, and quality.

This paper describes the GPU's Two-replacements Policy Island Model (TRPIM), which executes SM via warps. The warp granularity implies a sufficient number of individuals and the size of each individual, most of which are unaffected by correct mapping. GPU warps control the organized population into islands. However, the other genetic parameters such as tournament selection, uniform crossover, bit flip mutation, and replacement are employed to create the entire island with the two replacements methods. Every evolutionary operation beyond islands to the generational process requires synchronization barriers. The global memory stores sub-populations, offspring populations, fitness values, and data exchange among islands. The following are the categories for this paper's contributions:

1. The island model divides the population into multiple sub-populations to accelerate and execute each population's evolutionary process simultaneously. However, the sub-population distribution on the GPU is critical in time consumption. Instead of a block-based island, we propose a warp-based island mapping mechanism that maps a sub-population to a warp and stores all information in the shared memory. The warp based island reads the instructions one time, and all threads in the warp execute them parallel, which removes the thread divergence.
2. For implementing the island model on GPU, before the replacement, the algorithm has to synchronize and wait for the done of crossover and mutation. To reduce the time consumption of waiting duration, we propose a Two-replacements Policy Island Mode to replace the worse individuals with better ones asynchronously and synchronously, respectively.
3. Present a general framework of the Two-replacements Policy Island Model (TRPIM) on GPU and empirically evaluate it on the 0–1 knapsack problem.

The remainder of the paper is presented under the following sections: Sect. 2 details our new algorithm's methods. Section 3 contains the results of the experimental investigation. Finally, in Sect. 6, This paper is summarized and discusses future work.

2   Methodology

First, the warp based island structure for the island model on GPU is presented. Then, we describe the other components of TRPIM, including two-way replacements, migration policy, and other genetic operators.

2.1   Warp Based Island Structure for Island Model on GPU

In TRPIM design, the warps mapping structure is shown in Fig. 1. There are n Streaming-Multiprocessors (SMs), and the cores of an SM divides into r blocks.

A design structure consists of r blocks within n SM. There are k warps in a block, each used as a sub-population. After f_s generations, k islands exchange information in r blocks in global memory. Matrix threads are mapped to the y dimension of individual size, and the x dimension runs along the k warps is constructed. In the warp, each thread executes the same instruction simultaneously.

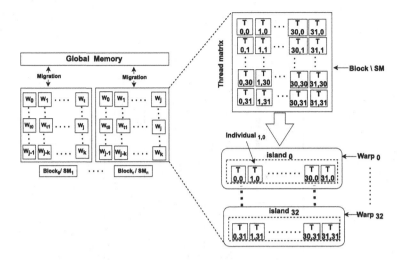

Fig. 1. General framework of warp based island mechanism on GPU

2.2   Two-Replacements Policy in Warp Based Island

Each warp in TRPIM represents one island. Each island is processed by genetic operators such as Tournament selection, uniform crossover, bit flip mutation, replacement, and migration. To ensure the accuracy of our design TRPIM algorithm, we used a two-replacement policy. First, the replacement policy applies to one thread to find a better fitness value for one individual solution. In this case, an individual would do better to replace the worst value in a thread. This will happen in each generation.

In the second replacement, we will select the best fitness values of solutions for the individual from each thread as a new offspring. And replace the worse values of an individual in the island/sub-population synchronously, as shown in Fig. 2. We will choose the best fitness values of solutions for the individual in the second replacement from each island as illustrated in Fig. 2. The R_1 is the first replacement policy for one thread that replaces worse individuals with better ones asynchronously. The stopping criteria of R_1 are 0.1 s. R_2 represents the second replacement that applies to the individual island synchronously after every 100 generations. And get the new offspring of the individuals of an island.

2.3   Migration Policy

Migration is an essential part of the island model. If migration didn't happen, the islands would merely be a collection of isolated individuals. The migratory mechanism would allow isolated islands to exchange their best individuals. In the island model, the frequency and interval of migration are essential variables.

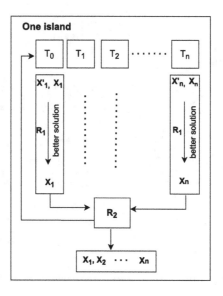

Fig. 2. Illustration of two replacements in island on GPU

Maximum warps in SM require more than one block to be active. As shown in Fig. 3, in every round of fs generations, the k island saves the best individual in global memory. These individuals are executed using four threads with copy operations in global memory. These four threads will select the best individual for the second block warps to migration. In each generation, the k islands write information to global memory to choose the best individual.

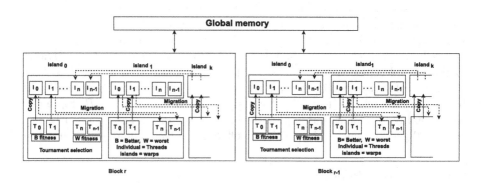

Fig. 3. Migration policy with global memory

2.4   Genetic Operators

Tournament Selection: The typical tournament selection [1,17] randomly chooses n individuals with replacement from the current sub-population size N into a tournament of size c and selects the optimal fitness value for crossover. We employ a warp-level parallel reduction strategy to maximize threads utilization for tournament selection to conduct tournaments with n individuals.

Crossover: Uniform crossover [11] ensures that the bits from two individual p_a and p_b are combined. The selecting uniform random numbers between 0 to 1 bit-wise procedures are needed. And executes the process of swapping bits in the individuals p_a and p_b to create the new offspring. In our algorithm, every warp reads individual p_a and p_b in chunks of i integer components. Each thread initially creates a z bit random number for the crossover mask.

With the EA approach, large instances size i uses to tackle the knapsack problem. Conditional code is eliminated from the crossover to save evaluation time. In this case, the thread divergence is not occurring.

Mutation: Mutation operator at the bit level, randomly flipping bits inside the current island. Every thread generates z bit random numbers with the mutation probability rate. However, the bit-wise operation is performed on the offspring to change the a element values. The bit-flip mutation operator is extensively used on binary string representations. This operator changes the value of each bit in a solution (from 0 to 1) with probability p. The p is an operator parameter. This parameter's recommended value is $p = 1/w$, where w is the length of the binary string.

3   Experimental Study

To evaluate the performance of TRPIM, we analyze the execution time and effectiveness of solution quality on knapsack problem instances. In the experimental setting, we explain the parameters of GPU architecture and EA parameters values used to execute the parallel evolutionary process. Then next, we describe the solution quality, speedup and the effectiveness of two-replacement policy in TRPIM.

3.1   Experimental Settings

We used the GeForce GTX 1080 Ti card for experiments with 28 SMs, and each SM has a 128 CUDA core. The experiments on CPU were executed on an Intel(R) Core(TM) i3-4170 CPU @ 3.70 GHz. The parameters of the GPU device are listed in Table 1.

Based on our TRPIM evaluation, the island model and evolutionary algorithm parameters are listed in Table 2. The evolutionary algorithm parameters

Table 1. GeForce GTX 1080 Ti

SMs	28
Total CUDA core	3584
Maximum threads per block	1024
Maximum threads per multiprocessor	2048
Maximum shared memory per block	49152

have a fixed size and rate for all experiments: The uniform crossover has the probability of 0.07, and the bit-flip mutation rate is 0.01. The migration size is 4 individuals in migration policy, and the migration frequency is 100. The bi-tonic sort is used to sort the population to determine which individuals are best suited for migration. We used the number of islands as 8, 16, 32, 64, 128, 256, 512. Each island has 32 individuals. The maximum number of generations is 10,000.

Table 2. The island model evolutionary algorithm parameters setting based on GPU

Parameters	Values
Number of individuals	32
Number of islands	4, 8, 16, 32, 64, 128, 256, and 512
Crossover type	Uniform
Crossover probability	0.07
Mutation type	Bit-flip
Mutation probability	0.01
Migration size	4
Migration frequency	100

All experimental results time costs and solution quality are the average cost of 20 runs. And the execution time is in milliseconds.

3.2 Benchmark Problem

A knapsack benchmark problem [10,12] with different items is selected to evaluate our TRPIM.

$$\text{maximize} \sum_{i=1}^{n} v_i x_i \tag{1}$$

$$\text{s.t.} \quad \sum_{i=1}^{n} w_i x_i \leqslant C, \quad x_i \in \{0, 1\} \tag{2}$$

We have n kinds of items in the knapsack problem, 1 through n. Each kind of item i has a value v_i and a weight w_i along with a maximum weight capacity C.

3.3 Solution Quality of TRPIM

In TRPIM on GPU, We analyze the solution quality of the two-replacement policy on knapsack problems. In this experiment, we take 32 islands, and one island has 32 threads. The knapsack problem is constructed at random, with item sizes in the range from 10000, 15000, 20000, 25000, 30000, 35000, 40000, and 45000. After every 100 generations, the best fitness values from one island are copied from global memory and replaced with worse values of individuals with neighbour islands on GPU. In without TRP, except the two replacements policy, the island model on GPU would have the same parameter values as given in Table 2. In Table 3 the TRP means two replacements policy. These 8 instances of problems are solved 20 times, with a maximum generation size of 10,000. The elements of individual instances are the same as the number of items in the knapsack problem. The average execution time and fitness values of the solution are listed in Table 3.

Table 3. Solution quality comparison with/without two replacements policy of TRPIM

Items (size)	Without TRP		With TRP		
	Time (ms)	Best solutions	Time (ms)	Best solution	Improvement
10000	1598	87695	1097	92925	+5230
15000	2287	128954	1659	136154	+7500
20000	3225	159415	2253	175866	+16451
25000	3601	188462	2947	217565	+29103
30000	4896	225018	3508	259318	+34300
35000	5220	253911	4184	301267	+47356
40000	5989	292117	4601	341277	+49160
45000	7126	331258	5404	382150	+50892

As shown in Table 3, in experiments without TRP on GPU, we gain the best fitness values in solution as 87695, 128954, 159415, 188462, 225018, 253911, 292117, and 331258. Respectively, 8 problem instances the average execution time 1598 ms, 2287 ms, 3225 ms, 3601 ms, 4896 ms, 5220 ms, 5989 ms, and 7126 ms. Next, we test the results with TRP on GPU, the 8 instances have gained the better fitness values of the solution as compared to without TRP. With TRP the improvement in solutions are 5230, 7500, 16451, 29103, 34300, 47356, 49160, and 50892. And the average time, respectively all instances are 1097 ms, 1659 ms, 2253 ms, 2947 ms, 3508 ms, 4184 ms, 4601 ms, 5404 ms. These results show that the version of TRPIM on GPU with a two-replacements policy affects the solution quality and average run-time.

To measure the solution quality of TRPIM on GPU vs CPU on Knapsack problem, we use 8 problem instances as shown in Table 4. The knapsack problem has instances 10000, 15000, 20000, 25000, 30000, 35000, 40000, and 45000. Every 100 generations, the best fitness values on one island are copied to global memory and then replaced by the worst values on neighbour islands on the GPU. A CPU island model would have the same parameter values as Table 2 except for the two-layered replacement policy. An average of 20 runs is used to determine solution quality.

Table 4. The effectiveness solution quality of TRPIM

Problem instances	CPU		GPU	
	Time (ms)	*Best solutions*	*Time (ms)*	*Best solutions*
10000	178162	81154	1097	92925 (+11771)
15000	260524	120125	1659	136154 (+16029)
20000	295151	159381	2253	175866 (+16485)
25000	360177	198699	2947	217565 (+18866)
30000	426998	236441	3508	259318 (+22877)
35000	508972	277826	4184	301267 (+23441)
40000	551220	312964	4601	341277 (+28313)
45000	–	NA	5404	382150

Using the 8 problem instances, we test the GPU version's solution quality. The bits of individual instances are the same as the number of items, the z bit indicating the presence of i item in the knapsack. The assigning of 32 islands has 32 individuals in respect of each island. The experiment shows that the GPU has better results with 10000, 15000, 25000, 30000, 35000, 40000, and 45000 items and gain the improvements of the best fitness values of solution respectively 8 problem instances are 11771, 16029, 16485, 18866, 22877, 23441, 28313, 382150. As we assign many problem instances, the CPU time increases more than GPU. The CPU version gain the better fitness values of solutions are 81145, 120125, 159381, 198699, 236441, 277826, 312964. When we chose the problem instances 45000, the experiment on CPU didn't gain the best fitness values. However, in the experiments with problem instances 45000 on GPU, our fitness values improved correspondingly at 382050 and took an average of 5404 ms.

3.4 Speedup

We conducted experiments with a two-replacement policy to evaluate the speed with different islands on GPU. The 8, 16, 32, 64, 128, 256, and 512 are the number of islands with 32 individuals as given in Table 5, are executed with problem instances 10000. The algorithm will terminate if it finds a better solution from the maximum number of generation 10000 when testing the speedup time of TRPIM. The island model on the CPU has the same parameter settings except for the two replacements policy.

Table 5. The average time with knapsack problem instances size 10000 on GPU vs CPU

Number of islands	*CPU*	GPU	
	Time (ms)	*Time (ms)*	*Speedup*
8	43698	1106	39.5x
16	86103	1142	75.3x
32	173938	1161	149.8x
64	345142	1252	275.6x
128	688854	1953	325.7x
256	1370257	3322	412.4x
512	2936476	7029	417.7x

As shown in Table 5 two-replacements policy gain the speedup time as 39.5, 75.3, 149.8, 275.6, 325.7, 412.4, 417.7, respectively the number of islands 8, 16, 32, 64, 128, 256, 512. When increasing the number of islands, the speedup time also increases. The CPU time increases more than GPU when the number of islands increases. With 512 islands, the TRPIM speeds up 417.7x time faster than the CPU version.

3.5   The Effectiveness of Two-replacements Policy

Finally, in Table 6, we investigate the efficacy of TRPIM with problem instances 20000 and 35000. And the number of islands 64, 128, and 256 is assigned for this experiment. The algorithm terminated when it found the best individuals from the maximum number of generations as 10000, which performed an average time cost of 20 runs.

Table 6. The average time and solution quality of TRPIM on knapsack problem instances

Number of islands	Problem instances	CPU		GPU	
		Time (ms)	*Best solutions*	*Time (ms)*	*Best solution*
64	20000	666183	200363	1014	233638 (+33275)
	35000	961501	350082	2671	399709 (+49627)
128	20000	1367273	200103	1322	220448 (+20345)
	35000	2058922	346124	2059	391533 (+45409)
256	20000	2310954	200982	1649	222771 (+21789)
	35000	4060857	349478	3288	379551 (+30073)

The number of islands 16 in respectively problem instances 20000 gives the improved the fitness values of solution 33275 and with 35000 problem instances have the solution quality as 49627. In the next experiment, we select the 128 islands respectively 20000 items and get a better solution as 20345. And get a solution with 35000 problem instances size. We choose 256 islands with 20000 problem instances and get the efficiency 21789 times better, and with 35000 problem instances gain the improved fitness values of solution 30073.

3.6 Conclusion

In this paper, we proposed a new model named TRPIM on GPU. We empirically conducted the island based warp mechanism and two-replacement policy in TRPIM with the migration via global memory on the knapsack problem to evaluate the performance. The island models divide populations into many sub-populations for a parallel evolutionary process to speed up the execution of the evolutionary algorithm. In an island-based mapping mechanism, a sub-population is mapped to warps rather than GPU blocks. We demonstrated that our design TRPIM gains 382150 average improvements of the solution to the knapsack problem, with a speedup of over by 417.7 ms times faster.

In the future, we intend to develop more effective methods for optimizing the TRPIM, such as improving the fitness evolution, managing memory to accelerate the island model, utilizing shared memory, and designing a new model for large problem instances without crossover parameters, and so on.

References

1. Fang, Y., Li, J.: A review of tournament selection in genetic programming. In: Cai, Z., Hu, C., Kang, Z., Liu, Y. (eds.) ISICA 2010. LNCS, vol. 6382, pp. 181–192. Springer, Heidelberg (2010). https://doi.org/10.1007/978-3-642-16493-4_19
2. Guide, D.: CUDA C programming guide. NVIDIA, July 29, 31 (2013)
3. Janssen, D.M., Pullan, W., Liew, A.W.C.: Graphics processing unit acceleration of the island model genetic algorithm using the CUDA programming platform. Concurr. Comput.: Pract. Exp. **34**(2), e6286 (2022). https://doi.org/10.1002/cpe.6286
4. Janssen, D.M., Liew, A.W.C.: Acceleration of genetic algorithm on GPU CUDA platform. In: 2019 20th International Conference on Parallel and Distributed Computing, Applications and Technologies (PDCAT), pp. 208–213. IEEE (2019). https://doi.org/10.1109/PDCAT46702.2019.00047
5. Li, C.C., Lin, C.H., Liu, J.C.: Parallel genetic algorithms on the graphics processing units using island model and simulated annealing. Adv. Mech. Eng. **9**(7), 1687814017707413 (2017). https://doi.org/10.1177/1687814017707413
6. Luong, T.V., Melab, N., Talbi, E.G.: GPU-based island model for evolutionary algorithms. In: Proceedings of the 12th Annual Conference on Genetic and Evolutionary Computation, pp. 1089–1096 (2010). https://doi.org/10.1145/1830483.1830685
7. Lynn, N., Ali, M.Z., Suganthan, P.N.: Population topologies for particle swarm optimization and differential evolution. Swarm Evol. Comput. **39**, 24–35 (2018). https://doi.org/10.1016/j.swevo.2017.11.002
8. NVIDIA, C.: CUDA C best practices guide v. 4.0 (2011)

9. Pospichal, P., Jaros, J.: GPU-based acceleration of the genetic algorithm. GECCO Compet. (2009). https://doi.org/10.1145/1830483.1830685

10. Shaheen, A., Sleit, A.: Comparing between different approaches to solve the 0/1 knapsack problem. Int. J. Comput. Sci. Netw. Secur. (IJCSNS) **16**(7), 1 (2016)

11. Singh, G., Gupta, N.: A study of crossover operators in genetic algorithms. In: Khosravy, M., Gupta, N., Patel, N. (eds.) Frontiers in Nature-Inspired Industrial Optimization. STNC, pp. 17–32. Springer, Singapore (2022). https://doi.org/10.1007/978-981-16-3128-3_2

12. Singh, R.P.: Solving 0–1 knapsack problem using genetic algorithms. In: 2011 IEEE 3rd International Conference on Communication Software and Networks, pp. 591–595. IEEE (2011). https://doi.org/10.1109/ICCSN.2011.6013975

13. Skolicki, Z., De Jong, K.: Improving evolutionary algorithms with multi-representation island models. In: Yao, X., et al. (eds.) PPSN 2004. LNCS, vol. 3242, pp. 420–429. Springer, Heidelberg (2004). https://doi.org/10.1007/978-3-540-30217-9_43

14. Slowik, A., Kwasnicka, H.: Evolutionary algorithms and their applications to engineering problems. Neural Comput. Appl. **32**(16), 12363–12379 (2020). https://doi.org/10.1007/s00521-020-04832-8

15. Sun, X., Chou, P., Wu, C.C., Chen, L.R.: Quality-oriented study on mapping island model genetic algorithm onto CUDA GPU. Symmetry **11**(3), 318 (2019). https://doi.org/10.3390/sym11030318

16. Sun, X., Lai, L.F., Chou, P., Chen, L.R., Wu, C.C.: On GPU implementation of the island model genetic algorithm for solving the unequal area facility layout problem. Appl. Sci. **8**(9), 1604 (2018). https://doi.org/10.3390/app8091604

17. Xie, H., Zhang, M.: Parent selection pressure auto-tuning for tournament selection in genetic programming. IEEE Trans. Evol. Comput. **17**(1), 1–19 (2012). https://doi.org/10.1109/TEVC.2011.2182652

LACNNER: Lexicon-Aware Character Representation for Chinese Nested Named Entity Recognition

Zhikun Yang[1], Shumin Shi[1,2]($\boxtimes$), Junyu Tian[1], En Li[1], and Heyan Huang[1,2]

[1] School of Computer Science and Technology, Beijing Institute of Technology, Beijing, China
{yzk,bjssm,tjy,ln,hhy63}@bit.edu.cn

[2] Beijing Engineering Research Center of High Volume Language Information Processing and Cloud Computing Applications, Beijing, China

Abstract. Named Entity Recognition (NER) is one of fundamental researches in natural language processing. Chinese nested-NER is even more challenging. Recently, studies on NER have generally focused on the extraction of flat structures by sequence annotation strategy while ignoring nested structures. In this paper, we propose a novel model, named LACNNER, that utilizing lexicon-aware character representation for Chinese nested NER. We select the typical character-level framework to overcome error propagation problem caused by incorrect word separation. Considering the situation that Chinese words always contain much richer semantic information than single characters do, it firstly obtains more significant matching words through external lexicon in our LACNNER model, and then generates lexicon-aware character representations that make full use of word-level knowledge for nested named entity. We also evaluate the effectiveness of LACNNER by taking ACE-2005-Zh dataset as a benchmark. The experimental results fully verified the positive effect of incorporating lexicon-aware character-representation in recognition of Chinese nested entity structure.

Keywords: Chinese nested NER · Character embedding · Information extraction

1 Introduction

As one of the fundamental researches in natural language processing, Named Entity Recognition (NER) aims to extract text spans representing a certain entity. NER includes two parts: entity mention extraction and entity mention classification. According to the text span boundary, entities can be divided into flat structured entities, where there is no inclusion between the word boundaries of any two entities mentioned in the sentence. However, a lot of NER studies generally extract flat structured entities while ignoring the nested ones. Figure 1 shows some examples of flat entity structures and nested entity structures. The

© Springer Nature Switzerland AG 2022
Y. Tan et al. (Eds.): ICSI 2022, LNCS 13345, pp. 254–264, 2022.
https://doi.org/10.1007/978-3-031-09726-3_23

inner entity usually reflects some property of the external entity. For example, a Location (LOC) or a Geo-Political Entity (GPE) nested within an organizational entity may indicate the location or attribution of the organization like the second case in the Fig. 1, so it is also necessary to identify nested named entities.

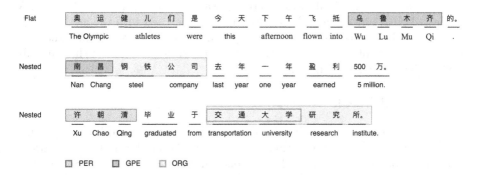

Fig. 1. Flat and nested structure of the entity. Person (PER), Geo-Political Entity (GPE), and Organization (ORG) types are indicated by different text background colors. The two entities of the first sentence belong to flat structure, and the entities of the last two sentences exist ORG nested GPE and ORG nested ORG, respectively.

The sequence annotation extraction architecture applied to flat NER is no longer applicable to nested NER. Several previous studies have proposed to decode nested entities using special structures such as hypergraphs or syntactic parsers, but these models suffer from structural ambiguity and complex manual design in the inference phase. Some other studies extract nested entities by hierarchical extraction from inside to outside or from outside to inside, and the extraction of the current layer depends on the extraction results of the previous layer, which is prone to error propagation issues.

In Chinese, a word is generally composed of more than one character. Considering that words contain richer semantic information than characters in Chinese, the character-level model does not capture the whole information of the sentence. Simply building a word-level model which uses divided words as input would lead to the propagation of segmentation error. So it is challenging to extract nested entities in Chinese. In this paper, we propose the model called LAC-NNER (Lexicon-Aware Character Representation for Chinese Nested Named Entity Recognition) to solve this problem. In detail, LACNNER uses a multi-headed self-attentive mechanism to encode both characters and word embeddings to obtain a representation of each character with word-level knowledge and then applies a biaffine classifier to decode the entity type of each text span, which can extract both flat and nested entity mentions. The words are obtained by matching the original text with an external lexicon. The position relationship between the matching words and the characters of the original text is reflected by the position embedding.

In summary, our main contributions are as follows: 1) We propose a simple but effective model to extract entity mentions with nested structures. 2) Taking into account the semantic gap between words and characters, we incorporate word-level information into the character-level model. 3) We validate the effectiveness of our model on the ACE2005-Zh dataset and obtain 1.48% gains.

The paper is organized in the following way. Section 2 presents related work on nested NER and Chinese NER. Section 3 is concerned with the methodology of LACNNER. Section 4 describes the experiments on ACE2005-Zh and analyzes the effectiveness of LACNNER. Section 5 summarizes the focus of the study as well as the experimental findings.

2   Related Work

In early studies, NER mainly focused on flat entities instead of nested entities, thus the sequence annotation framework is the most traditional solution. But this approach is not suitable for identifying nested structures. Some studies extract nested entities by constructing special data structures. Finkel et al. extract nested entities by constructing a constituent syntactic parser [5]. Some works propose to represent the nested structure by a hypergraph [11,17,19,26], but illegal structures and structural ambiguities need to be avoided when modeling hypergraphs, which can make modeling more complicated. Fu et al. use Tree-CRF to decode the nested structure [7], and Wang et al. converts a sentence into a tree structure and identifies nested entities by adjusting the structure of the tree through state transfer operations [26]. Lin et al. proposed the concept of anchor words and then identified the anchor words before calculating the complete entity boundaries [15].

There are also some studies that use a layered extraction approach to identify nested entities hierarchically from inside to outside or from outside to inside using sequence annotation. Ju et al. and Li et al. identifies and classifies them from inside to outside by dynamically stacking the extraction layers [10,12]. Fisher et al. uses layered merging to combine tokens belonging to the same entity. In addition to introducing error propagation issue [6], the layered structure also tends to prevent the model from capturing the entity relationships between different layers. Wang et al. designed a character-level normal pyramid and inverse pyramid structure with bidirectional interactions between neighboring layers [27], but the model is complex and inefficient.

A few works recognize nested entities by extracting and classifying text spans of arbitrary length within a sentence. Sohrab et al. and Yu et al. enumerates all the text spans in a sentence and classifying the text spans [21,29]. There is also some work based on the pipeline approach [20,24,28,31], where the boundaries of the entity mentions are first extracted and then classified. The pipeline method inevitably introduces error propagation issue. Other studies transform NER into other NLP task forms. Strakova et al. uses sequential-to-sequence architecture to generate all labels of tokens [22], and Li et al. proposes that MRC model can also be used to solve NER [14], and some priori knowledge can be introduced based on the design of the query.

Chinese NER using the character-level model is better than the word-level, but extracting entity words needs to be sensitive to the boundaries of words while the character-level model is missing the lexical information. Zhang et al. was the first to propose that text and matching words connected can form a lattice structure [30], and incorporating word-level information to the model requires the ability to encode this structure. Some researchers uses Lattice-LSTM, GNNs, and CNNs as feature extractors [8,9,23,30]. Li et al. flattens the lattice structure into a sequential structure using Transformer layer to enhance the computational speed of the model [13]. Another method is fuse lexical features by training token embeddings with lexical information rather than by modifying the model structure [4,16,18].

3   Approach

We mainly focus on nested NER in Chinsese text, and propose a kind of model, named LACNNER, its overall structure is shown in Fig. 2. The model firstly generates lexicon-aware character representations that contain word-level knowledge, and then concatenates them with the features extracted by the BiLSTM. Finally, the concatenation vectors are fed into a biaffine classifier to obtain entity classification results.

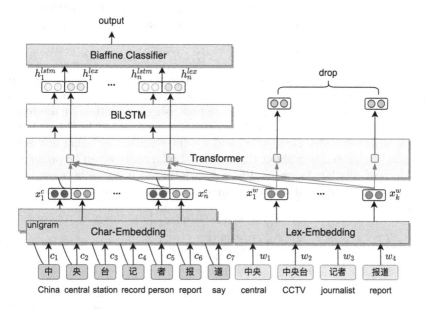

Fig. 2. The overall architecture of LACNNER. The solid arrows indicate the data flow direction and the light gray arrows indicate the self-attentive mechanism. (Color figure online)

Formally, we denote the inputs to the model as $s = \{c_1, c_2, \ldots, c_n\}$, where c_i denotes the i_{th} character of sentence s, and n is the length of s. We use $\mathcal{L}$ to represent the external lexicon. The words obtained by matching each sentence s in the corpus with $\mathcal{L}$ are represented by $w_s = \{w_1, w_2, \ldots, w_k\}$, where w_j denotes the j_{th} matched words and there are k words in total.

3.1 Embedding Layer

Character unigram embeddings, character bigram embeddings and words embeddings are pretrained on a large Chinese corpus to represent characters, two characters and words as static vectors. For the unigram embedding and bigram embedding of c_i we denote x_i^{uni} and x_i^{bi}, respectively. Similarly, the embedding of w_j is written as x_j^w. For c_i we use the concatenation of x_i^{uni} and x_i^{bi} as the final character embedding x_i^c:

$$x_i^c = [x_i^{uni}; x_i^{bi}] \tag{1}$$

3.2 Character Representation

The BiLSTM encoder can model the long-range dependencies to extract dynamic features containing contextual information. Taking $X^c = \{x_1^c, x_2^c, \ldots, x_n^c\}$ as input, we use the concatenation of the BiLSTM's hidden states in both directions as the character representation containing contextual information of c_i:

$$h_i^{lstm} = [\overrightarrow{LSTM}(x_1^c, \ldots, x_i^c); \overleftarrow{LSTM}(x_i^c, \ldots, x_n^c)] \tag{2}$$

Inspired by the FLAT model of Li et al., we can calculate lexicon-aware character representations by using an external lexicon and encode the interactions between matching words and the original text using a self-attentive mechanism. Set x_i, x_j to be any two spans in $X = \{x_1^c, \ldots, x_n^c, x_1^w, \ldots, x_k^w\}$. There are three positional relationships between x_i and x_j: intersection, inclusion and separation, and the information about the positional relationships can be preserved by four kinds of relative distances [13]. d_{ij}^{hh} denotes the distance between head of any two spans x_i, x_j, and the other three have similar meanings. The formulas are as follows:

$$\begin{aligned} d_{ij}^{hh} &= head[i] - head[j] \\ d_{ij}^{ht} &= head[i] - tail[j] \\ d_{ij}^{th} &= tail[i] - head[j] \\ d_{ij}^{tt} &= tail[i] - tail[j] \end{aligned} \tag{3}$$

where $head[i]$ and $tail[i]$ denote the head and tail position of span x_i. The final relative position encoding of spans is a non-linear transformation of the four distances [13]:

$$R_{ij} = ReLU(W_r[P_{d_{ij}^{hh}}; P_{d_{ij}^{th}}; P_{d_{ij}^{ht}}; P_{d_{ij}^{tt}}]) \tag{4}$$

where W_r is a learnable parameter, and the calculation of P_d is calculated using the same periodic trigonometric function as Transformer [25].

In this paper, we use a variant of the self-attentive mechanism that can encode relative positions to capture the dependencies of matching words with text. [2], so that the model learns the position relationship between the matching words and the original text. The difference with Vanilla Transformer lies in the calculation of the attention weight matrix:

$$A_{i,j} = W_q^T X_i^T X_j W_{k,X} + W_q^T X_i^T R_{ij} W_{k,R} + u^T X_j W_{k,X} + v^T R_{ij} W_{k,R} \quad (5)$$

where $W_q, W_{k,R}, W_{k,E}, u, v$ are learnable parameters.

Transformer outputs a representation of each character as well as the matching word. After the self-attentive interaction, the output carry the knowledge of the matching words, thus we drop the word representations and keep the part of the character representations denoted as h^{lex} for the subsequent computation of the biaffine classifier.

3.3   Biaffine Classifier

Two character representations, h^{lstm} and h^{lex}, carrying different information are concatenated as inputs to the biaffine classifier. The biaffine classifier calculates a scoring tensor for any span in s. The scoring tensor for a $span_{ij}$ starting at index i and ending at index j is denoted as $r_m(ij)$, which is c-dimensional. The number of entity categories is noted as c, which includes "non-entity" type. The biaffine classifier is calculated as follows:

$$
\begin{aligned}
h_s(i) &= FFNN_s([h_i^{lstm}; h_i^{lex}]) \\
h_e(j) &= FFNN_e([h_j^{lstm}; h_j^{lex}]) \\
r_m(ij) &= h_s(i)^T U_m h_e(j) + W_m[h_s(i); h_e(j)] + b_m
\end{aligned}
\quad (6)
$$

where U is a high-dimensional $d \times c_e \times d$ learnable parameter (d denotes the feature dimension of h_s and h_e), W is a $2d \times c_e$ learnable parameter, and b_m is a bias. The category with the highest score given by the classifier is the predicted classification of the model for $span_{ij}$:

$$y'(ij) = argmax \ r_m(ij) \quad (7)$$

The loss function for training is softmax cross entropy.

4   Experiment

4.1   Data Set

To validate our approach on Chinese nested NER, we take the ACE 2005 Mandarin Chinese dataset (ACE-2005-Zh) as a benchmark. It is an available Chinese nested NER dataset with high annotation quality. We split the dataset into 80%, 10%, and 10% for the train, development, and test set respectively. The statistics of ACE-2005-Zh are presented in Table 1.

Table 1. The statistics of ACE-2005-Zh.

Attribute	Train	Dev	Test	Total
Documents	507	63	63	633
Sentences	6450	755	814	8019
Words	205010	24493	26077	255580
Entity mentions	26965	3370	3268	33603
Nested entity mentions	44.57%	48.64%	43.15%	44.84%

In our experiments, character unigram embeddings and bigram embeddings are trained by zhang et al. on the Chinese Giga-word using word2vec. The lexicon embeddings also trained by this team on automatically segmented CTB6.0 [30]. We consider the words encoded by lexicon embeddings as a base external lexicon $\mathcal{L}_{base}$, which contains about 600,000 words. We also filtered the core concept words and entity words of OpenEntitiy [1] from $\mathcal{L}_{base}$ to get a streamlined version lexicon $\mathcal{L}_{small}$ with only 75,000 words.

4.2 Evaluation Metrics

We use the classical evaluation metrics in classification tasks, that is Precision (P), Recall (R) and $F1$, to measure the effectiveness of LACNNER. P focuses on measuring samples predicted to be positive classes, while R puts more emphasis on the accuracy of true positive samples. $F1$ is the harmonic average of P and R. The evaluation metrics are calculated as follows:

$$P = \frac{TP}{TP+FP}$$
$$R = \frac{TP}{TP+FN} \tag{8}$$
$$F1 = 2 * \frac{P*R}{P+R}$$

where TP, FP, TN, and FN are used to indicate True Positive, False Positive, True Negative and True Negative, respectively.

4.3 Hyperparameters

In the training phase, we use BERTAdam as an optimizer, with a learning rate of $5e^{-4}$. We set the dropout rate of the unigram and bigram character embedding layers to 0.15, and the dropout rate of the lexicon character embedding layer to 0.4. For the encoder, we use a single-layer Transformer with 4 heads and a two-layer BiLSTM with 288 units in the hidden layer. In the Biaffine Classifier, the feature dimension is reduced to 150. The model is trained for 60 iterations with an early stop mechanism to prevent overfitting of the model training. The above hyperparameters are obtained by the grid search algorithm to achieve the best combination of parameters on the validation dataset.

4.4   Results and Analysis

We use Yu's model (LSTM-Biaffine) [29] as the baseline, since it is one of the most typical algorithms for solving nested NER, and has outstanding performance in the English dataset. LSTM-Biaffine feeds the character embedding into a BiLSTM and finally to a Biaffine Classifier. We experimented on both $\mathcal{L}_{base}$ and $\mathcal{L}_{small}$ lexicon, and both obtain higher F1 scores than baseline, which shows that a model that excels in English datasets also struggles with Chinese datasets. Models incorporating lexicon-aware character representation are more suitable for the Chinese text (Table 2).

Table 2. Experimental results on ACE-2005-Zh.

Model	P	R	F1
LSTM-Biaffine	71.62	70.78	71.20
van-Transformer	71.33	69.75	70.53
LACNNER ($\mathcal{L}_{base}$)	74.10	**71.30**	72.68
LACNNER ($\mathcal{L}_{small}$)	**77.36**	68.61	**72.73**
LSTM-Biaffine +BERT	76.64	**82.30**	79.37
van-Transformer +BERT	72.08	77.41	74.65
LACNNER+BERT ($\mathcal{L}_{base}$)	79.97	79.90	79.94
LACNNER+BERT ($\mathcal{L}_{small}$)	**81.15**	79.14	**80.13**

Meanwhile, we construct the van-Transformer model, whose self-attentive mechanism uses the Vanilla Transformer that cannot encode relative position information. The rest of the structure is consistent with our model. Van-Transformer is inferior to LSTM-Biaffine indicating that the Vanilla Transformer not only does not obtain effective information but even brings the noise to the model.

We found that smaller lexicon does not affect the final results if it is of high quality. OpenEntity ensures the confidence of $\mathcal{L}_{small}$ so that despite the tenfold difference in size between $\mathcal{L}_{small}$ and $\mathcal{L}_{base}$, the results are still comparable.

Pre-trained language models can bring more semantic-level information, so experiments were also done to replace the static word vectors with dynamic word vectors obtained from BERT (base) [3]. The F1 score improved significantly. The conclusions remain consistent with the above (Table 3).

Table 3. Results of ablation experiments.

Settings	P	R	F1
Default	74.10	71.30	72.68
w/o context representation	62.12	48.98	54.78
w/o lexicon-aware character representation	68.85	71.00	69.91
w/o matched words input	73.34	69.12	71.17

Ablation experiments show that fusing word-level information using a self-attentive mechanism that can encode relative location information can effectively improve the effectiveness of model nested named entity recognition. Contextual features encoded using BiLSTM are still the predominant way to encode information. The different degrees of F1 decline also indicate that the two representations capture complementary information.

5 Conclusion

In this paper, we put forward a novel model named LACNNER for recognizing Chinese nested entities. We both consider the lack of important word-level information in the existing character-level models and the problem of word separation errors in the word-level ones. In LACNNER, we firstly find out the matching words related to text from lexicon, then obtain the dependency between text and matching words by self-attention mechanism to get lexicon-aware character representation, which be treated as a complementary feature to extract nested entities. Finally, we implement a biaffine classifier to aggregate such representations. Experiments show that LACNNER outperforms the character-level models on ACE-2005-Zh dataset. Furthermore, ablation tests also demonstrate the significance of lexicon-aware character representation for Chinese Nested NER as we expected.

Acknowledgements. This work is supported by the National Natural Science Foundation of China (Grant No. 61732005, No. 61671064) and National Key Research & Development Program (Grant No. 2018YFC0831700).

References

1. Choi, E., Levy, O., Choi, Y., Zettlemoyer, L.: Ultra-fine entity typing. In: Proceedings of the 56th Annual Meeting of the Association for Computational Linguistics (Volume 1: Long Papers), pp. 87–96 (2018)
2. Dai, Z., Yang, Z., Yang, Y., Carbonell, J., Le, Q.V., Salakhutdinov, R.: Transformer-XL: attentive language models beyond a fixed-length context. arXiv preprint arXiv:1901.02860 (2019)
3. Devlin, J., Chang, M.W., Lee, K., Toutanova, K.: BERT: pre-training of deep bidirectional transformers for language understanding. arXiv preprint arXiv:1810.04805 (2018)

4. Ding, R., Xie, P., Zhang, X., Lu, W., Li, L., Si, L.: A neural multi-digraph model for chinese ner with gazetteers. In: Proceedings of the 57th Annual Meeting of the Association for Computational Linguistics, pp. 1462–1467 (2019)
5. Finkel, J.R., Manning, C.D.: Nested named entity recognition. In: Proceedings of the 2009 Conference on Empirical Methods in Natural Language Processing, pp. 141–150 (2009)
6. Fisher, J., Vlachos, A.: Merge and label: a novel neural network architecture for nested NER. arXiv preprint arXiv:1907.00464 (2019)
7. Fu, Y., Tan, C., Chen, M., Huang, S., Huang, F.: Nested named entity recognition with partially-observed TreeCRFs. In: Proceedings of the AAAI Conference on Artificial Intelligence, pp. 2–9 (2021)
8. Gui, T., Ma, R., Zhang, Q., Zhao, L., Jiang, Y.G., Huang, X.: CNN-based Chinese NER with lexicon rethinking. In: IJCAI, pp. 4982–4988 (2019)
9. Gui, T., et al.: A lexicon-based graph neural network for Chinese NER. In: Proceedings of the 2019 Conference on Empirical Methods in Natural Language Processing and the 9th International Joint Conference on Natural Language Processing (EMNLP-IJCNLP), pp. 1040–1050 (2019)
10. Ju, M., Miwa, M., Ananiadou, S.: A neural layered model for nested named entity recognition. In: Proceedings of the 2018 Conference of the North American Chapter of the Association for Computational Linguistics: Human Language Technologies, Volume 1 (Long Papers), pp. 1446–1459 (2018)
11. Katiyar, A., Cardie, C.: Nested named entity recognition revisited. In: Proceedings of the 2018 Conference of the North American Chapter of the Association for Computational Linguistics: Human Language Technologies, vol. 1 (2018)
12. Li, H., Xu, H., Qian, L., Zhou, G.: Multi-layer joint learning of Chinese nested named entity recognition based on self-attention mechanism. In: Zhu, X., Zhang, M., Hong, Yu., He, R. (eds.) NLPCC 2020. LNCS (LNAI), vol. 12431, pp. 144–155. Springer, Cham (2020). https://doi.org/10.1007/978-3-030-60457-8_12
13. Li, X., Yan, H., Qiu, X., Huang, X.: FLAT: Chinese NER using flat-lattice transformer. arXiv preprint arXiv:2004.11795 (2020)
14. Li, X., Feng, J., Meng, Y., Han, Q., Wu, F., Li, J.: A unified MRC framework for named entity recognition. arXiv preprint arXiv:1910.11476 (2019)
15. Lin, H., Lu, Y., Han, X., Sun, L.: Sequence-to-nuggets: nested entity mention detection via anchor-region networks. arXiv preprint arXiv:1906.03783 (2019)
16. Liu, W., Xu, T., Xu, Q., Song, J., Zu, Y.: An encoding strategy based word-character LSTM for Chinese NER. In: Proceedings of the 2019 Conference of the North American Chapter of the Association for Computational Linguistics: Human Language Technologies, Volume 1 (Long and Short Papers), pp. 2379–2389 (2019)
17. Lu, W., Roth, D.: Joint mention extraction and classification with mention hypergraphs. In: Proceedings of the 2015 Conference on Empirical Methods in Natural Language Processing, pp. 857–867 (2015)
18. Ma, R., Peng, M., Zhang, Q., Huang, X.: Simplify the usage of lexicon in Chinese NER. arXiv preprint arXiv:1908.05969 (2019)
19. Muis, A.O., Lu, W.: Labeling gaps between words: recognizing overlapping mentions with mention separators. arXiv preprint arXiv:1810.09073 (2018)
20. Shen, Y., Ma, X., Tan, Z., Zhang, S., Wang, W., Lu, W.: Locate and label: a two-stage identifier for nested named entity recognition. arXiv preprint arXiv:2105.06804 (2021)
21. Sohrab, M.G., Miwa, M.: Deep exhaustive model for nested named entity recognition. In: Proceedings of the 2018 Conference on Empirical Methods in Natural Language Processing, pp. 2843–2849 (2018)

22. Straková, J., Straka, M., Hajič, J.: Neural architectures for nested NER through linearization. arXiv preprint arXiv:1908.06926 (2019)

23. Sui, D., Chen, Y., Liu, K., Zhao, J., Liu, S.: Leverage lexical knowledge for Chinese named entity recognition via collaborative graph network. In: Proceedings of the 2019 Conference on Empirical Methods in Natural Language Processing and the 9th International Joint Conference on Natural Language Processing (EMNLP-IJCNLP), pp. 3830–3840 (2019)

24. Tan, C., Qiu, W., Chen, M., Wang, R., Huang, F.: Boundary enhanced neural span classification for nested named entity recognition. In: Proceedings of the AAAI Conference on Artificial Intelligence, vol. 34, pp. 9016–9023 (2020)

25. Vaswani, A., et al.: Attention is all you need. In: Advances in Neural Information Processing Systems, vol. 30 (2017)

26. Wang, B., Lu, W., Wang, Y., Jin, H.: A neural transition-based model for nested mention recognition. arXiv preprint arXiv:1810.01808 (2018)

27. Wang, J., Shou, L., Chen, K., Chen, G.: Pyramid: a layered model for nested named entity recognition. In: Proceedings of the 58th Annual Meeting of the Association for Computational Linguistics, pp. 5918–5928 (2020)

28. Wang, Y., Li, Y., Tong, H., Zhu, Z.: HIT: nested named entity recognition via head-tail pair and token interaction. In: Proceedings of the 2020 Conference on Empirical Methods in Natural Language Processing (EMNLP), pp. 6027–6036 (2020)

29. Yu, J., Bohnet, B., Poesio, M.: Named entity recognition as dependency parsing. arXiv preprint arXiv:2005.07150 (2020)

30. Zhang, Y., Yang, J.: Chinese NER using lattice LSTM. arXiv preprint arXiv:1805.02023 (2018)

31. Zheng, C., Cai, Y., Xu, J., Leung, H., Xu, G.: A boundary-aware neural model for nested named entity recognition. In: Proceedings of the 2019 Conference on Empirical Methods in Natural Language Processing and the 9th International Joint Conference on Natural Language Processing (EMNLP-IJCNLP). Association for Computational Linguistics (2019)

Simple Flow-Based Contrastive Learning for BERT Sentence Representations

Ziyi Tian[1]📷, Qun Liu[1(✉)], Maotao Liu[1], and Wei Deng[2]

[1] Chongqing Key Laboratory of Computational Intelligence, Chongqing University of Posts and Telecommunications, Chongqing 400065, China
{s200201068,s200231202}@stu.cqupt.edu.cn, liuqun@cqupt.edu.cn
[2] Center of Statistical Research, Southwestern University of Finance and Economics, Chengdu 611130, Sichuan, China
dengwei@swufe.edu.cn

Abstract. Natural language processing is a significant branch of machine learning, and pre-trained models such as BERT have been widely used in it. Previous research has shown that sentence embeddings from pre-trained language models without fine-tune have difficulty in capturing their exact semantics. The ambiguous semantics leads to poor performance on semantic text similarity (STS) tasks. However, fine-tune tends to skew the model toward high-frequency distributions due to the heterogeneous nature of word frequency and word sense distributions. Therefore, fine-tune is not a optimal choice. To address this issue, we propose an unsupervised flow-based contrastive learning model. The model maps sentence embedding distributions to smooth and isotropic Gaussian distributions, thus mitigating the impact caused by irregular word frequency distributions. To evaluate the performance of our model, we use an industry-recognized method that outperforms competing baselines in different sentence-related tasks.

Keywords: Flow model · Contrastive learning · Deep learning

1 Introduction

Extracting high-quality sentence representations facilitates a wide range of natural language processing (NLP) tasks. In particular, good sentence representations facilitate a wide range of downstream tasks. Recently, pre-trained language models like BERT [1] and its variants [1,2] have been widely used as natural language representations. Great success has been achieved in many NLP tasks by

This work is supported by the key cooperation project of Chongqing municipal education commission (HZ2021008), partly funded by the State Key Program of National Nature Science Foundation of China (61936001), National Nature Science Foundation of China (61772096), the Key Research and Development Program of Chongqing (cstc2017zdcy-zdyfx0091) and the Key Research and Development Program on AI of Chongqing (cstc2017rgzn-zdyfx0022).

ⓒ Springer Nature Switzerland AG 2022
Y. Tan et al. (Eds.): ICSI 2022, LNCS 13345, pp. 265–275, 2022.
https://doi.org/10.1007/978-3-031-09726-3_24

fine-tuning BERT and its variant. However, the unevenness of word frequency distribution in the dataset makes it difficult to capture the underlying semantics of sentence well in the task of STS.

Previous work [3] has demonstrated that due to word frequency, not all words are isotropic in the latent space but tend to degenerate and be distributed into a narrow cone, largely limiting word embeddings' expressive power. Li et al. [4] further investigates two problems with the BERT sentence embedding space, namely word frequency shifted embedding space and sparse dispersion of low frequency words, which leads to difficulties in using BERT sentence embeddings directly through simple similarity measures such as dot product or cosine similarity.

To alleviate the above problems, we propose a flow model introducing contrastive learning, namely ContFlow. Specifically, our model consists of two main parts, firstly, the mapping process of the traditional flow model. The anisotropic sentence embedding distribution obtained from the BERT encoder is mapped to an isotropic Gaussian distribution, and maximum likelihood is used for optimization. The second is the contrastive learning module, which constructs positive and negative sample pairs by sentence meanings and optimizes the sentence vectors using contrastive loss. The two modules work together to obtain the final sentence representation.

We evaluate the performance of BERT-ContFlow on STS datasets and utilize an general model evaluation method for seven STS tasks. Experimental results show that our method generally improves model performance and outperforms competing baselines that do not use fine-tune to construct BERT sentence embeddings in context. Our model metrics are improved by a maximum of 5.25% on STS dataset, by an average of 2.8% on STS benchmarks and by an average of 1.05% on various sentence migration tasks. Our contributions are summarized as follows.

- Analyze the reasons for the poor performance of BERT sentence embedding.
- A flow model using contrastive learning is proposed to alleviate the uneven word frequency distribution in the dataset.
- Experimental results on seven STS tasks show that our method significantly improves model performance.

2   Related Work

2.1   Sentence Representation

Several works have been studied to understand sentences representation learning. Iyyyer et al. [9] introduces deep averaging networks, which provide unweighted averages of word vectors through multiple hidden layers before classification. The original BERT and its variants are designed to fine-tune each downstream task to achieve its best performance. In the process of fine-tune, as being proposed by Devlin et al. [1], the predefined tokens embedded from the last layer of the encoder (also called [CLS]) are considered as representations of the input sequence. This simple and effective approach is feasible because [CLS] serves as the only communication gate between the pre-trained encoders and the task-specific layer that captures the overall information during BERT supervised fine-tuning. However, in cases where the labeled dataset is unavailable or is small, it is not clear what the best strategy is for deriving sentence embeddings from BERT. Reimers and Gurevych [2] experimentally demonstrate that in such cases, an averaging pooling at the last layer of BERT would be superior to [CLS] embeddings, and use siamese networks to fine-tune the BERT.

2.2   Fine-Tune BERT

Pre-trained sentence embeddings models like BERT have succeeded in the natural language processing community. Researchers find that sentence embeddings without fine-tune struggle to accurately capture the semantic information of sentences. Li et al. [4] argues that BERT sentence embeddings in latent space have the characteristics of anisotropy. They suggest using flow model to map the BERT sentence embedding distribution into Gaussian distribution. Su et al. [5] finds that the whitening operation also enhances the isotropy of the sentence representation. Moreover, the models' training speed is improved due to the reduced dimensionality of the sentences. The other works also use supervised datasets for sentence representation learning. Conneau et al. [12] finds that using sizeable natural language inference datasets is capable of obtaining higher quality sentence embeddings. And a general sentence representation with supervised training using the Stanford natural language inference dataset demonstrates its superiority.

2.3   Contrastive Representation Learning

Contrastive learning has long been a more scientific modeling approach and has been widely used in many fields. The aim of contrastive learning is to learn effective representations by constructing pairs of positive and negative samples, pulling semantically similar neighbors together and separating non-neighbors [14]. Kim et al. [6] proposes a contrastive learning approach using siamese network architecture that allows BERT to utilize its own information to construct positive and negative sentence pairs to obtain higher quality sentence embeddings without introducing any external resources. Yan et al. [15] also adopts the same siamese network architecture as SBERT and introduces the ideology of contrastive learning. The difference is that ConSERT constructs positive and negative sentence pairs by four different data enhancement methods (Adversarial Attack, Token Shuffling, Cutoff, Dropout). Gao et al. [7] proposes a supervised training approach and an unsupervised training approach respectively. The unsupervised SimCSE predicts the input sentences themselves through applying different dropout masks. The supervised SimCSE uses the NLI dataset and treats the entailment pairs as positive and the contradiction pairs and other instances within the batch as negative pairs.

3   Method

In this section, we present a sentence representation learning method named BERT-ContFlow. The target is to map the initial sentence vector output from the BERT encoder onto a Gaussian distribution via flow model.

We first present the overall framework of BERT-ContFlow and then describe the specific roles of each modules in the model.

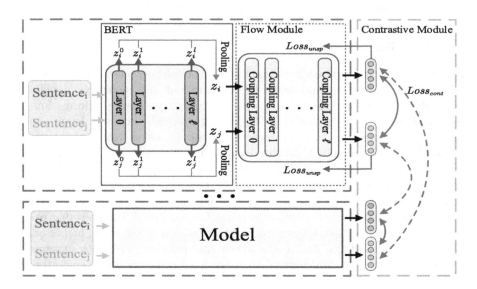

Fig. 1. The schematic diagram of the Bert-ContFlow method. Sentence pairs Sentence$_i$ and Sentence$_j$ are used as inputs. The initial representation vectors z_i, z_j are obtained by pre-training BERT. The flow model is trained to map z_i, z_j to the desired vector space and the process is optimized by maximum likelihood and contrastive loss.

3.1   General Framework

Our model uses the structure of siamese networks [2], as shown in Fig. 1. There are three main components in our framework. Firstly, A BERT encoder that computes each input sentence representation. Secondly, the Flow part, which maps the sentence representation from BERT into a Gaussian distribution, optimizing the problem of sentence representation anisotropy. Finally, the flow model has a contrastive loss layer. The distance of positive sentences reduces while negative sentences increases over the latent space.

3.2   BERT Layer Combination

The pooling operation [1] is the key step in BERT. Referring to the work of BERT-FLOW [4], we adopt a first-last averaging pooling strategy, i.e., averaging the sentence embeddings obtained by the vector representations of layer 1 and 12 of the BERT encoder(denoted by first-last-avg).

3.3   Flow-Based Model

We convert anisotropic sentence embedding distributions to smooth and isotropic Gaussian distributions via Flow-based model [4]. Model implements a reversible transformation from a change in the distribution of the latent space $\mathcal{H}$ to the distribution of the observed space $\mathcal{Z}$. The formula for the model is defined as:

$$h \sim \pi(h), z = f(h). \tag{1}$$

where $h \sim \pi(h)$ denotes the prior distribution, z is the sentence embedding from BERT, and f symbolizes the invertible mapping. Assuming that $\pi(h)$ approximately obeys a uniform distribution on $[h', h' + \Delta h]$ and that $p(z)$ also approximately obeys a uniform distribution on $[z', z' + \Delta z]$, since the sampling probabilities are consistent. When Δz and Δh are tiny, we have:

$$p(z') = \pi(h')|\frac{dh}{dz}|. \tag{2}$$

By the inverse of the Jacobian determinant, we obtain:

$$p(z') = \pi(h')|\det(J_{f^{-1}})|. \tag{3}$$

where $J_{f^{-1}}$ denotes the Jacobi determinant of the f^{-1} operation, f^{-1} represents the transformation from representation vectors z to h and det denotes the determinant value of the matrix. To implement the transformation between the sentence embedding distribution of BERT to the standard Gaussian distribution, we solve for f^{-1} by maximum likelihood as follows:

$$f^{-1} = \arg\max \sum \log p(z). \tag{4}$$

Note that only flow parameters are optimized during training, while the BERT parameters remain unchanged. In our model, the losses in the flow model are as follows:

$$L_{unsp} = \log \pi(f_\phi^{-1}(z)) + \log |\det \frac{\partial f_\phi^{-1}(z)}{\partial z}|. \tag{5}$$

where $f_\phi^{-1}(z)$ is the sentence vector obtained by passing z through flow model, and π is the prior probability distribution.

3.4   Contrastive Loss Layer

Our model makes similar instances closer together in the latent space and different samples further apart in the latent space by introducing the idea of contrastive learning. In our model, we follow the contrastive framework of SimCSE [7], where contrastive learning is guided by constructing positive and negative sentence pairs $D = \{(x_i, x_i^+, x_i^-)\}_{i=1}^{i=m}$. For n batches of sentence pairs, the training objective can be formulated as:

$$L = -\log \frac{e^{sim(h_i, h_i^+)/\tau}}{\sum_{i=1}^{N} e^{sim(h_i, h_j)/\tau}}. \tag{6}$$

where h_i, h_j^+ denote the embedding vectors of positive sentence pairs x_i, x_j^+ generated by the model, *sim* represents the similarity calculation function, and τ is the temperature hyperparameter. For a batch of sentence pairs from the NLI dataset, we consider each premise in the sentence pair as the main sentence, the hypothesis necessarily related to it as the positive example, and the other hypotheses in the batch as the negative example. We input the main clause and its affirmative and negative cases into the BERT encoder, respectively, and pass the resulting sentence embeddings to flow model. We optimize the parameters of flow model by computing the supervised contrastive loss of a batch of sentence pairs from the final sentence embeddings output by flow model as follows:

$$L_{cont} = -\log \frac{e^{sim(f_\phi^{-1}(z_i), f_\phi^{-1}(z_j^+))/\tau}}{\sum_{i=1}^{N}(e^{sim(f_\phi^{-1}(z_i), f_\phi^{-1}(z_j^+)/\tau} + e^{sim(f_\phi^{-1}(z_i), f_\phi^{-1}(z_j^-)/\tau})}. \qquad (7)$$

where z_i is the sentence embedding obtained by passing data x_i through the BERT, and h_i is the sentence embedding obtained by passing z_i through flow model. The loss of the whole model consists of flow model loss and contrastive loss as follows:

$$L = \lambda L_{unsp} + (1 - \lambda) L_{cont}. \qquad (8)$$

where λ is the hyperparameter, L_{unsp} and L_{cont} are the losses of the flow and contrastive models, respectively.

4   Experiments

To verify the effectiveness of the proposed method, we conduct experiments on various tasks related to semantic STS in a variety of configurations.

4.1   Setups

Dataset. For training the model, we use the NLI dataset, including SNLI dataset [16] and the MNLI dataset [17]. It labels sentence pairs using implicative, neutral and contradictory tags. In the NLI dataset, each premise is presented at least three times and is assumed to be different.

For evaluating the model, we use seven datasets, i.e., the STS benchmark [8], the Sickness Correlation (SICK-R) dataset [10] and the STS tasks (STS12-STS16) from 2012–2016 [11]. These datasets are obtained through the SentEval toolkit [12].

Baselines. We first choose a non-BERT method as a baseline: Avg.Glove [13]. In addition, various BERT sentence embedding models were introduced as baselines, including Avg.BERT [2], BERT-CLS [2], BERT-FLA [2], BERT-Flow [4], BERT-Whitening [5], InferSent-Glove [12], SBERT [2].

Evaluation. We calculate the cosine similarity between the final embedding vectors of the sentence pairs. Then we evaluate the model's performance by using the Spearman correlation coefficient of cosine similarity and the labels.

Implementation Details. We mainly compare the BERT-Flow [4] and the BERT-Whitening [5] models with the maximum sequence length set to 128, and we basically agree with their experimental settings and notations. Flow-based maximization likelihood and contrastive loss are only used to update the reversible mapping for the proposed method, while the BERT model parameters remain unchanged. we use the AdamW optimizer and set the learning rate to 1e-3.

4.2   STS Results

The hyperparameters λ and τ in our model are 0.1 and 0.3. First-Last-Average is chosen as the pooling method for the experiment.

Results Based on Original BERT. As shown in Table 1, the original BERT (BERT$_{base}$ and BERT$_{large}$) do not outperform Glove embedding. Our method raises 3.41–5.25% and 2.84–5.06% on BERT$_{base}$ and BERT$_{large}$ respectively.

Results Based on Fine-Tuned BERT with NLI. In Table 2, BERT$_{base}$ and BERT$_{large}$ are fine-tuned on the NLI dataset by the method in [2]. We can observe that our BERT-ContFlow outperforms BERT-Flow and BERT-Whitening on STS tasks. These experimental results suggest that our method can further improve the performance of SBERT even though it is trained on the NLI dataset that was prepared under the supervision.

4.3   Visualization Results

We visualize some variants of the BERT sentence representations to demonstrate that our approach is effective. Specifically, We sample 20 positive pairs and 20 negative pairs from the STS-B dataset. Then we compute their vectors and plot them in 2D with the help of t-SNE algorithm. Figure 2 confirms that BERT$_{base}$-ContFlow encourages the BERT$_{base}$ sentences embeddings to be more consistent with their positive pairs while remaining relatively far from their negative pairs.

Table 1. Results based on original BERT. The results are shown as the spearman correlation coefficient as $\rho \times 100$ between cosine similarity and original labels.

Model	STS-12	STS-13	STS-14	STS-15	STS-16	STS-B	SICK-R
Avg.GloVe	55.14	70.66	59.73	68.25	63.66	58.02	53.76
Avg.BERT	38.78	57.98	57.98	63.15	61.06	46.35	58.40
BERT-CLS	20.16	30.01	20.09	36.88	38.03	16.5	42.63
BERT$_{base}$-FLA	57.84	61.95	62.48	70.95	69.81	59.04	63.75
BERT$_{large}$-FLA	57.68	61.37	61.02	68.04	70.32	59.56	60.22
BERT$_{base}$-Flow	59.54	64.69	64.66	72.92	71.84	58.56	65.44
BERT$_{base}$-Whitening	61.69	65.70	66.02	75.11	73.11	68.19	63.60
BERT$_{base}$-ContFlow	**63.69**	**68.83**	**68.66**	**76.14**	**74.77**	**73.44**	**68.85**
BERT$_{large}$-Flow	61.72	66.05	66.34	74.87	74.74	68.09	64.62
BERT$_{large}$-Whitening	62.54	67.31	67.12	75.00	76.29	68.54	62.40
BERT$_{large}$-ContFlow	**63.72**	**68.53**	**68.80**	**75.60**	**77.31**	**72.91**	**67.46**

Table 2. Results based on fine-tuned BERT with NLI. The results are shown as the spearman correlation coefficient as $\rho \times 100$ between cosine similarity and original labels.

Model	STS-12	STS-13	STS-14	STS-15	STS-16	STS-B	SICK-R
InferSent - Glove	52.86	66.75	62.15	72.77	66.86	68.03	65.65
SBERT$_{base}$-NLI	70.97	76.53	73.19	79.09	74.30	77.03	72.91
SRoBERTa$_{base}$-NLI	71.54	72.49	70.80	78.74	73.69	77.77	74.46
SBERT$_{large}$-NLI	72.27	78.46	74.90	80.99	76.25	79.23	73.75
SRoBERTa$_{large}$-NLI	74.53	77.00	73.18	81.85	76.82	79.10	74.29
BERT$_{base}$-NLI-Flow	67.75	76.73	75.53	80.63	77.58	79.10	78.03
BERT$_{base}$-NLI-Whitening	69.11	75.79	75.76	82.31	79.61	78.66	76.33
BERT$_{base}$-NLI-ContFlow	**69.57**	**77.51**	**77.94**	**82.91**	**79.88**	**81.50**	**78.51**
BERT$_{large}$-NLI-Flow	69.61	79.45	77.56	82.48	79.36	79.89	77.73
BERT$_{large}$-NLI-Whitening	70.41	76.78	76.88	82.84	81.19	79.55	75.93
BERT$_{large}$-NLI-ContFlow	**72.15**	**80.24**	**79.20**	**84.06**	**81.51**	**83.11**	**78.20**

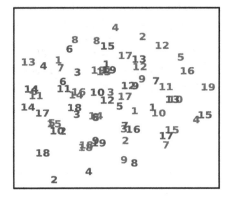
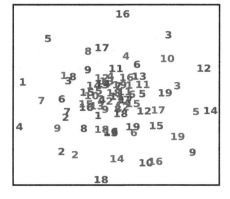

Fig. 2. Sentence representation visualization. (Left) Embeddings from the original BERT$_{base}$. (Right) Embeddings from the BERT$_{base}$-ContFlow model. Red numbers correspond to positive sentence pairs and blue to negative pairs. (Color figure online)

4.4 Ablation Studies

This section performs ablation experiments on two hyperparameters and pooling methods of our sentence embedding method to better understand their relative importance.

Table 3. Results of ablation experiments with different λ on the STS task.

λ	STS-12	STS-13	STS-14	STS-15	STS-16	STS-B	SICK-R	Avg.
0.1	**63.69**	**68.83**	**68.66**	**76.14**	**74.77**	**73.44**	**68.85**	**70.62**
0.3	62.76	67.24	67.30	76.03	73.68	71.17	66.22	69.20
0.5	62.24	66.63	66.64	75.62	73.55	69.59	65.06	68.47
0.7	62.08	66.29	66.62	75.16	73.27	68.78	64.57	68.11
0.9	62.21	66.56	66.42	75.03	73.26	68.56	64.21	68.04

Table 4. Results of ablation experiments with different τ on the STS task.

τ	STS-12	STS-13	STS-14	STS-15	STS-16	STS-B	SICK-R	Avg.
0.1	63.01	67.74	67.92	76.19	73.84	72.36	67.28	69.76
0.3	**63.69**	**68.83**	**68.66**	76.14	**74.77**	**73.44**	**68.85**	**70.62**
0.5	62.97	68.43	68.65	76.29	74.62	72.79	68.65	70.34
0.7	63.09	68.21	68.42	76.23	74.62	72.70	68.47	70.25
0.9	63.20	67.96	68.47	**76.38**	74.22	72.54	68.05	70.12

Effect of Lambda. We explore the effect of λ (the weight of the supervised contrastive loss term) on the resulting sentence embeddings (Table 3). For the STS task, the optimal λ value for the model is 0.1. Increasing λ from 0.3 to 0.9 reduces the quality of the generated sentence embeddings, suggesting that the combination of maximum likelihood and supervised contrastive loss lead to better sentence embeddings than individual losses.

Effect of Temperature. We explore the effect of temperature (used to calculate the sentence pair similarity magnitude) on the sentence embedding model (Table 4). The smaller the temperature value, the more the model focuses on the negative examples with higher similarity. For the STS task, the optimal τ value for the model is 0.3. Decreasing or increasing the τ value diminishes the quality of the generated sentence embeddings.

Table 5. Results of different pooling strategies on the STS task.

Method	CLS.	Last-avg.	Last2avg.	First-last-avg.
BERT$_{base}$-ContFlow	61.23	66.67	64.88	**70.63**
BERT$_{large}$-ContFlow	57.38	65.13	63.33	**70.62**
BERT$_{base}$-NLI-ContFlow	77.42	77.24	75.35	**78.26**
BERT$_{large}$-NLI-ContFlow	78.24	78.57	77.97	**79.78**

Effect of Pooling Strategies. We explore the effect of pooling strategies (CLS., Last-avg., Last2avg. and First-last-avg.) on the sentence embedding model (Table 5). Experiments results show that averaging the first layer and the last layer of BERT model sentence embeddings (denoted by first-last-avg) consistently produces better results than the other pooling strategies, so we choose first-last-avg as the default configuration.

5   Conclusion

This paper presents BERT-ContFlow, a contrastive learning framework for transferring sentence representations to downstream tasks. Experimental results on seven semantic similarity benchmark datasets show that our approach can enjoy the benefits of combining contrastive learning and flow models. The visualization and ablation experiments demonstrate the effectiveness of our proposed approach. We hope that our work will provide a new perspective for future research on sentence representation migration.

References

1. Devlin, J., Chang, M.W., Lee, K., Toutanova, K.: BERT: pre-training of deep bidirectional transformers for language understanding. In: Proceedings of the 2019 Conference of the North American Chapter of the Association for Computational Linguistics: Human Language Technologies, pp. 4171–4186. Association for Computational Linguistics (2019)
2. Reimers, N., Gurevych, I.: Sentence-BERT: sentence embeddings using siamese BERT-networks. In: Proceedings of the 2019 Conference on Empirical Methods in Natural Language Processing and the 9th International Joint Conference on Natural Language Processing, pp. 3980–3990. Association for Computational Linguistics (2019)
3. Gao, J., He, D., Tan, X., Qin, T., Wang, L.W., Liu, T.Y.: Representation degeneration problem in training natural language generation models. In: 7th International Conference on Learning Representations. OpenReview.net (2019)
4. Li, B.H., Zhou, H., He, J.X., Wang, M.X., Yang, Y.M., Li, L.: On the sentence embeddings from pre-trained language models. In: Proceedings of the 2020 Conference on Empirical Methods in Natural Language Processing, pp. 9119–9130. Association for Computational Linguistics (2020)
5. Su, J.L., Cao, J.R., Liu, W.J., Ou, Y.Y.W.: Whitening sentence representations for better semantics and faster retrieval. arXiv preprint arXiv:2103.15316 (2021)

6. Kim, T., Yoo, K.M., Lee, S.: Self-guided contrastive learning for BERT sentence representations. In: Proceedings of the 59th Annual Meeting of the Association for Computational Linguistics and the 11th International Joint Conference on Natural Language Processing, pp. 2528–2540. Association for Computational Linguistics (2021)
7. Gao, T.Y., Yao, X.C., Chen, D.Q.: SimCSE: simple contrastive learning of sentence embeddings. In: Proceedings of the 2021 Conference on Empirical Methods in Natural Language Processing, pp. 6894–6910. Association for Computational Linguistics (2021)
8. Cer, D., Diab, M., Agirre, E., Lopez-Gazpio, I., Specia, L.: SemEval-2017 task 1: semantic textual similarity multilingual and crosslingual focused evaluation. In: Proceedings of the 11th International Workshop on Semantic Evaluation, pp. 1–14. Association for Computational Linguistics (2017)
9. Iyyer, M., Manjunatha, V., Boyd-Graber, J., Daumé III, H.: Deep unordered composition rivals syntactic methods for text classification. In: Proceedings of the 53rd Annual Meeting of the Association for Computational Linguistics and the 7th International Joint Conference on Natural Language Processing, pp. 1681–1691. The Association for Computer Linguistics (2015)
10. Marelli, M., Menini, S., Baroni, M., Bentivogli, L., Bernardi, R., Zamparelli, R.: A SICK cure for the evaluation of compositional distributional semantic models. In: Proceedings of the Ninth International Conference on Language Resources and Evaluation, pp. 216–223. European Language Resources Association (2014)
11. Agirre, E., et al.: SemEval-2015 task 2: semantic textual similarity, English, Spanish and pilot on interpretability. In: Proceedings of the 9th International Workshop on Semantic Evaluation, pp. 252–263. The Association for Computer Linguistics (2015)
12. Conneau, A., Kiela, D., Schwenk, H., Barrault, L., Bordes, A.: Supervised learning of universal sentence representations from natural language inference data. In: Proceedings of the 2017 Conference on Empirical Methods in Natural Language Processing, pp. 670–680. Association for Computational Linguistics (2017)
13. Pennington, J., Socher, R., Manning, C.D.: GloVe: global vectors for word representation. In: Proceedings of the 2014 Conference on Empirical Methods in Natural Language Processing, pp. 1532–1543. ACL (2014)
14. Hadsell, R., Chopra, S., LeCun, Y.: Dimensionality reduction by learning an invariant mapping. In: 2006 IEEE Computer Society Conference on Computer Vision and Pattern Recognition, pp. 1735–1742. IEEE Computer Society (2006)
15. Yan, Y.M., Li, R.M., Wang, S.R., Zhang, F.Z., Wu, W., Xu, W.R.: ConSERT: a contrastive framework for self-supervised sentence representation transfer. In: Proceedings of the 59th Annual Meeting of the Association for Computational Linguistics and the 11th International Joint Conference on Natural Language Processing, pp. 5065–5075. Association for Computational Linguistics (2021)
16. Bowman, S.R., Angeli, G., Potts, C., Manning, C.D.: A large annotated corpus for learning natural language inference. In: Proceedings of the 2015 Conference on Empirical Methods in Natural Language Processing, pp. 632–642. The Association for Computational Linguistics (2015)
17. Williams, A., Nangia, N., Bowman, S.R.: A broad-coverage challenge corpus for sentence understanding through inference. In: Proceedings of the 2018 Conference of the North American Chapter of the Association for Computational Linguistics: Human Language Technologies, pp. 1112–1122. Association for Computational Linguistics (2018)

Data Mining

Feature Selection for EEG Data Classification with Weka

Marina Murtazina$^{(\boxtimes)}$ 🆔 and Tatiana Avdeenko 🆔

Novosibirsk State Technical University, Karla Marksa Avenue 20, 630073 Novosibirsk, Russia
`murtazina@corp.nstu.ru`

Abstract. The paper explores the application of feature selection techniques for the brain activity classifying patterns task. The study aim is to compare the machine learning algorithms results depending on the chosen feature selection technique. As an example for analysis, the task of classifying of open-eyes and closed-eyes resting states according to EEG data was chosen. For the experiment, EEG records of the resting states from the data set "EEG Motor Movement/Imagery" were used. Features in the time and frequency domains for 19 electrodes corresponding to the 10–20 system were extracted from the EEG records presented in the EDF format. Python was used to form the feature matrix and convert it to ARFF format. The resulting dataset contains 209 features for classification and a target feature. In the experimental part of the work, the classification results are compared before and after feature selection. The experiment examined 10 Weka attribute evaluators.

Keywords: EEG · Feature selection · Weka · Classification · Feature extraction

1 Introduction

Currently, the classification of mental states according to electroencephalogram (EEG) data is important for solving problems in such areas as medicine, healthcare, education, robotics, etc. [1, 2]. The task of classifying mental states according to EEG data is extremely difficult in view of the large number of features that can be obtained from EEG data, which can be noisy and also have a large dimension. Feature selection is a critical step in solving classification problems. Feature selection is a procedure for selecting a subset of features from the feature space needed to build a model. In the feature selection process, irrelevant, noisy, and redundant features are removed. This makes it possible to increase and decrease the computational cost of training classifiers and improve their accuracy. Experimental results show that the accuracy of classifying mental states according to EEG data increases when feature selection techniques such as scatter matrices [3], linear regression [4], combinations of linear regression with genetic algorithms [5], mutual information metrics [6], the ReliefF algorithm [7], etc.

The data analysis and machine learning software Weka includes a number of attribute estimators to select the optimal set of features. This toolkit is used, among other things, in the selection of features extracted from physiological signals. In [8], to select features from physiological signals, the CfsSubsetEval attribute estimator from the Weka

© Springer Nature Switzerland AG 2022
Y. Tan et al. (Eds.): ICSI 2022, LNCS 13345, pp. 279–288, 2022.
https://doi.org/10.1007/978-3-031-09726-3_25

package with the Logistic classifier was used to evaluate a set of attributes. In [9], the problem of long learning time and complex calculations based on features extracted from physiological signals was discussed. Feature selection is carried out using the attribute evaluators CfsSubsetEval, InfoGainAttributeEval, ClassifierSubsetEva from the Weka package. The best result was obtained by the ClassifierSubsetEva estimator for the SMO classifier.

One of the urgent tasks among the classification of mental states according to EEG data is the classification of states of rest with open and closed eyes according to EEG data. In [10], the attribute estimator CfsSubsetEval was used to search for the optimal set of features (frequency components of delta waves and alpha waves) for eye state recognition. The work [11] considers the issues of selecting features for predicting the state of the eyes using the EEG. To select features, 5 attribute evaluators from the Weka package were used. After feature selection, 21 classification algorithms were investigated. The best result was obtained on a set of features obtained using ReliefF.

The purpose of this study is to analyze the possibilities of applying the feature selection techniques included in the Weka tool to search for a subset of features in the time and frequency domains from which mental states can be identified. As an example for analysis, a classification of rest states with open and closed eyes was chosen.

The work is organized as follows. Section 1 substantiates the relevance of the research topic. Section 2 contains the theoretical background necessary for understanding the work. Section 3 describes the materials and research methods. Section 4 presents the results of the experiment. Section 5 summarizes the work done.

2   Theoretical Background

2.1   EEG

Currently, EEG is a widely used method for monitoring brain activity. EEG, as a method for studying brain activity, has almost a century and a half history. The first electrophysiological measurements date back to the last quarter of the 19th century. At that time, the English physician R. Cato conducted the first experiments to measure the electrical activity of the cerebral hemispheres of rabbits and monkeys. During the experiments, electrical activity was measured on the naked brains of animals. In 1924, the German neurologist G. Berger proved that weak electrical currents generated in the brain can be recorded without opening the skull, and began to use the electroencephalography method to record the activity of the human cerebral cortex. In the course of his experiments, G. Berger established that the EEG changes depending on the functional state of the brain [12]. Since then, EEG recording has been carried out using electrodes placed on the scalp.

By the middle of the 20st century, the EEG method has been actively used by many scientists, but not always in the same way, which has led to a number of inconsistencies in research. Therefore, in 1947, at the first international congress of EEG researchers, the need to create a standard electrode positioning system was realized [13]. In 1958, the International Federation of Electroencephalography and Clinical Neurophysiology adopted a standard that defines the scheme for applying electrodes, called the 10–20 system [12, 13]. The international 10–20 system is the standard system for positioning

19 scalp electrodes. The spatial resolution of the 10–20 system is the 10–10 system, which uses more electrodes. Four electrodes have been renamed in the 10–10 system. On Fig. 1 shows the location of the electrodes according to the 10–20 and 10–10 systems. Renamed electrodes are highlighted in red.

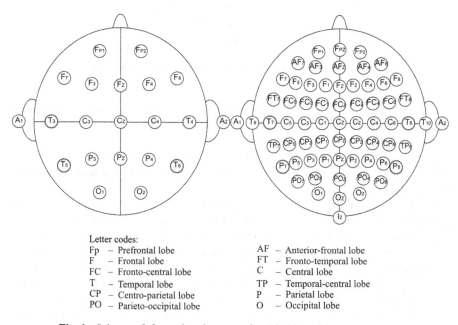

Letter codes:
Fp	– Prefrontal lobe		AF	– Anterior-frontal lobe
F	– Frontal lobe		FT	– Fronto-temporal lobe
FC	– Fronto-central lobe		C	– Central lobe
T	– Temporal lobe		TP	– Temporal-central lobe
CP	– Centro-parietal lobe		P	– Parietal lobe
PO	– Parieto-occipital lobe		O	– Occipital lobe

Fig. 1. Scheme of electrodes placement for '10–20' and '10–10' systems.

Each electrode is identified by a combination of a letter code and a number. The letter code indicates the region of the brain. Odd numbers are used to designate electrodes located above the left brain, odd numbers - above the right hemisphere of the brain. The letter "z" is used instead of the number in the identifier of the electrodes located above the midline.

EEG enables analyzing the neural activity of the brain during wakefulness and sleep. In the waking state, the brain can be at rest (with eyes open or closed), in an active state, or in a cognitive state [14].

2.2 Building a Classifying Model Based on EEG Data

The classifying model predicts the belonging of objects to a particular class of a categorical variable, depending on a set of features. The process of building a classifying model based on EEG data can be divided into two stages: forming a data sample for training a classifier, training a classifying model (see Fig. 2).

EEG data collection is a process that records the electrical activity caused by the firing of neurons in the brain. The result of this process is a set of electroencephalogram records stored in one of the standard EEG data storage formats (EDF, BDF, GDF, EEG, etc.).

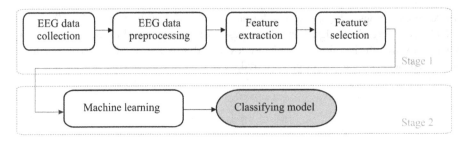

Fig. 2. Stages of building a classifying model based on EEG data.

During pre-processing of EEG data, artifacts are removed. The purpose of removing artifacts is to separate data that are true neural signals from extraneous noise that arose during the registration of electrical activity.

As features for EEG data, indicators from areas related to time series analysis can be used, such as power spectral density from signal processing, entropy from information theory, and so on [15].

Feature selection is a process in which a subset of features is extracted from the extracted feature space, sufficient to solve the classification problem. Feature selection approaches can be divided into filter methods, wrapper methods, and embedded methods. The methods of the first group measure the degree of significance of each feature without taking into account the specific algorithms that will be used to train the classifying model. The methods of this group are based on probability theory and statistical approaches. Filtering methods enables ranking features by relevance, assessing the correlation degree of each dependent variable with the target variable. Methods of the second group involve taking into account the classification algorithm when assessing the significance of features. There are three approaches in this category: forward, backward, and stepwise selection. Methods of the third group do not separate feature selection and classifier training The main method of the last category is the regularization (for example, Lasso or Ridge) [16–18]. The result of the feature extraction process is a dataset for the feature selection. In the process of feature selection, a subset of features is formed on which the classifier is trained.

Further, to train classifiers with a data set containing features obtained from EEG signals, software designed for machine learning can be used. Currently, there are many platforms that enable quickly performing training and quality assessment of classifying models. Among them are the analytical platforms KNIME and Weka. To train and test a classifying model in KNIME, it is necessary to build a separate workflow for each classifier. In the Weka Program Analyzer, it is possible to explore the operation of classification algorithms by selecting an algorithm and setting its parameters through a system of dialog boxes. The latter makes the Weka Analyzer a popular tool among researchers.

3 Materials and Methods

3.1 Formation of the Features Matrix from EEG Records

For the experiment, the data set "EEG Motor Movement/Imagery" [19] was used. This data set is hosted in the PhysioNet data warehouse in the public domain [20]. The data set includes EEG recordings of 109 subjects in EDF format. EEG was recorded from 64 electrodes placed according to the international 10–10 system. For each subject, the data set includes EEG recordings at rest with eyes open and closed, and EEG recordings in which the subjects performed various motor and imaginary tasks.

For EEG at resting states, a matrix of features extracted using the eeglib Python library for 19 electrodes corresponding to the international 10–20 system was formed from the EEG Motor Movement/Imagery dataset. The feature matrix has dimensions of 218 rows × 210 columns. Each subject is represented in it by two instances of data that correspond to states of rest with open and closed eyes. For each data instance, the following features of the time and frequency domains are extracted: the Hjort parameters (activity, mobility, complexity) and the power of four frequency bands (delta, theta, alpha, beta) calculated using Welch Power Spectral Density (PSD) and Discrete Fourier Transform (DFT). The sequence numbers of the extracted features are shown in Fig. 3.

Feature	Channel name																		
	Fp1	Fp2	F7	F3	Fz	F4	F8	T7	C3	Cz	C4	T8	P7	P3	Pz	P4	P8	O1	O2
Hjorth activity	1	2	3	4	5	6	7	8	9	10	11	12	13	14	15	16	17	18	19
Hjorth mobility	20	21	22	23	24	25	26	27	28	29	30	31	32	33	34	35	36	37	38
Hjorth complexity	39	40	41	42	43	44	45	46	47	48	49	50	51	52	53	54	55	56	57
delta band power (PSD)	58	62	66	70	74	78	82	86	90	94	98	102	106	110	114	118	122	126	130
theta band power (PSD)	59	63	67	71	75	79	83	87	91	95	99	103	107	111	115	119	123	127	131
alpha band power (PSD)	60	64	68	72	76	80	84	88	92	96	100	104	108	112	116	120	124	128	132
beta band power (PSD)	61	65	69	73	77	81	85	89	93	97	101	105	109	113	117	121	125	129	133
delta band power (DFT)	134	138	142	146	150	154	158	162	166	170	174	178	182	186	190	194	198	202	206
theta band power (DFT)	135	139	143	147	151	155	159	163	167	171	175	179	183	187	191	195	199	203	207
alpha band power (DFT)	136	140	144	148	152	156	160	164	168	172	176	180	184	188	192	196	200	204	208
beta band power (DFT)	137	141	145	149	153	157	161	165	169	173	177	181	185	189	193	197	201	205	209

Fig. 3. Ordinal numbers of extracted features.

Python was used to form the feature matrix from EEG records and convert it to ARFF (Attribute-Relation File Format) format.

3.2 Feature Selection in Weka

Feature selection in Weka uses a combination of attribute evaluator and search method. The attribute evaluator is a feature quality estimation algorithm. The search method is a method for finding the optimal feature space. Search methods define how attributes are looked up. The individual attribute or attribute set selected by the search method is passed as input to the attribute evaluator. Table 1 shows a list of the attribute evaluators used in this work, as well as compatible search methods.

Table 1. Weka attribute evaluators [21].

Attribute evaluator name	Algorithm	Search method
CfsSubsetEval	Feature subset selection based on correlation	BestFirst, GreedyStepwise
ClassifierAttributeEval	Feature subset selection by using a user-specified classifier	Ranker
ClassifierSubsetEval	Feature subset selection using the training data or a separate hold out testing set	BestFirst, GreedyStepwise
CorrelationAttributeEval	Feature subset selection using Pearson's correlation	Ranker
GainRatioAttributeEval	Feature subset selection by measuring the gain ratio	Ranker
InfoGainAttributeEval	Feature subset selection using information gain	Ranker
OneRAttributeEval	Feature subset selection using the OneR classifier	Ranker
ReliefFAttributeEval	Feature subset selection using the Relief algorithm	Ranker
SymmetricalUncertAttributeEval	Feature subset selection using the symmetrical uncertainty with respect to the class	Ranker
WrapperSubsetEval	Feature subset selection using a learning scheme	Ranker

4   Experimental Results

As part of the experiment, 14 sets of features were generated using the combination of Attribute evaluator and Search method. When using attribute evaluators that require a classification algorithm, the Logistic and SMO algorithms were selected from the same "functions" group. The results of feature selection are presented in Table 2.

Table 2. Feature selection results.

ID	Feature selection methodology	Feature set
FS1	CfsSubsetEval and BestFirst (or GreedyStepwise)	24, 58, 59, 62, 63, 82, 124, 128, 134, 138, 204, 205, 208
FS2	CorrelationAttributeEval and Ranker >0.4	134, 138, 208, 204, 135, 139, 158, 200, 154, 142, 146, 196, 150, 192, 188, 58, 172
FS3	GainRatioAttributeEval and Ranker >0.2	58, 128, 204, 62, 132, 208, 138, 124, 134, 59, 82, 63, 200, 108
FS4	InfoGainAttributeEval and Ranker >0.2	62, 134, 138, 58, 204, 63, 128, 132, 135, 208, 139, 154, 124, 59, 78
FS5	OneRAttributeEval and Ranker >70	62, 134, 139, 204, 128, 138, 58, 132, 63, 208, 124, 120
FS6	ReliefFAttributeEval and Ranker >0.02	134, 138, 135, 139, 158, 208, 204, 146, 154, 150, 142, 58, 63, 59, 62, 159, 200, 196, 172, 184, 188
FS7	SymmetricalUncertAttributeEval and Ranker >0.2	62, 58, 138, 134, 204, 128, 132, 208, 63, 124, 59, 139, 154, 200
FS8	ClassifierAttributeEval (Logistic) and Ranker >0.2	138, 58, 134, 62, 204, 208, 132, 200, 128, 124, 63, 139, 135, 59, 196, 108, 120, 192, 96, 112, 172, 116
FS9	ClassifierAttributeEval (SMO) and Ranker >0.15	134, 138, 208, 204, 135, 196, 188, 152, 139, 158, 200, 156, 172, 192, 184, 168, 154, 148, 142
FS10	ClassifierSubsetEval (Logistic) and BestFirst	58, 62, 65, 74, 90, 96, 108, 120, 132
FS11	ClassifierSubsetEval (SMO) and BestFirst (or GreedyStepwise)	25, 110, 134, 204
FS12	ClassifierAttributeEval (Logistic) and BestFirst	58,65,66,112,128,134,140,166,189,201
FS13	ClassifierAttributeEval (Logistic) and GreedyStepwise	58,65,66,112,128,134,140,189
FS14	ClassifierAttributeEval (SMO) and BestFirst (or GreedyStepwise)	36,117,125,134,159,191,204

When using the two search methods BestFirst and GreedyStepwise, the same feature set was obtained in a number of cases. This applies to feature sets FS1, FS11, FS14.

To evaluate the result of feature selection, the following algorithms were used: Logistic and SMO, as well as the RandomForest algorithm from the "trees" group. Figure 4 compares the accuracy of these classifiers on the full feature set and on 14 subsets of this feature set.

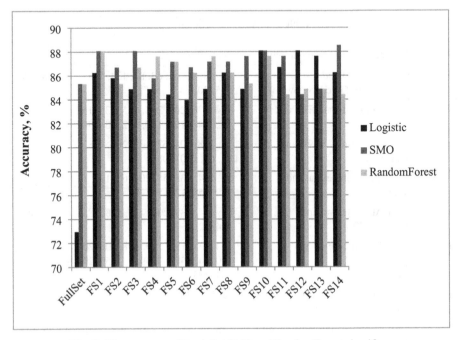

Fig. 4. The accuracy of Logistic, SMO and RandomForest classifiers.

When using attribute evaluators that use filtering methods for all classifiers, there was an improvement in the accuracy estimate compared to the full feature set. Feature selection is considered successful if the dimension of the feature space is reduced, and the accuracy of the classifier has become higher or has not changed. The feature selection methods used to obtain the feature sets FS1, FS3, FS4, FS7, FS10 proved to be successful for all classification algorithms. At the same time, the best result was obtained on the FS10 and FS10 set for the Logistic algorithm (the classifier accuracy changed from 72.94% to 88.07%). On average, the classification accuracy for the full feature set was 81.19%. The average classification accuracy increased for all 14 feature sets. For the FS10 set, the increase in the average classification accuracy was 6.72%.

5 Conclusion

This paper explores the issues of feature selection for solving the problem of classifying rest states. For the experiment, a set of features extracted from EEG signals in the time and frequency domains was used. In the experimental part of the work, 10 attribute evaluators of the Weka library were studied in combination with search methods to solve the problem of extracting features extracted from EEG signals. The experiment shows that when using attribute evaluators using wrapper methods, the accuracy of the corresponding classifier improves. When using the Logistic algorithm, the classification accuracy increases by 15.14%. The further direction of work will be directed to the study of the dependence of the result of the work of attribute evaluators on their settings.

Acknowledgment. The research is supported by Ministry of Science and Higher Education of Russian Federation (project No. FSUN-2020–0009).

References

1. Bird, J.J., Manso, L.J., Ribeiro, E.P., Ekárt, A., Faria, D.R.: A study on mental state classification using EEG-based brain-machine interface. In: 2018 International Conference on Intelligent Systems (IS), pp. 795–800 (2018)
2. Edla, D.R., Mangalorekar, K., Dhavalikar, G., Dodia, S.: Classification of EEG data for human mental state analysis using random forest classifier. Procedia Comput. Sci. **132**, 1523–1532 (2018)
3. Gupta, A., Agrawal, R.K.: Relevant feature selection from EEG signal for mental task classification. In: Tan, P.-N., Chawla, S., Ho, C.K., Bailey, J. (eds.) PAKDD 2012. LNCS (LNAI), vol. 7302, pp. 431–442. Springer, Heidelberg (2012). https://doi.org/10.1007/978-3-642-30220-6_36
4. Timofeeva, A.Y., Murtazina, M.S.: Feature selection for EEG data based on logistic regression. In: 2021 XV International Scientific-Technical Conference on Actual Problems of Electronic Instrument Engineering (APEIE), pp. 604–609 (2021)
5. Becerra-Sánchez, P., Reyes-Munoz, A., Guerrero-Ibañez, A.: Feature selection model based on EEG signals for assessing the cognitive workload in drivers. Sensors **20**, 5881 (2020)
6. Deligani, R.J., Borgheai, S.B., McLinden, J., Shahriari, Y.: Multimodal fusion of EEG-fNIRS: a mutual information-based hybrid classification framework. Biomed. Opt. Express **12**, 1635–1650 (2021)
7. Zhang, Y., Cheng, C., Chen, T.: Multi-channel physiological signal emotion recognition based on ReliefF feature selection. In: 2019 IEEE 25th International Conference on Parallel and Distributed Systems (ICPADS), pp. 725–730 (2019)
8. Escudero, J., Ifeachor, E., Fernández, A., López-Ibor, J.J., Hornero, R.: Changes in the MEG background activity in patients with positive symptoms of schizophrenia: spectral analysis and impact of age. Physiol. Meas. **34**(2), 265–279 (2013)
9. Ghaderi, A., Frounchi, J., Farnam, A.: Machine learning-based signal processing using physiological signals for stress detection. In: 2015 22nd Iranian Conference on Biomedical Engineering (ICBME), pp. 93–98 (2015)
10. Zhou, Z., Li, P., Liu, J., Dong, W.: A novel real-time EEG based eye state recognition system. In: Liu, X., Cheng, D., Jinfeng, L. (eds.) ChinaCom 2018. LNICSSITE, vol. 262, pp. 175–183. Springer, Cham (2019). https://doi.org/10.1007/978-3-030-06161-6_17
11. Uwisengeyimana, J.D., AlSalihy, N.K., Ibrikci, T.: Statistical performance effect of feature selection techniques on eye state prediction using EEG. Int. J. Stat. Med. Res. **5**, 224–230 (2016)
12. Teplan, M.: Fundamentals of EEG measurement. Meas. Sci. Rev. **2**(2), 1–11 (2002)

13. Klem, G., Lüders, H., Jasper, H., Elger, C.: The ten-twenty electrode system of the International federation. The international federation of clinical neurophysiology. Electroencephalogr. Clin. Neurophysiol. Suppl. **52**, 3–6 (1999)
14. Mazher, M., Faye, I., Qayyum, A., Malik, A.S.: Classification of resting and cognitive states using EEG-based feature extraction and connectivity approach. In: 2018 IEEE-EMBS Conference on Biomedical Engineering and Sciences (IECBES), pp. 184–188 (2018)
15. Murtazina, M., Avdeenko, T.: Applying classification algorithms to identify brain activity patterns. In: Tan, Y., Shi, Y. (eds.) ICSI 2021. LNCS, vol. 12690, pp. 452–461. Springer, Cham (2021). https://doi.org/10.1007/978-3-030-78811-7_42
16. Hag, A., et al.: Enhancing EEG-based mental stress state recognition using an improved hybrid feature selection algorithm. Sensors **21**, 8370 (2021)
17. Jiang, K., Tang, J., Wang, Y., Qiu, C., Zhang, Y., Lin, C.: EEG feature selection via stacked deep embedded regression with joint sparsity. Front. Neurosci. **14**, 829 (2020)
18. Kumar, C.A., Sooraj, M.P., Ramakrishnan, S.: A comparative performance evaluation of supervised feature selection algorithms on microarray datasets. Procedia Comput. Sci. **115**, 209–217 (2017)
19. Schalk, G., et al.: BCI2000: A general-purpose Brain-Computer Interface (BCI) system. IEEE Trans. Biomed. Eng. **51**(6), 1034–1043 (2004)
20. Goldberger, A., et al.: PhysioBank, PhysioToolkit, and PhysioNet: components of a new research resource for complex physiologic signals. Circulation **101**(23), e215–e220 (2000)
21. Bouckaert, R.R., et al.: WEKA manual for version 3-8-3. University of Waikato, Hamilton (2018)

Modified Correlation-Based Feature Selection for Intelligence Estimation Based on Resting State EEG Data

Tatiana Avdeenko, Anastasiia Timofeeva[✉], and Marina Murtazina

Novosibirsk State Technical University, 20, Karl Marx Avenue, Novosibirsk, Russia
{avdeenko,a.timofeeva}@corp.nstu.ru

Abstract. The effectiveness of the correlation-based method (CFS) for feature selection based on electroencephalogram (EEG) data of the resting state for the purpose of intelligence assessment is investigated. A modification of the CFS is proposed, which makes it possible to vary the cardinality of a subset of selected features using a hyperparameter. A practical example of the analysis of the relationship between the intelligence quotient (IQ), the age of subjects, the features extracted from EEG data, and the effects of their interaction is considered. A comparison of the genetic algorithm and the forward selection was made to find the optimal subset of features within the modified CFS. It was found that it is quite sufficient to use the method of forward selection. Using the nested cross-validation procedure, it was shown that the modified approach gives a lower mean absolute error compared to usual CFS, as well as building a stepwise regression by the forward selection method based on the Bayesian information criterion (BIC). In terms of the mean absolute error, the modified CFS is close to the least absolute shrinkage and selection operator (LASSO) and the improved algorithm Bolasso-S.

Keywords: Correlation feature selection · IQ · EEG · Resting state · Estimation · Cross-validation

1 Introduction

It is now well known that human cognitive activity is accompanied by electrical oscillations of the brain. In recent decades, various methods have been developed and used to measure this electrical activity and explore its relationship with brain function. The most effective measures include fMRI, EEG, fNIR, PET, MEG [1]. Among them, electroencephalogram (EEG) is the oldest and most common technique for measuring brain activity. The cost of EEG measurements is low enough that university laboratories can easily afford.

In connection with the development of methods for measuring brain activity, there is a high interest in studying the relationship between brain activity patterns and human cognitive abilities when solving specially organized tasks in accordance with the neural efficiency hypothesis of intelligence [2]. According to this hypothesis, brains of more

© Springer Nature Switzerland AG 2022
Y. Tan et al. (Eds.): ICSI 2022, LNCS 13345, pp. 289–300, 2022.
https://doi.org/10.1007/978-3-031-09726-3_26

intelligent individuals work more efficiently when engaged in cognitive task performance as compared to those of less intelligent ones.

For example, some authors used EEG to study various factors, for example, different levels of cognitive load during solving various arithmetic problems [3]. The paper [4] explores the possibilities of assessing intelligence using data from various brain mapping systems, including EEG data.

Thus, the neural efficiency hypothesis predicts that the level of cognitive abilities would be correlated to brain activity during cognitive load. However, it is still unclear whether the brain activity at rest can be a good predictor of individual differences in intelligence. It has been proposed that the most informative way to investigate resting state activity is the network neuroscience approach.

However, research on the relationship between the brain resting state activity and level of intelligence is inconsistent. In some papers the brain resting state characteristics correlated to intelligence [5] and non-verbal intelligence [6]. However, a recent study [7] failed to find any significant links between measures of the brain resting state dynamics and several widely used intelligence measures.

Thus, an urgent task is to find relationships between EEG indicators of the brain resting state and the results of intelligence tests. One of the important steps in solving this problem can be finding a set of the most informative EEG indicators (frequency bands, electrodes) for assessing intelligence.

After feature extraction, a high-dimensional EEG feature space is usually obtained. The choice of suitable features for the available set of EEG signals is a complex task. It must be reduced to increase the interpretability of the results and reduce the risk of overfitting the model. Usually, to solve this problem [4], such methods of feature extraction are used as Linear Discriminant Analysis (LDA) and Principal Component Analysis (PCA).

LDA is a supervised algorithm, that is, it uses information about the response, PCA is based only on the analysis of the correlation matrix of input features, that is, an unsupervised algorithm. Both approaches build a new feature space based on some combination of the available features. This is not always convenient either from the point of view of interpreting the constructed model or from the point of view of design of subsequent experiments (reducing the number of channels for further EEG studies). Therefore, this article explores various feature selection methods.

The difficulty of EEG data analysis is that the extracted features are highly correlated, so fast and well-scalable one-dimensional methods for feature selection are not suitable here. They evaluate features individually, so the resulting set includes many redundant, highly correlated features.

Multidimensional methods take these relationships into account and try to exclude not only irrelevant (not affecting the response), but also redundant features. Most often, feature selection is carried out in conjunction with the construction of predictive models using built-in methods such as LASSO regression [8, 9]. It gives a sparse solution, including only essential features, which, however, is very sensitive to the regularization parameter.

In addition, in the problems of building regressions, stepwise procedures are often used [10], namely, the methods of forward selection and backward elimination. However, such greedy search algorithms are not guaranteed to achieve a global optimum.

Finally, another class of feature selection methods is filtering methods. Among multivariate methods, the approach based on correlations is well known [11]. This approach is proposed for solving classification problems. Its applicability to feature selection for IQ estimation from EEG data has not yet been studied. It is this problem that our article is devoted to, in which a more flexible modification of the feature selection method based on correlations is proposed.

2 Theoretical Background

2.1 Modified Correlation Feature Selection

The Correlation-based Feature Selection (CFS) was proposed in [11]. Features are chosen to provide the highest correlation with the response and the weakest relationship between the features themselves. This solves the following optimization problem:

$$\frac{\sum\limits_{i \in S_k} R_i}{\sqrt{k + 2 \sum\limits_{i,j \in S_k, i \neq j} r_{ij}}} \to \max_{S_k}, \tag{1}$$

where R_i is the absolute value of the correlation coefficient between the i-th feature and the response, r_{ij} is the absolute value of the correlation coefficient between the i-th and j-th feature, S_k is a subset of k features.

Unlike other approaches to the model structure selection, the CFS method does not allow varying the complexity of the model. For example, when using information criteria, one can choose between the Akaike (AIC), which gives the most complex model, the Bayesian (BIC), which selects fewer input features, and another combination of residual sum of squares and number of model parameters. A similar effect is achieved when using regularization methods by varying the regularization parameter. On the one hand, this creates additional difficulties, since it is required to choose an appropriate value of the regularization parameter (to choose between the AIC and BIC criteria). On the other hand, it gives the approaches additional flexibility.

Probably, not for all practical problems, a fixed ratio between the complexity of the model (the number of features k) and the correlation with the response (the numerator of the ratio (1)) will give the best result. In this regard, it is proposed to modify problem (1) as follows:

$$\frac{\sum\limits_{i \in S_k} R_i}{\left(k + 2 \sum\limits_{i,j \in S_k, i \neq j} r_{ij}\right)^c} \to \max_{S_k}, \tag{2}$$

where the constant c allows you to vary the complexity of the model. For $c = 1/2$, we obtain the well-known problem (1). If $c < 1/2$, then the number of selected features will be greater than when $c = 1/2$. For $c > 1/2$, on the contrary, we obtain a smaller cardinality of the subset of selected features. Further, in the course of experiments on real data, we will test the performance of the proposed modification.

2.2 Heuristic Search Algorithms

The problem (2) is a non-linear integer programming problem. The search for the global optimum by enumeration of all combinations leads to an NP-complete problem. In practice, procedures stepwise variable selection have become widespread, which reduce the number of calculations, but do not ensure the achievement of the optimal set of input variables due to "greedy" strategies. Such algorithms are considered heuristic since they are not guaranteed to be accurate or optimal, but they are sufficient to solve the problem.

Many popular feature selection methods use hill climbing search, a mathematical optimization technique that belongs to the local search algorithm family. The algorithm is iterative. It starts with an arbitrary (random) subset of features and then tries to find the best solution by stepwise changing one of the elements of the (selection or elimination of one of the features). If at the current step it is possible to improve the solution, then at the next step the subset of features is again corrected to obtain a new solution. The steps are repeated until no improvement can be found at some step. Since the initial subset is chosen randomly, then, generally speaking, the algorithm can give different results when it is restarted, especially in the case of large data arrays, on which the function from problem (2) has many local optima.

A special case of hill climbing search algorithms are forward selection and backward elimination methods [12]. In contrast to the general case, where the initial subset is chosen randomly, these methods assume an exactly specified starting solution. For this reason, they always give the same result when restarted.

The forward selection procedure starts with a single feature candidate as the desired subset. This candidate is chosen as a solution to problem (2) for $k = 1$. This corresponds to the maximum correlation with the response. Further, in the loop over i, a new subset is formed by adding another i-th candidate to the already selected $(i-1)$ feature candidates so that the function from (2) reaches the maximum over all possible candidates added at the i-th iteration. The following rule is used as a stopping criterion: if at the current iteration it is not possible to improve the value of the objective function compared to the previous iteration, then the loop breaks.

Another heuristic approach to solving the optimization problem (2) is genetic algorithms (GA). They are based on procedures analogous to the genetic processes of biological organisms. The advantage of GA is that these methods are robust and can successfully address a wide range of problem areas, including those that are difficult to solve with other methods. Genetic algorithms are not guaranteed to find a globally optimal solution, but they are generally good at finding "acceptably good" solutions "acceptably fast".

Using the genetic algorithm in [13] a selection method based on Pearson's correlation coefficients is proposed. During the experiments, it was shown that the proposed method can be appropriate for improving the efficiency of feature selection.

Authors of [14] used a combination of the CFS method and a genetic algorithm to select the optimal subset of traits for the recognition of apple leaf diseases. As a result, it is shown that this method of selection with support vector machines (SVM) classification has an advantage compared to segmentation based on k-means and classification using neural networks, as well as the extraction of color features of tomato leaf diseases. However, the results of the genetic algorithm have not been compared with simpler procedures, for example, the forward selection.

3 Research

3.1 Data Processing Software

The de facto standard for EEG recording in commercial equipment and in research projects is now the EDF format [15]. Python libraries such as eeglib [16] was used to extract features from EEG records. For each instance of data on 19 electrodes, frequency domains features are extracted, supported by the eeglib: Power of four frequency bands (delta, theta, alpha, beta) calculated using discrete Fourier transformation (DFT), Power of four frequency bands (delta, theta, alpha, beta) calculated using the power spectral density (PSD) estimated through the Welch's method. So the feature matrix contains 152 columns.

The constructed feature matrix was used in the selection procedures. All calculations were performed in the R environment using the libraries: MASS for choosing a model in a stepwise algorithm, GA [17] for applying of genetic algorithm methods. The problem of subset selection can be naturally treated by GAs using a binary string, with 1 indicating the presence of a predictor and 0 its absence from a given candidate subset. The fitness function to be maximized is defined in (2). The maximum number of iterations to run before the GA search is halted is set to 1000. The population size is set to 50. The probability of crossover between pairs of chromosomes is set to 0.8. The probability of mutation in a parent chromosome is set to 0.1. The number of best fitness individuals to survive at each generation is set to the top 5% individuals.

3.2 Data Description and Preprocessing

The data of EEG records of 107 subjects at rest were analyzed, as well as the values of their IQ2 component of the intelligence structure according to the Amthauer method, which shows the ability to abstract. The sample varied by gender (1 - male, 2 - female) and age. Table 1 shows a contingency table. It can be seen that subjects over 25 years of age are rare.

As for the distribution of the IQ2 indicator, in group 2 it is more normal than in group 1, which is clearly seen from Fig. 1, *a*.

For these reasons, it was decided to limit the sample to only individuals with sex 2 and less than 25 years of age. As a result, 79 individuals remained. Individuals were grouped into age groups: 17 years old, 18 years old, 19 years old and older. It turned out that the distribution of IQ2 is not the same depending on age. This is shown in Fig. 1, *b*.

From the EEG data, the band power spectrums calculated using DFT and PSD were extracted from various channels and frequency ranges. There are 152 features in total. Since the records had different durations (minimum −2 min, maximum −6), the features were extracted for each minute. Observations for different minutes for one subject were considered as repeated. As a result, the sample size was 222.

Table 1. Sample distribution of subjects by gender and age.

Gender	Age							
	17	18	19	20	21	26	31	41
1	3	7	8	6	1	0	0	0
2	15	40	20	3	1	1	1	1

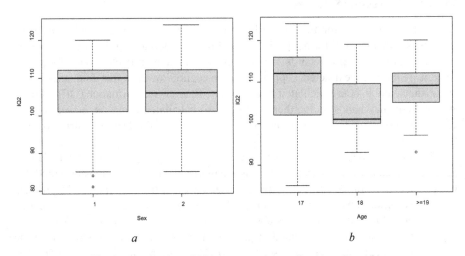

a   b

Fig. 1. Distribution of IQ2 in groups by gender (a) and age (b)

3.3 Estimated Model

During the analysis, it was found that the effect of the interaction of age and EEG features has a significant effect on IQ2. Therefore, the following model was taken as the basis:

$$y_i = \theta_0 + \sum_{j=1}^{J} \theta_j z_{ij} + \sum_{l=1}^{k} \alpha_l x_{il} + \sum_{l=1}^{k} \sum_{j=1}^{J} \gamma_{lj} x_{il} z_{ij} + \varepsilon_i, \tag{3}$$

where y_i are the IQ2 values for the i-th observation, z_{ij} is the i-th value of the binary variable reflecting the subject's belonging to group j by age, $J = 2$ taking into account the reduction (the smallest group is excluded -17 years old), x_{il} is the value of the l-th feature according to EEG data for of the i-th observation, k is the number of features taken from the EEG data, ε_i is a random error, $\theta_0, ..., \theta_J, \alpha_1, ..., \alpha_L, \gamma_{11}, ..., \gamma_{LJ}$ are the parameters to be estimated.

The model (3) is estimated using regression methods (least squares). The age factor must be included in the model. When EEG factors are included in the model, the effect of their interaction with age is also added. The selection of EEG factors is based on the usual and modified CFS method.

When choosing a correlation measure in relations (1) and (2), it was taken into account that the input features from the EEG data are quantitative, in addition, the response (IQ2), although it takes discrete values (integer numbers), but in a fairly wide range, so it can also be considered as a quantitative indicator. For these reasons, it is acceptable to use the usual Pearson correlation coefficient or the Spearman correlation coefficient (as done below), which provides greater resistance to outliers. When calculating correlations in relations (1) and (2), other factors included in the model (age and interaction effects) are also taken into account.

3.4 Results

First, forward selection and the genetic algorithm for the usual CFS method with objective function (1) were compared on the entire sample. It turned out that forward selection gives the best solution: the value of the objective function was 0.3269, and for the genetic algorithm it was equal to 0.3268. Figure 2 shows a plot of best and average values of the objective function at each step of the GA search. As a result of forward selection, 8 features were selected, and 12 features were selected based on the genetic algorithm. Thus, the use of a genetic algorithm leads to obvious overfitting.

The forward selection first selects the theta band power for channel T4, the delta band power for channel O1, the beta band power for channel T4, the beta band power for channel F4, calculated using DFT. Interestingly, the penultimate feature includes the beta band power for channel T4, calculated using PSD, which is highly correlated with the beta band power for channel T4, calculated using DFT (Spearman correlation coefficient is equal to 0.9914). This indicates that the usual CFS method is prone to overfitting and needs to be more flexible in order to be able to vary the complexity of the model.

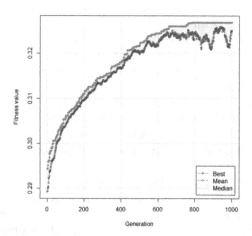

Fig. 2. Plot of best and average fitness values at each step of the GA search

Next, a comparison was made between the forward selection algorithm and the genetic algorithm when varying the values of the parameter c in the objective function (2) from 0.43 to 0.58 with a step of 0.005. In most cases, both algorithms gave the same results. The Table 2 shows those cases when the optimal values of the objective function differed. In general, the differences are not very large. At the same time, the forward selection algorithm has advantages in terms of execution time, does not require setting tuning parameters, and gives a deterministic result. Therefore, it was concluded that the direct selection algorithm is more suitable for solving the optimization problem under consideration. Further, only this algorithm is used.

Table 2. Comparison of forward selection algorithm and genetic algorithm.

c	Objective value from GA	Objective value from forward selection	The difference
0.440	0.575819	0.575818	1.13E–06
0.445	0.544239	0.544238	9.01E–07
0.460	0.462200	0.462199	1.10E–06
0.470	0.417635	0.417636	−2.53E–07
0.490	0.349778	0.349921	−1.42E–04
0.500	0.326816	0.326930	−1.14E–04
0.505	0.317640	0.317648	−7.12E–06
0.535	0.279142	0.279194	−5.22E–05

At the next stage, we checked how the variation of the parameter c in the modified CFS method affects the number of selected features. The value of parameter c varied from 0.43 to 0.58 with a step of 0.005. The graph of the obtained values of the number of selected features k is shown in Fig. 3.

It turned out that with the smallest value c, all features were selected, with the largest value c, the algorithm returns an empty set. At $c = 0.51$, the first 4 features listed above are added. In general, with a reasonable variation of the parameter, it is possible to choose the structure of the model that satisfies the researcher.

Let's check how prone to overfitting is the model built by the usual CFS method compared to the one modified using cross-validation. For this, different cross-validation procedures were used, as they gave slightly different results. The error in all cases was calculated as the average module of deviations of the observed response values from the predicted ones.

The simplest is the leave-one-out cross validation (LOOCV). Here the sample was drawn from subjects (rather than observations). The advantage of LOOCV is that it is a deterministic outcome, but only one subject remains per test sample. In addition, a 5-fold cross-validation with 100 repetitions was used. At each repetition, the sample of subjects was randomly shuffled. Finally, another approach is the Monte Carlo cross validation (MCCV), which involves random selection of subjects into training and test samples. The test sample was 20%. 500 repetitions were made. The disadvantage of this procedure is that the test samples may randomly overlap, and some subjects may not be included in the test sample at all.

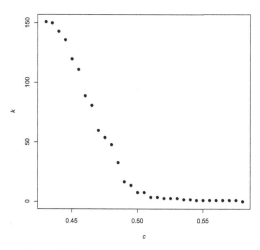

Fig. 3. Dependence of the number of selected features on the parameter c

Figure 4 shows the mean values of the absolute deviations of the observed response values from the predicted ones when the values of the parameter c vary from 0.49 to 0.58 with a step of 0.005.

However, in practice there is a problem of choosing the value of the parameter c. Hyperparameter selection should be separated from the model training procedure. Therefore, to evaluate the performance of the proposed approach, it is more correct to use nested cross-validation. To do this, the sample is divided into three parts: training, validation and test. Here they were divided in the ratio of 60%, 20% and 20%, respectively. On the training sample, features are selected and the parameters of the model (3) are estimated for various values of the parameter c. The test sample is used to estimate the prediction error for various values of the parameter c and to choose the optimal value of the parameter c that provides the smallest average error. Further, this optimal value is used in the procedure for selecting and estimating a regression model based on a sample, including training and validation together. Based on the obtained regression estimates, a prediction is built for the test sample. It is important that here the test sample is not involved in any way when choosing the parameter c, while in the cross-validation procedures described above, conclusions about the best or worse value of the parameter c were made based on the prediction error on the test sample. The procedure is based on MCCV, that is, the division into three parts was carried out randomly.

The inner cross-validation loop included 100 repetitions. At each iteration, the average absolute deviations of the predicted response values from the observed values was calculated for each value of the parameter c. Further, the error was averaged over 100 repetitions. The outer cross-validation loop also included 100 repetitions. It accumulated prediction errors for all test samples. As a result, the average of the absolute values of these errors was calculated.

For comparison, the performance of the usual CFS method (at $c = 0.5$), the forward selection method based on the BIC criterion, the LASSO and the Bolasso-S [18] were tested on the same test samples for 100 repetitions. In this case, the training sample

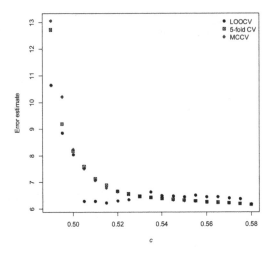

Fig. 4. Dependence of the error on the c parameter for various cross-validation procedures

included the rest of the observations, that is, the training and validation samples. The average of the absolute values of the errors are given in the Table 3. The standard errors of the mean are given in parentheses. In addition, at each iteration of the outer cross-validation loop, the number of EEG features included in the model was saved. The median values obtained over 100 repetitions are also shown in the Table 3. In parentheses are the first and third quartiles of the number of EEG features. It should be noted that when selecting based on the BIC, LASSO, Bolasso-S, the effects of the interaction of EEG factors and age did not necessarily correspond to the main effects of EEG factors. Therefore, when calculating the number of features, the number of main effects of features was taken from the EEG data in model (3).

Table 3. Comparison of feature selection methods using nested cross-validation.

Method	Mean absolute error	Number of EEG features
CFS	8.055 (0.173)	8 (7; 9)
modified CFS	6.109 (0.073)	1 (0; 1)
BIC (forward)	8.667 (0.370)	11 (6; 17)
LASSO	6.181 (0.065)	1 (1; 2)
Bolasso-S	6.165 (0.062)	0 (0; 0)

According to the results of nested cross-validation (Table 3), the proposed modification of the CFS method with the optimal selection of the parameter c in the objective function (2) gives the smallest prediction error. In the experiments, in 70% of cases, one feature was selected from the EEG data. If we use the proposed approach to feature selection to the entire original data, then this feature is the theta band power calculated using DFT for channel T4.

If we limit to choosing only one single feature, that is, with a fixed $k = 1$ in the model structure (3), then we can select a feature that gives the least error during the cross-validation procedure. For this, a 5-block cross-validation with 100 repetitions was chosen. It turned out that the beta band power calculated using DFT for channel F4 would be the best choice. The mean absolute error was 5.524 compared to 5.986 for the theta band power calculated using DFT for channel T4. Thus, the CFS method gives 1.08 times the worst result.

Finally, the obtained results are compared with the other regularized procedures: principal component regression (PCR) and Ridge regression. The number of components and the regularization parameter are chosen using nested cross-validation. The Ridge regression gives a mean absolute error equal to 5.918, the PCR gives the smallest mean absolute error 5.745. In all repetitions, one principal component is extracted.

4 Conclusion

The Correlation-based Feature Selection does not guarantee optimal feature selection in terms of the smallest error of IQ on test samples. However, by the proposed modification, it was possible to significantly improve the selection procedure: to make it more flexible. This made it possible to significantly reduce the error on test samples, thereby reducing the risk of model overfitting. As result, in terms of the mean absolute error calculated using nested cross-validation, the modified CFS is close to popular feature selection methods such as LASSO and Bolasso-S.

From the point of view of the task of estimating IQ using EEG data, the other regularized techniques, for example the PCR, give a smaller prediction error. This can be explained by the high correlation of the extracted features from close channels and frequency ranges. Therefore, in the future, the connectivity patterns of the brain resting state activity will be used to estimate the intelligence quotient.

Acknowledgments. The research is supported by Ministry of Science and Higher Education of Russian Federation (project No. FSUN-2020–0009).

References

1. Forsythe, C., Liao, H., Trumbo, M., Cardona-Rivera, R.E.: Cognitive neuroscience of human systems. Work and Everyday Life. CRC Press: Taylor&Frencis Group (2015)
2. Haier, R.J., Siegel, B., Tang, C., Abel, L., Buchsbaum, M.S.: Intelligence and changes in regional cerebral glucose metabolic rate following learning. Intelligence **16**, 415–426 (1992)
3. Zarjam, P., et al.: Estimating cognitive workload using wavelet entropy-based features during an arithmetic task. Comput. Biol. Med. **43**, 2186–2195 (2013)
4. Firooz, S., Setarehdan, S.K.: IQ estimation by means of EEG-fNIRS recordings during a logical-mathematical intelligence test. Comput. Biol. Med. **110**, 218–226 (2019)
5. Langer, N., Pedroni, A., Gianotti, L.R.R., Hänggi, J., Knoch, D., Jäncke, L.: Functional brain network efficiency predicts intelligence. Hum Brain Map **33**, 1393–1406 (2012)
6. Zakharov, I., Tabueva, A., Adamovich, T., Kovas, Y., Malykh, S.: Alpha band Resting-State EEG connectivity is associated with non-verbal intelligence: front. Hum. Neurosci. (2020)

7. Kruschwitz, J.D., Waller, L., Daedelow, L.S., Walter, H., Veer, I.M.: General, crystallized and fluid intelligence are not associated with functional global network efficiency: a replication study with the human connectome project 1200 data set. Neuroimage **171**, 323–331 (2018)

8. Sagaert, Y.R., Aghezzaf, E.H., Kourentzes, N., Desmet, B.: Tactical sales forecasting using a very large set of macroeconomic indicators. Eur. J. Oper. Res. **264**(2), 558–569 (2018)

9. Tibshirani, R.: Regression shrinkage and selection via the lasso. J. R. Stat. Soc. Ser. B (Methodol.) **58**(1), 267–288 (1996)

10. Zhang, Z.: Variable selection with stepwise and best subset approaches. Ann. Transl. Med. **4**, 136 (2016)

11. Hall, M.A.: Correlation-based feature selection for machine learning. Ph.D. thesis. University of Waikato, Hamilton (1999)

12. Sutter, J. M., Kalivas, J. H.: Comparison of forward selection, backward elimination, and generalized simulated annealing for variable selection: Microchemical journal, vol. 47(1–2), pp. 60–66 (1993)

13. Saidi, R., Bouaguel, W., Essoussi, N.: Hybrid feature selection method based on the genetic algorithm and pearson correlation coefficient. In: Hassanien, Aboul Ella (ed.) Machine Learning Paradigms: Theory and Application. SCI, vol. 801, pp. 3–24. Springer, Cham (2019). https://doi.org/10.1007/978-3-030-02357-7_1

14. Chuanlei, Z., et al.: Apple leaf disease identification using genetic algorithm and correlation based feature selection method. Int. J. Agric. Biol. Eng. **10**(2), 74–83 (2017)

15. Gershon, A., Devulapalli, P., Zonjy, B., Ghosh, K., Tatsuoka, C., Sahoo, S.S.: Computing functional brain connectivity in neurological disorders: efficient processing and retrieval of electrophysiological signal data. AMIA Jt Summits Transl. Sci. Proc. **2019**, 107–116 (2019)

16. Bao, F.S., Liu, X., Zhang, C.: PyEEG: an open source Python module for EEG/MEG feature extraction. Comput. Intell. Neurosci. 2011 (2011). art. 406391

17. Scrucca, L.: GA: a package for genetic algorithms in R. J. Stat. Softw. **53**, 1–37 (2013)

18. Bach, F.R.: Bolasso: model consistent lasso estimation through the bootstrap. In: Proceedings of the 25th international conference on Machine learning, pp. 33–40. Helsinki, Finland (2008)

Feature Weighting on EEG Signal by Artificial Bee Colony for Classification of Motor Imaginary Tasks

Demison Rolins de Souza Alves[1,2,4(⊠)] (ID), Otávio Noura Teixeira[2,3,4] (ID), and Cleison Daniel Silva[2,3] (ID)

[1] Electrical Engineering Post Graduation Program of the Federal University of Pará (UFPA), Belém, PA, Brazil
demison.alves@itec.ufpa.br
[2] Applied Computing Post Graduation Program of the Federal University of Pará (UFPA), Tucurui, PA, Brazil
{otaviont,cleison}@ufpa.br
[3] Tucuruí University Campus Federal University of Pará (UFPA), Tucurui, PA, Brazil
[4] Research Group on Evolutionary Algorithms (UFPA), Belém, PA, Brazil

Abstract. In this article, we propose an alternative approach to define degrees of freedom to improve feature extraction and signal classification for EEG-based brain-machine interface systems. The idea is to use the EEG signal information in the frequency domain and thus weight the sampled covariance matrices to highlight features. To perform this step, we use an auxiliary diagonal matrix in which the diagonal entries are parameterized by two Gaussian membership functions. The parameters are determined by the Artificial Bee Colony Algorithm (ABC). The classification of signals is performed by the Minimum Distance to the Riemannian Mean (MDRM) algorithm and the experiments use the dataset IIa of the IV international competition of brain-computer interfaces (BCI), in which the proposed approach is compared with the approach of the state of art. The results suggest that the proposed method can be used to extract significant features and improve the classification of motor imagery tasks.

Keywords: Brain-machine interface · Artificial Bee Colony · Riemann geometry · Covariance matrices

1 Introduction

The Brain-computer interfaces (BCI) are robust systems for interpreting brain electrical activity enabling a communication pathway without any muscle interactions [14]. In BCI systems the user's thoughts can be interpreted and used to give action to another device, however, the applications of BCI systems are not limited to this, we can observe several problem resolutions in the medical area

© Springer Nature Switzerland AG 2022
Y. Tan et al. (Eds.): ICSI 2022, LNCS 13345, pp. 301–310, 2022.
https://doi.org/10.1007/978-3-031-09726-3_27

for the prevention smoking, alcoholism [1]. Normally, for the acquisition of electrical brain signals, the electroencephalogram (EEG) is the selected biomedical reading technique as it is robust and mainly non-invasive [12]. A classic use of BCI systems is for the rehabilitation, commonly, the brain pattern for this task is that of motor imagery (MI) e.g. motor imagination of the hands, feet and tongue [13,14].

Unfortunately, the electrical brain activity obtained contains noise [12] that can reduce system performance, therefore, is necessary to apply robust preprocessing techniques, such as temporal filters, to the raw signal [13]. Currently, approaches using classifiers based on Riemannian geometry such as the Minimum Distance to Riemanian Mean (MDRM) are state-of-the-art for solving many problems in BCI systems [9]. Flexibility is an important aspect in BCI systems, as brain electrical activity has a high variance from person to person, so ideally the system should have high flexibility to maintain acceptable performance among a group of individuals [11].

In this paper, we improve feature extraction and classification steps for EEG-based brain-machine interface systems increasing your flexibility. The approach proposed by [10] is solved by the Artificial Bee Colony (ABC) plus two Gaussian membership functions to highlight discriminating features, frequencies, of the EEG signal. This approach uses the information in the frequency domain of the filtered EEG signals introducing degrees of freedom to the signal that builds the covariance matrices by a linear transformation to finally, classify the EEG signal by the MDRM Algorithm.

In Sect. 2, we discuss the decomposition of the EEG signals using sinusoidal components and their interpretation for feature weighting and classification using Riemann distance. In Sect. 3, we describe the ABC. In Sect. 4, the experiment. Finally, we provide conclusions about the developed method.

2   Brain-Computer Interface Approach

In BCI systems, normally, a chain of procedures is performed following these steps: preprocessing, feature extraction, feature selection and classification. In this section we will present our approach to optimizing the classification of EEG signals.

2.1   Preprocessing by Signal Decomposition Using a Sinusoidal Basis

In this approach described by [10], there is an alternative to classic temporal filters. For this, a projection of the EEG signal is performed by a base of orthogonal signals formed by sinusoids. The methodology is interesting for three reasons: emulating bandpass filtering, easy access to the spectral content of EEG signals and reduced computational cost compared to classical approaches. We can describe it by starting the representation an EEG signal segment as a matrix $Z \in \mathbb{R}^{p \times q}$, where p describes the number of sensors and q the number of samples recorded from a sampling rate f_s. The signal is projected in the base defined by $2m$ distinct sinusoidal signals for sine and cosine belonging to the set of frequencies of interest $\mathbf{F} = \{f_1, ..., f_k, ..., f_m\}$, with $k = 1, 2, ..., m.$, such that $f_l \geq 0$; $f_u \leq \frac{f_s}{2}$; $f_l \leq f_k \leq f_u$ and finally $\Delta_f = |f_k - f_{k+1}| = \frac{f_s}{q}$. Where f_l is the minimum frequency; f_u is the maximum frequency and Δ_f is the minimum interval between frequencies.

A minimization problem is solved to have the new representation of the signal in the frequency domain filtered in the desired frequency range by the matrix $\Gamma \in \mathbb{R}^{p \times 2m}$ where the spectral content is explicit. In each column, i and $i + m$, with $i = 1, 2, ..., m$ corresponds to the same frequency in F representing the filtered Z segment and in the frequency domain. In this work, the spectral information of the signal by Γ is used in the next steps.

2.2   Feature Extraction by Parameterized Covariance Matrices

Feature extraction is an important step to highlight features to improve the classification step. In [10], the EEG signal covariance matrices are parameterized in the frequency domain. This parameterization is given by:

$$cov\{\Gamma\} = \Gamma W \Gamma^T, W \succeq 0, \tag{1}$$

where $cov\{\Gamma\} \in \mathbb{R}^{p \times p}$ is the covariance matrix of Γ constructed with an auxiliary matrix. The matrix $W \in \mathbb{R}^{2m \times 2m}$ has a constraint of being semi-definite positive symmetric and can be used to weight the frequency information contained in Γ to highlight discriminant information.

2.3   Defining the W Weight Matrix

An efficient way to define W with a semi-defined positive symmetric matrix is to use it as a positive diagonal matrix, however, it can still be computationally expensive and more susceptible to overfitting. A way around this is to adopt two Gaussian membership functions to define the elements of the diagonal matrix.

The gaussian membership function is popularly presented in fuzzy systems. In this work, we reduce the computational cost of the process to optimize the parameters of W adopting two strategies: the classic in Eq. (2) and the modified by Eq. (3):

$$y(w, c, \sigma) = e^{-\frac{1}{2} \frac{(w-c)^2}{\sigma^2}}, \tag{2}$$

where w is the frequency; c is the mean and σ the standard deviation.

$$k(w, c_1, c_2, \sigma_1, \sigma_2) = \max(e^{-\frac{1}{2} \frac{(w-c_1)^2}{\sigma_1^2}}, e^{-\frac{1}{2} \frac{(w-c_2)^2}{\sigma_2^2}}), \tag{3}$$

where w is the frequency; c_1 is mean 1; c_2 mean 2; σ_1 standard deviation 1 and σ_2 standard deviation 2.

Each element of W is defined by Eq. (2) or Eq. (3), note on the diagonal of W have $2m$ parameters, however, the sine and cosine values are associated with the same frequency in F so the weights will be equal, for example, using Eq. (2) we can define the matrix W following this rules: $A = \text{concatenation}(F, F)$, $W = diag(y(a_1), y(a_2), ..., y(a_{2m}))$. The parameters of mean and standard deviation used to construct those functions are defined by ABC which will be discussed in the next section.

2.4   Classification by Riemann Distance

Proposed by [2] the MDRM uses covariance matrices as features for classification, this technique uses a riemannian distance metric obtained directly from the geometric space of the covariance matrices. The modeling of the metric for the minimum distance to the mean is somewhat generic, but using it through a Riemannian metric is interesting because of the intricate properties of covariance matrices, as they are part of the Riemannian manifold of definite positive symmetric matrices.

The riemann distance is expressed by the length of the geodesic which is the smallest possible curve to connect two points in a Riemannian manifold by:

$$\delta_R^2(P_1, P_2) = \| log(P_1^{-1}P_2) \|_F^2 = \sum_{i=1}^{p} log^2 \lambda_i, \tag{4}$$

where P_1 and P_2 are any two points (covariance matrices) and $\|\ \|_F$ the frobenius norm; λ_i the eigenvalues of the matrix $P_1^{-1}P_2$.

MDRM needs classification reference points obtained through the mean geometry of the covariance matrices by:

$$M_g^* = \operatorname*{argmin}_{M_g} \sum_{i=1}^{j} \delta_R^2(M_g, cov\{\Gamma_i\}), \tag{5}$$

where j is the number of examples, for each class we have a mean matrix, for example, assuming there are two classes Υ and Ψ the matrices $M_{g\Upsilon}$ and $M_{g\Psi}$ are reference points used to identify a segment of the signal class. To define whether a segment of the signal Ω belongs to class Υ or Ψ, just calculate the Riemann distance by Eq. (4), between the means and Ω if it is closer to $M_{g\Upsilon}$ then $\Omega \in \Upsilon$ otherwise $\Omega \in \Psi$.

It is interesting to highlight that MDRM is a parameter-free algorithm and the classic approach only needs temporal filters, so the methodology adopted in this work proposed by [10] is a way to make it more flexible.

3   Artificial Bee Colony

The Artificial Bee Colony (ABC) proposed by [6] is a classic algorithm in swarm intelligence performing an abstraction of the process for searching and selecting honey bees forage. For the construction of the algorithm, it is necessary to use four phases: initialization, employed bees, onlooker bees, scout bees. In the initialization, the solutions are generated randomly and the others are performed iteratively refining the solutions. To control the Algorithm it needs three parameters: maximum number of cycles (MCN), number of solutions (SN) and a threshold for solutions trapped in local traps $(LIMIT)$ following the description of [5–8] we can describe the algorithm in the next subsections.

3.1 Initialization

The process for building the initial population of solutions is defined uniformly randomly in the search space, which can be written as:

$$x_{ij} = l_j + U(0,1)(u_j - l_j), \tag{6}$$

where j is the j-th variable of decision of i-th solution; $1 \leq i \leq SN$; $1 \leq j \leq D$ and (u,l) are the upper and lower limits of the problem. Then the limit values for each solution are set to 0.

3.2 Employed Bees Phase

The first group, employed bees, of the algorithm are used to search for food sources (solutions) in space based on their memory and visual information creating new solutions v_i for each solution x_i , $i = 1, 2, ..., SN$ modifying only one dimension in space of decision variables given by:

$$v_{ij} = x_{ij} + \phi(x_{ij} - x_{kj}), \tag{7}$$

where $k \neq i$ is a randomly chosen index and ϕ is a uniform distributed random number between $[-1, 1]$. Then, for each solution generated, a greedy strategy is applied between v_i and x_i, if the solution v_i is better then it will replace the solution x_i and the limit for x_i will be reset to 0, otherwise it will be discarded and the limit for x_i will increase by 1.

3.3 Onlooker Bees Phase

The second group, onlooker's bees, use information from food sources to choose and explore it based on objective function values. For this, a selection based on the classic roulette wheel method is applied. The probability p_i can be defined as:

$$p_i = \frac{fit_i}{\sum\limits_{n=1}^{SN} fit_n}, \tag{8}$$

where fit_i is proportional to the value of the objective function of solution x_i. Then, the same process described above for employed bees is applied.

3.4 Scout Bees Phase

The third group, scout bees, are applied as a mechanism to escape local traps. In the optimization process, each solution has a (limit counter) to check if there has been an improvement. If the solution with the highest limit value has exceeded or equaled the value of the LIMIT parameter, it will be restarted by Eq. (6).

4 Experiment

4.1 Dataset

Dataset 2a from the IV international competition of brain-computer interface systems presented by [3] is used with EEG signals from 9 subjects, each performed four distinct cognitive tasks, left hand movement imagination (class 1), right hand (class 2), feet (class 3) and tongue (class 4). The number of EEG electrodes placed was equal to 22 described by a sampling rate 250 Hz, in this dataset there is a division between training (**1**) and test (**2**) set, for which each has 72 tests for each imaginary motor task described above.

In this work, we will only use data from the imagery of the right and left hands, making use of two scenarios A: training with set 1 and testing with 2 and B: training with set 2 and testing with 1 for each subject.

4.2 Experiment Settings

Preprocessing of EEG Data. The window extracted starts at 0.5 to 2.5 s after the cue; totaling 500 samples; $f_l = 8$ Hz and $f_u = 30$ Hz; spacing between frequencies $\Delta_f = 0.5$; m = 45 distinct frequencies of the F set of frequencies $\{8, 8.5, 9, ..., 29.5, 30\}$ according to guidelines of [10].

Artificial Bee Colony Algorithm Parameters

- **Dimensions:** the number of dimensions is proportional to the strategy needed, for example, to construct the Gaussian presented in Eq. (2) we need two parameters, while for Eq. (3) we need four. The two strategies are represented, respectively, by ABC_{2d} and ABC_{4d}.
- **Search Space:** the range defined for each decision variable was stipulated in a real value for the means of $[f_1, f_m]$ and standard deviation of $[1,10]$, however, the mean values are rounded in construction of the Gaussian for the closest number belonging to the set of frequencies F.
- **Runs:** 10 independent executions were performed for each subject in scenarios A and B. For the same subject in scenarios A and B, the initial solutions were the same (same seed)
- **SN:** number of solutions is equal to 10.
- **LIMIT:** is equal to SN * dimension.
- **MCN:** 50 iterations.

4.3   Objective Function and Wrapper Strategy

As an objective function, we use the stratified 10-Fold cross-validation accuracy linked to the Wrapper strategy which is an approach to receive feedback from the learning algorithm. Then, iteratively, the features are weighted. In Fig. 1, we can see how the Wrapper is happening: iteratively W matrix configurations are evaluated and selected by applying the proposed ABC_{2d} or ABC_{4d} strategies after the stopping criterion (MCN) is reached by ABC, we use the best W matrix found to train (using the full training base) and finally the evaluation for the test set is performed.

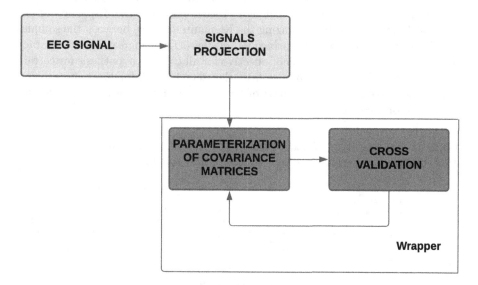

Fig. 1. Strategy Wrapper applied to classify EEG signals; In blue the iterative search process for parameter optimization. (Color figure online)

4.4   Numerical Results

The Tables 1 and 2 show classification rates in scenarios A and B using descriptive statistics of cross-validation results for strategies ABC_{2d} and ABC_{4d}. The data sets A6, A7, A8, A9, B3, B4, B5, B6, B8 obtained a good estimate of the results for the test set.

In Tables 4 and 5 we can see the results obtained for the test set. Adopting four strategies: W = ABC_{2d}, W = ABC_{4d}, W = I (Identity, original spectral information) and the results of state-of-the-art approach by MDRM of [4]. The strategies proposed in this work obtained equivalent or superior results for the data set: A3, A5, A6, A7, A8, A9, B1, B2, B3, B4, B5, B6, B7, B8, B9. It is worth highlighting the performance of A7, B5, B7 where there were significant gains in accuracy.

The two-tailed Wilcoxon Signed Rank test was applied between the two proposals ABC_{2d} and ABC_{4d} adopting a significance level for $\alpha = 0.05$ look at Table 3 for data sets A3, A5, A7, A9, B2, B6, B7, B8 the null hypothesis was rejected, so there is a significant difference between the two approaches. In this sets $W = ABC_{4d}$ presented the best accuracy performance for five cases A5, A7, A9, B2 and B7. This might suggest that $W = ABC_{4d}$ is a better strategy to define degrees of freedom. The results obtained are competitive with the classical method by Riemannian Geometry (MDRM) showing an overall superior average performance of 77.57% and 77.86% respectively by $W = ABC_{2d}$ and $W = ABC_{4d}$ versus 76.43% by [4] and 75.49% by $W = I$.

The approach used in this work adjusting the position in space of the geometric mean of the covariance matrices of each class and the dispersion of the samples. In this way, can minimizing the Riemann distance between the samples and the geometric mean of its corresponding class. Thus, increasing the accuracy of the system. However, it was not effective for all cases, a hypothesis for certain subjects is the variance of samples from datasets 1 and 2 for the same subject. This could justify the effectiveness of the proposal for scenario A and not for scenario B or vice versa.

Table 1. Accuracy results obtained from the Artificial Bee Colony Algorithm using $W = ABC_{2d}$ in scenarios **A** and **B** for the *Stratified 10-fold cross-validation*

$W = ABC_{2d}$	A1	A2	A3	A4	A5	A6	A7	A8	A9	B1	B2	B3	B4	B5	B6	B7	B8	B9
Mean	95.07	62.87	99.33	74.53	79.74	74.61	81.24	97.19	93.10	95.76	65.90	97.95	75.23	72.27	68.75	93.10	97.95	90.83
Median	95.14	62.62	99.33	74.76	79.81	74.38	81.24	97.19	93.10	95.76	65.90	97.95	75.19	72.29	68.62	93.05	97.95	90.95
Std	0.23	0.36	0.02	0.30	0.21	0.33	0.00	0.00	0.00	0.02	0.00	0.00	0.17	0.02	0.28	0.22	0.00	0.26
Best	95.14	63.29	99.33	74.76	79.81	75.00	81.24	97.19	93.10	95.76	65.90	97.95	75.71	72.29	69.29	93.71	97.95	91.00
Worst	94.43	62.57	99.29	74.19	79.14	74.33	81.24	97.19	93.10	95.71	65.90	97.95	75.14	72.24	68.62	93.00	97.95	90.33

Table 2. Accuracy results obtained from the Artificial Bee Colony algorithm using $W = ABC_{4d}$ in scenarios **A** and **B** for the *Stratified 10-fold cross-validation*

$W = ABC_{4d}$	A1	A2	A3	A4	A5	A6	A7	A8	A9	B1	B2	B3	B4	B5	B6	B7	B8	B9
Mean	96.73	63.73	99.36	75.53	79.33	74.59	81.38	97.26	93.65	96.24	66.02	98.02	76.59	73.70	70.19	94.36	98.35	91.13
Median	96.52	63.95	99.29	75.57	79.14	74.43	81.29	97.19	93.76	96.43	65.93	97.95	76.52	73.64	70.05	94.33	98.62	91.00
Std	1.38	0.34	0.40	0.70	0.77	0.33	0.61	0.23	0.63	0.33	1.63	0.21	0.74	0.86	0.72	0.56	0.34	0.28
Best	98.67	63.95	100.00	76.90	80.43	75.10	81.95	97.90	94.48	96.52	68.86	98.62	77.90	75.10	71.43	95.05	98.62	91.67
Worst	95.05	63.24	98.62	74.86	77.76	74.33	79.90	97.19	92.43	95.76	64.00	97.95	75.14	72.48	69.33	93.62	97.95	91.00

Table 3. Wilcoxon Signed-Rank test results for $W = ABC_{2d}$ versus $W = ABC_{4d}$ in scenarios **A** and **B** for test set

	A1	A2	A3	A4	A5	A6	A7	A8	A9	B1	B2	B3	B4	B5	B6	B7	B8	B9
p-value	0.13	0.12	**0.04**	0.09	**0.01**	0.51	**0.04**	0.65	**0.00**	0.40	**0.03**	0.40	0.08	0.06	**0.01**	**0.00**	**0.04**	0.37

Table 4. Classification results of the EEG signals in terms of accuracy for the IIa data of the IV Competition of BCI systems (Scenario A) for four approaches: $W = I$; $W = ABC_{2d}$; $W = ABC_{4d}$ and Reference [4]

	A1	A2	A3	A4	A5	A6	A7	A8	A9
$W = ABC_{2d}$	88.96 ± 0.22	57.57 ± 1.92	93.68 ± 0.22	70.07 ± 1.89	66.67 ± 0.00	73.47 ± 0.64	82.43 ± 0.57	**96.60 ± 0.22**	90.28 ± 0.00
$W = ABC_{4d}$	90.00 ± 1.97	59.17 ± 2.22	92.85 ± 1.43	68.82 ± 1.01	**67.85 ± 0.93**	72.99 ± 1.06	**83.33 ± 1.22**	96.53 ± 0.57	**92.15 ± 0.57**
$W = I$	93.05	62.5	91.66	73.61	57.63	71.52	65.27	96.52	91.66
REF	**93.75**	**63.19**	**94.44**	**75.00**	63.19	71.53	72.92	96.53	91.67

Table 5. Classification results of the EEG signals in terms of accuracy for the IIa data of the IV Competition of BCI systems (Scenario B) for four approaches: $W = I$; $W = ABC_{2d}$; $W = ABC_{4d}$ and Reference [4]

	B1	B2	B3	B4	B5	B6	B7	B8	B9
$W = ABC_{2d}$	**78.40 ± 0.51**	53.06 ± 1.61	**93.12 ± 0.69**	**67.50 ± 0.91**	**72.15 ± 1.01**	65.49 ± 2.12	72.71 ± 2.91	93.40 ± 0.37	80.76 ± 1.31
$W = ABC_{4d}$	78.33 ± 3.61	**55.62 ± 3.82**	92.57 ± 1.88	66.32 ± 1.15	70.00 ± 3.38	62.15 ± 1.77	**79.65 ± 1.09**	92.57 ± 0.93	80.62 ± 0.83
$W = I$	75.0	50.00	90.27	63.19	65.27	63.88	74.30	93.05	80.55
REF	74.31	50.00	88.89	65.28	63.89	61.81	73.61	**94.44**	**81.25**

5 Conclusion

In this paper, we define degrees of freedom to improve feature extraction and classification steps for EEG-based brain-machine interface systems. These parameters are used to weight the spectral information of the EEG signal to try to extract discriminant features to optimize the classification step. We adopt the use of the Artificial Bee Colony Algorithm to construct two Gaussian membership functions that determine the diagonal of the auxiliary parameter matrix W.

The statistical results suggest that both approaches $W = ABC_{2d}$ and $W = ABC_{4d}$ are competitive with the classical approach of [4] and $W = I$ with considerable accuracy gains for certain subjects. The proposed method can be considered as an alternative, especially when the state-of-the-art approach by MDRM method does not present an accuracy performance above 90%. Thus, it is interesting to search alternatives to increase the robustness of the BCI system.

Future works consider investigating the performance of the presented method for multi-class classification and exploring new ways to determine the W parameter matrix to optimize the EEG signal classification task.

Acknowledgment. This work is supported by PROPESP/UFPA (PAPQ).

References

1. Abdulkader, S.N., Atia, A., Mostafa, M.S.M.: Brain computer interfacing: applications and challenges. Egypt. Inform. J. **16**(2), 213–230 (2015). https://www.sciencedirect.com/science/article/pii/S1110866515000237
2. Barachant, A., Bonnet, S., Congedo, M., Jutten, C.: Multiclass brain-computer interface classification by Riemannian geometry. IEEE Trans. Biomed. Eng. **59**(4), 920–928 (2011)

3. Brunner, C., Leeb, R., Müller-Putz, G., Schlögl, A., Pfurtscheller, G.: BCI competition 2008-Graz data set a. Institute for Knowledge Discovery (Laboratory of Brain-Computer Interfaces), Graz University of Technology, vol. 16, pp. 1–6 (2008)
4. Congedo, M., Barachant, A., Andreev, A.: A new generation of brain-computer interface based on Riemannian geometry. CoRR abs/1310.8115 (2013). http://arxiv.org/abs/1310.8115
5. Karaboga, D., Basturk, B.: On the performance of artificial bee colony (ABC) algorithm. Appl. Soft Comput. 8(1), 687–697 (2008). https://doi.org/10.1016/j.asoc.2007.05.007, https://www.sciencedirect.com/science/article/pii/S1568494607000531
6. Karaboga, D.: An idea based on honey bee swarm for numerical optimization. Technical report, Technical report-tr06, Erciyes university, engineering faculty, computer ... (2005)
7. Karaboga, D., Akay, B.: A comparative study of artificial bee colony algorithm. Appl. Math. Comput. 214(1), 108–132 (2009). https://doi.org/10.1016/j.amc.2009.03.090, https://www.sciencedirect.com/science/article/pii/S0096300309002860
8. Karaboga, D., Basturk, B.: A powerful and efficient algorithm for numerical function optimization: artificial bee colony (ABC) algorithm. J. Glob. Optim. 39(3), 459–471 (2007). https://doi.org/10.1007/s10898-007-9149-x
9. Lotte, F., et al.: A review of classification algorithms for EEG-based brain-computer interfaces: a 10 year update. J. Neural Eng. 15(3), 031005 (2018)
10. Silva, C., Duarte, R., Goulart, R., Trofino, A.: Towards a LMI approach to feature extraction improvements and classification by Riemann distance. In: 2016 12th IEEE International Conference on Control and Automation (ICCA) pp. 990–995, June 2016. https://doi.org/10.1109/ICCA.2016.7505409
11. Silva, C., Duarte, R., Trofino, A.: Feature extraction improvements using an LMI approach and Riemannian geometry tools: an application to BCI. In: 2016 IEEE Conference on Control Applications (CCA), pp. 966–971, September 2016. https://doi.org/10.1109/CCA.2016.7587938
12. Teplan, M., et al.: Fundamentals of EEG measurement. Meas. Sci. Rev. 2(2), 1–11 (2002)
13. Wolpaw, J., Wolpaw, E.W.: Brain-Computer Interfaces: Principles and Practice. OUP, USA (2012)
14. Wolpaw, J.R., Birbaumer, N., McFarland, D.J., Pfurtscheller, G., Vaughan, T.M.: Brain-computer interfaces for communication and control. Clin. Neurophysiol. 113(6), 767–791 (2002)

A Lung Segmentation Method Based on an Improved Convex Hull Algorithm Combined with Non-uniform Rational B-Sample

Xianghang Shi[1,4], Jing Liu[2], Jingzhou Xu[3], and Mingli Lu[4(✉)]

[1] School of Mechanical Engineering, Yancheng Institute of Technology,
Yancheng, People's Republic of China
[2] The Affiliated Infectious Diseases Hospital of Soochow University, The Fifth
People's Hospital of Suzhou, Suzhou, Jiangsu, People's Republic of China
[3] Information and Computing Science School of Advanced Technology,
Xian Jiaotong-Liverpool University, Suzhou, China
[4] School of Electrical and Automatic Engineering, Changshu Institute of Technology,
Changshu, People's Republic of China
luml@cslg.edu.cn

Abstract. Currently, tuberculosis (TB) remains one of the major threats to people's health. Specifically, the problem of under-segmentation due to adhesion of pulmonary tuberculosis lesions to the pleura is a thorny problem in image segmentation. In this paper, An effective lung parenchyma patching method is proposed, which is composed of an improved convex hull algorithm with non-uniform rational B-splines. Our method is mainly divided into three parts. First, the temporal image processing method is used to preliminarily segment the lung parenchyma. Then, the lesion area was discriminated based on the convex hull algorithm and discrete point derivative frequency. Finally, the NURBS fitting method is introduced to complete the fitting of the defect contour. According to our experimental results, the completed lesion contour blends naturally with the original lung contour. Compared with some existing algorithms, our method performs better.

Keywords: Lung segmentation · Convex hull algorithm · Non-uniform ratioinal B-splines

1 Introduction

In this paper, we focus on the patching of the edges of pleural adhesions in TB lesions, followed by the completion of lung segmentation. However, these lesions are distributed at the edges of the lung and often show almost the same pixel values as the chest cavity [1], which makes it difficult to find pixel feature information that we can use.

© Springer Nature Switzerland AG 2022
Y. Tan et al. (Eds.): ICSI 2022, LNCS 13345, pp. 311–319, 2022.
https://doi.org/10.1007/978-3-031-09726-3_28

In order to solve the problem that similar pixels cannot be segment, it was proposed to patch the contour before lung segmentation. Ammi [2] proposed an automatic lung segmentation algorithm based on improved convex hull algorithm and mathematical morphology technique. Shen [3] used the bidirectional differential chain code combined with the support vector machine. The method complements the lung parenchyma. Liu [4] proposed a lung image segmentation algorithm based on random forest method and multi-scale edge detection technology.

The main motivation of this paper is to adopt three methods of temporal image processing, improved convex hull algorithm and NURBS fitting. Use these methods to gradually realize the process from lung parenchyma completion to segmentation. Thus, we make the following two contributions: 1) It is proposed to calculate the frequency of slope change based on equal interval sampling, so as to effectively judge whether the concave area is the area to be completed. 2) The NURBS fitting method is introduced to make the contour patching accurate and efficient.

The rest of the paper is organized as follows: Sect. 2 details our distinguishing and completion methods in the tuberculosis region. Section 3 provides our experimental results, and Sect. 4 summarizes our method about this paper.

2   Methods

As shown in Fig. 1. The overall framework of our work content can be divided into three parts. First, the CT images are preprocessed and the lung parenchyma is segmented. The second step is to determine the area that needs to be completed according to our proposed method and implement the completion. Finally, the complete lung parenchyma can be segmented by filling the lung mask.

Fig. 1. Overall process block diagram

2.1   Primary Segmentation of Lung Parenchyma

We use the superposition of multiple image processing methods. The main processing steps are summarized in Fig. 2. The rough steps are described as: Binarize the image using the maximum between-class variance method [5]. Elimination of vascular shadows in lung parenchyma using closure operation [6]. Extract the largest connected area [7]. Fill the thoracic region according to the flood fill method to get its mask [8]. Subtract the results of the above two steps. Set the connected area threshold to 800, this can remove small artifacts such as trachea and medical equipment. Perform a bitwise AND operation on the mask and the original image to obtain a preliminary lung parenchyma segmentation image.

According to the process flow of Fig. 2, we will show the results of each step as shown in Fig. 3. Follow the direction the arrow points, The methods of treatment are as follows: Original image input, Gaussian filtering, binarization, morphological operations, connected region selection, hole filling, mask acquisition, edge extraction, and lung parenchyma extraction.

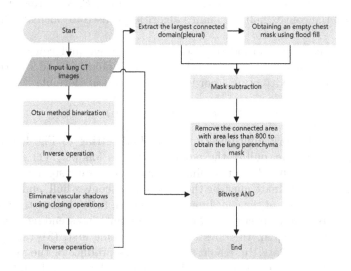

Fig. 2. Lung parenchyma initial segmentation steps

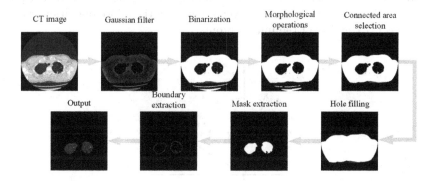

Fig. 3. Effect of the process of primary segmentation of the lung parenchyma

2.2 Lung Segmentation Method Based on Improved Convex Hull Algorithm and NURBS

This method will be described in the following way. Compute the convex hull of the lung from the set of lung contour edge points. Identify the mediastinal side and other recessed areas. Identify true and false lesions and solve the over-segmentation problem caused by false lesions. Locate the points on both sides of the area to be completed (lesions), Then use non-uniform rational B-splines to approximate the positioning points.

2.2.1   Find the Convex Hull of the Lung Parenchyma

Before calculating the convex hull, we need to use edge detection technology to find all the points on the lung contour and set them as P. After that, in order to quickly find the convex hull, Graham's algorithm is used to calculate the convex hull.

Graham scanning method to discriminate concave and convex points [9]: Suppose two points $p_0, p_1 \in P$, take the third point p_2 and determine the concavity and convexity of p_1 according to the following formula.

$$r = (p_2 - p_0) \times (p_1 - p_0) \tag{1}$$

where "$\times$" is the fork multiplication symbol. Referring to Fig. 4, the idea of calculating the convex hull of the point set P is as follows: (1) All points on the point set P are placed in a two-dimensional coordinate system, then the point with the smallest ordinate must be the point on the convex hull, and this point is specified as the origin of the coordinate, let it be p_0. (2) Calculate the arguments of other points with respect to p_0, arrange the points in ascending order. When the arguments are equal, the ones closest to p_0 come first. (3) Take p_2 as the current point, Calculate the result r of the cross product of the vectors $\overrightarrow{p_0 p_1}$, $\overrightarrow{p_0 p_2}$. If $r < 0$ then determine that p_1 is a point on the convex hull. Conversely, it is not a point on the convex hull. If $r = 0$, the point farther from the origin is considered to be the point on the convex hull. (4) Checking if each current point is the last element in the point set. Keep repeating the above operation, we will find the convex hull Q of the lung contour.

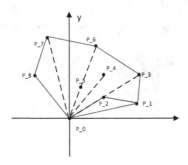

Fig. 4. Graham scan method to find convex hull

2.2.2 Eliminate Pseudo-lesions Based on Frequency of Slope Change

Combined with the description of Fig. 5. First, the purple line represents the convex hull contour of the left and right lung parenchyma. In general, the distance between the convex hull points at both ends of the mediastinum side C is the largest adjacent distance in Q, so we set the threshold dis according to prior knowledge to exclude the mediastinal side. However, the lung contour also contains true lesions A and pseudo lesions B. According to the analysis, the difference between true lesions and pseudo lesions lies in the complexity of the edges of the recessed areas. Therefore, we sample the edges of A and B with an interval of t and calculate the slope between their two adjacent points and count the frequency of positive and negative slope changes. The frequency of slope change between sampling points is defined as follows.

$$\omega = \begin{cases} \omega + 1 & \frac{s_{n-1}(y)-s_n(y)}{s_{n-1}(x)-s_n(x)} \times \frac{s_n(y)-s_{n+1}(y)}{s_n(x)-s_{n+1}(x)} < 0 \\ \omega & \frac{s_{n-1}(y)-s_n(y)}{s_{n-1}(x)-s_n(x)} \times \frac{s_n(y)-s_{n+1}(y)}{s_n(x)-s_{n+1}(x)} > 0 \end{cases} \tag{2}$$

$$S = \{s_1, s_2...s_n, s_{n+1}\}$$

Among them, ω is the frequency of slope change, $\omega_{initial} = 0$. If the slope values of $s_{n-1}s_n$ and s_ns_{n+1} have opposite signs, then $\omega = \omega + 1$. S is the lung parenchyma outline point set of the area to be determined, and s_i ($i \in \{1,2,3...n-1,n,n+1\}$) is the point on it.

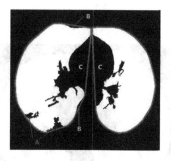

Fig. 5. Lung area description

2.2.3 Completing the Lung Parenchyma Based on NURBS

A NURBS curve of degree k is defined as [10]:

$$l(u) = \frac{\sum\limits_{i=0}^{n} N_{i,k}(u)\lambda_i d_i}{\sum\limits_{i=0}^{n} N_{i,k}(u)\lambda_i} \tag{3}$$

where $d_i (i = 0, 1, ..., n)$ is the control point of the curve. $N_{i,k}(u)$ is called a canonical B-spline basis function of degree k. It is determined by a non-decreasing parameter sequence $U : u_0 \leq u_1 \leq \cdots \leq u_{n+k+1}$ called the node vector u. $\lambda_i(i = 0, 1, ...n)$represents the weight of each control point

The basis functions in NURBS are evaluated using the Cox-de Boor[11] recursion algorithm as described in Eq. (4).

$$\begin{cases} N_{i,0}(u) = \begin{cases} 1 & if u_i \leq u \leq u_{i+1} \\ 0 & otherwise \end{cases} \\ N_{i,k}(u) = \frac{u-u_i}{u_{i+k}-u_i}N_{i,k-1}(u) + \frac{u_{i+k+1}-u}{u_{i+k+1}-u_{i+1}}N_{i+1,k-1}(u) \end{cases} \quad (4)$$

The formula shows that to obtain the i -th B-spline $N_{i,k}(u)$ of k -th power, a total of $k + 2$ nodes $u_i, u_{i+1}, ...u_{i+k+1}$ are required.

In order to reduce the influence of the control point weight on the curve, it is not discussed in the present invention, and all of them are set to 1 by default. So it is simplified to $l(u)$

$$l(u) = \sum_{i=0}^{n} N_{i,k}(u)d_i \quad (5)$$

As shown in Fig. 6, the red points represent the control points we located, The lung contour is fitted according to these points. The yellow curve represents the fitting result.

After the fitting is completed, using flood fill to fill in the hole area, so that lung mask is obtained. Lung parenchyma was extracted from the original image according to the mask.

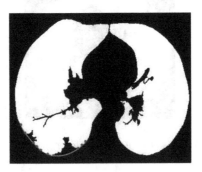

Fig. 6. The effect of NURBS fitting

3   Experimental Results

In this section, we will specifically discuss the implementation process to verify the effectiveness of our proposed method. The programming language is python 3.6 in the experiment. The experimental data were selected from CT scan images of lungs containing pleural adhesions of tuberculosis lesions. The images were 512 * 512 pixels. Our experimental parameters are as follows: $10 < dis < 60$ is the depressed region to be judged, $t = 5$ is the sampling interval of the depressed edge points, $\omega = 8$ indicates the frequency of positive and negative slope changes between the sampling points, and $k = 3$ indicates a degree of the curve fitted.

The experimental results are shown in Fig. 7. Lines 1 and 2 represent common and smaller types of lesions. Although the lesions in the third row seem to have regular features, their edges are also more tortuous at the pixel level. The fourth row belongs to the rare lesion location. The lesions in the fifth row have a larger area of damage.

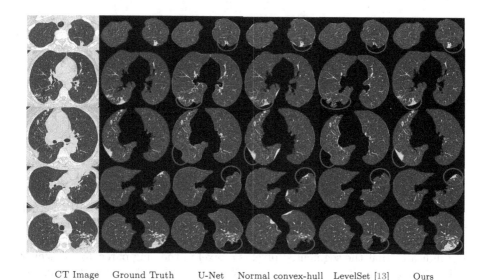

CT Image   Ground Truth   U-Net   Normal convex-hull   LevelSet [13]   Ours

Fig. 7. Visual comparison of lung segmentation results.

As far as this study is concerned, although U-net is an excellent image segmentation model, it is almost incapable of segmentation in areas with very close pixel values, resulting in a lot of under-segmentation. The normal convex hull algorithm often has the problem of over-segmentation, and the completed contours are all straight lines. This reduces the accuracy of the segmentation. The level set method mainly obtains the image boundary through continuous iteration, which causes the expansion curve to stop expanding in the lesion area. Our method can better distinguish true and false lesions, and this method makes the fitted contour more realistic and accurate.

We randomly selected 100 images of pulmonary tuberculosis lesions containing pleural adhesions as experimental data. Four methods were evaluated according to four evaluation indicators. The statistical object is the entire lung parenchyma, so the proportion of lesions is relatively small. The evaluation indicators are defined as formula (6). The statistical results are shown in Fig. 8.

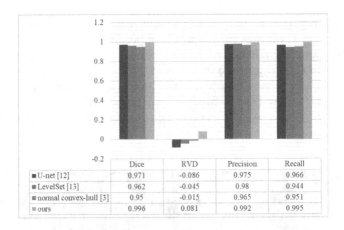

The table within the figure:

	Dice	RVD	Precision	Recall
■ U-net [12]	0.971	-0.086	0.975	0.966
■ LevelSet [13]	0.962	-0.045	0.98	0.944
■ normal convex-hull [3]	0.95	-0.015	0.965	0.951
■ ours	0.996	0.081	0.992	0.995

Fig. 8. Comparison results of different methods

$$Dice = \frac{2*(R_{seg} \cap R_{gt})}{R_{seg}+R_{gt}} \qquad RVD = \left(\frac{R_{seg}}{R_{gt}} - 1\right) * 100\%$$

$$Precision = \frac{TP}{TP+FP} \qquad Recall = \frac{TP}{TP+FN} \tag{6}$$

Among them, R_{gt} represents ground turth, and R_{seg} represents the actual segmentation result. Pixel-level statistics, TP indicates that the segmented pixel is a pixel in the lung parenchyma, which is actually on the lung parenchyma. FP indicates that the segmented pixel is a pixel in the lung parenchyma, which in fact is not in the lung parenchyma. FN indicates that the segmented pixel is not a pixel in the lung parenchyma, which is actually in the lung parenchyma.

4   Conclusion

Accurate lung parenchymal segmentation can provide strong support for tuberculosis lesion detection and disease diagnosis. In this paper, an improved convex hull algorithm combined with non-uniform rational B-sample method is proposed. Compared with other methods, this method reduces the adjustment of hyperparameters and enhances the ability of contour prediction. Using this method, the problem of missing lung parenchyma edges caused by pleural adhesion type TB lesions can be effectively solved. Thus, the complete lung parenchyma can be extracted. Experimental results show that our method has better superiority in lung segmentation.

Acknowledgement. This work was supported by National Natural Science Foundation of China (No. 61876024), and partly by the higher education colleges in Jiangsu province (No. 21KJA510003), and Suzhou municipal science and technology plan project (No. SYG202129), and Natural Science Research Fund for colleges and universities in Jiangsu Province (No. 17KJB510002).

References

1. Magnusson, K.E.G., Jaldén, J., Gilbert, P.M., Blau, H.M.: Global linking of cell tracks using the Viterbi algorithm. IEEE Trans. Med. Imaging **34**(4), 911–929 (2014)

2. Pulagam, A.R., Kande, G.B., Ede, V.K.R., Inampudi, R.B.: Automated lung segmentation from HRCT scans with diffuse parenchymal lung diseases. J. Digit. Imaging **29**(4), 507–519 (2016)

3. Shen, S., Bui, A.A.T., Cong, J., Hsu, W.: An automated lung segmentation approach using bidirectional chain codes to improve nodule detection accuracy. Comput. Biol. Med. **57**, 139–149 (2015)

4. Liu, C., Zhao, R., Pang, M.: Lung segmentation based on random forest and multiscale edge detection. IET Image Process. **13**(10), 1745–1754 (2019)

5. Somasundaram, K., Kalavathi, P.: Medical image binarization using square wave representation. In: Balasubramaniam, P. (ed.) ICLICC 2011. CCIS, vol. 140, pp. 152–158. Springer, Heidelberg (2011). https://doi.org/10.1007/978-3-642-19263-0_19

6. Chudasama, D., Patel, T., Joshi, S., Prajapati, G.I.: Image segmentation using morphological operations. Int. J. Comput. Appl. **117**(18) (2015)

7. Weng, M., Liu, Q., Guo, J.: MSER and connected domain analysis based algorithm for container code locating process. In: International Conference on Industrial Informatics-Computing Technology, pp. 83–86 (2017)

8. Bhargava, N., Trivedi, P., Toshniwal, A., Swarnkar, H.: Iterative region merging and object retrieval method using mean shift segmentation and flood fill algorithm. In: International Conference on Advances in Computing and Communications, pp. 157–160 (2013)

9. Graham, R.L.: An efficient algorithm for determining the convex hull of a finite planar set. Info. Pro. Lett. **1**, 132–133 (1972)

10. Bingol, O.R., Krishnamurthy, A.: NURBS-python: an open-source, object-oriented NURBS modeling framework in python. SoftwareX **9**, 85–94 (2019)

11. Beccari, C.V., Casciola, G.: A Cox-de Boor-type recurrence relation for C1 multidegree splines. Comput. Aided Geometr. Design **75**(10), 17–84 (2019)

12. Ronneberger, O., Fischer, P., Brox, T.: U-net: convolutional networks for biomedical image segmentation. In: Navab, N., Hornegger, J., Wells, W.M., Frangi, A.F. (eds.) MICCAI 2015. LNCS, vol. 9351, pp. 234–241. Springer, Cham (2015). https://doi.org/10.1007/978-3-319-24574-4_28

13. Li, C., Kao, C.-Y., Gore, J.C., Ding, Z.: Minimization of region-scalable fitting energy for image segmentation. IEEE Trans. Image Process. **17**(10), 1940–1949 (2008)

Similar Feature Extraction Network for Occluded Person Re-identification

Xiao Jiang, Ju Liu[✉], Yanyang Han, Lingchen Gu, and Xiaoxi Liu

School of Information Science and Engineering, Shandong University,
Qingdao 266237, Shandong, China
xjiang@mail.sdu.edu.cn, juliu@sdu.edu.cn

Abstract. Occluded person re-identification (Re-ID) is a challenging task in real-world scenarios due to the extensive conditions that persons are occluded by various obstacles. Although state-of-the-art methods with additional cues such as pose estimation and segmentation have achieved great success, they did not overcome data bias and the dependency on the accuracy of other detectors. In this paper, we propose a novel similar feature extraction network (SFE-Net) for occluded person Re-ID to address these issues. Firstly, we introduce the adaptive convolution method to separate the features of occluded and non-occluded regions, where local and global features are sufficiently used. We then apply adaptive aggregating parameters to find a better weighting strategy automatically. Finally, the transformer encoder architecture is utilized for generating discriminative features. Extensive experiments show SFE-Net outperforms state-of-the-art methods on both occluded and holistic datasets.

Keywords: Occluded person Re-ID · Adaptive convolution · Transformer encoder · Adaptive aggregating parameters

1 Introduction

Person Re-identification (Re-ID) aims at associating the same person in non-overlapping camera views. It is an attractive task in the computer vision field, which has various applications in video surveillance, pedestrian tracking, public security and smart city. Re-ID has achieved significant progress with the development of deep learning in recent years. However, in contrast to the existing holistic Re-ID methods [5,6,18], occluded person Re-ID is more practical and challenging in real-world scenarios due to the extensive conditions that persons are occluded by various obstacles (e.g., cars, trees and other persons) or not captured completely in the cameras.

One of the major challenges in the occluded person Re-ID is how to learn discriminative features from unoccluded regions. Recently, many methods [9,16] detect non-occluded body parts using person masks, human parsing or pose estimation models for accurate alignment of visible body parts. However, their

© Springer Nature Switzerland AG 2022
Y. Tan et al. (Eds.): ICSI 2022, LNCS 13345, pp. 320–330, 2022.
https://doi.org/10.1007/978-3-031-09726-3_29

accuracy strongly depends on the additional body part detectors and suffers from the noises of the datasets bias between the occluded Re-ID and them. The other methods focus on learning the body parts representations [7], which have a lower computational complexity but facing some difficulties that generating discriminative features and ensuring all of salient features available.

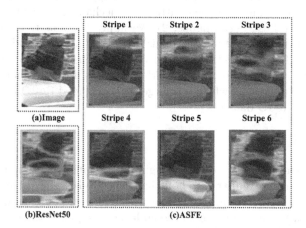

Fig. 1. Comparison of attention maps generated by similar adaptive features extracting module (ASFE) and ResNet50. (a) Occluded Re-ID suffers from occlusions in pedestrian images. (b) ResNet50 is not robust for occluded Re-ID. (c) Stripe 1–6 show the attention maps generated by the ASFE. The green boxes highlight the matching of body parts. The images with red box denote the occlusions (stripe5) or background noise (stripe6). (Color figure online)

In this paper, we propose a more salient and discriminative part-based framework named similar feature extraction network (SFE-Net). The framework include a similar adaptive features extracting module (ASFE), adaptive aggregating parameters (AAP), a transformer-based salience feature capture module (TSFC) leveraging the transformer encoder [15] and a global branch. Firstly, the ASFE separates the features of occluded and non-occluded regions by precisely extracting similar global features from the global feature. As shown in Fig. 1, we average global features into 6 stripes, and use them as convolution kernels to the convolution of global feature. Then we get 6 similar global features respectively. Some of these similar global features are the salient features of the person (such as stripe1, stripe2, stripe3, stripe4), the others are the occlusions (stripe5) or background noise (stripe6). A PCB-like [12] strategy is used to supervise the classification of each local convolutional feature for extracting the most discriminative pedestrian features without any additional body detectors. In this way, the little patches of the body can be represented more comprehensively through multiple stripes adaptive convolution. It overcomes the defects of the common CNN which can not distinguish occlusions also lead to omitting some discriminative details. Moreover, it is considered which of similar global features are more important. Therefore, we use the updatable adaptive parameters to optimize the features extracted by ASFE. Finally, the TSFC efficiently aggregates these similar global features for the more robust feature presentation performance thanks to transformer architecture.

The main contributions of this paper include: 1) We propose a novel Similar Feature Extraction Network to extract complete, unobstructed pedestrian features and utilize the multi-head attention mechanism to capture salient features for more robust performance; 2) We apply the adaptive convolution to the occluded person Re-ID for distinguishing the pedestrian and occlusion features. Only local and global features of occlusion are used without any additional body detectors; 3) Our approach outperforms state-of-the-art methods on both Occluded-DukeMTMC [9] which is a large-scale occluded Re-ID benchmark and holistic person Re-ID datasets.

2   Related Work

Occluded person Re-ID is more challenging due to incomplete information and spatial misalignment compared to holistic person Re-ID. Miao et al. [9] utilized a pose estimation method to guide the feature alignment. Wang et al. [16] involved an adaptive direction graph convolutional layer to suppress the meaningless features and a cross-graph embedded alignment layer to align two groups of local features. Different from the above methods which most rely heavily on additional tools (human parsing models or pose estimators) to bridge the gap between the occlusion and non-occlusion. We only use their own feature maps to solve this problem, which make the method simpler and more effective.

Transformer was first proposed in the field of natural language processing (NLP) by Vaswani et al. [15], and achieved great success with multi-head self-attention mechanism. Recently, researchers have started using it in various computer vision tasks. Li et al. [7] proposed the first work that exploiting the transformer encoder-decoder architecture for occluded person Re-ID. Transformer has an advantage in capture long-range dependencies. That is the reason we use transformer encoder in our proposed approach to capture salience feature from the similar global features, which contain complete, unobstructed pedestrian information for better feature representation learning.

Adaptive convolution is a novel approach to the matching task of two images, proposed by Liao et al. [8]. One of the two images is reshaped as a 1 × 1 convolution kernel, and the convolution is adopted on the other image with this convolution kernel. Since the channels of the two images are normalized by L2, convolution actually measures the cosine similarity of each location of the two features. In addition, the convolution kernel is adaptively constructed according to the image content. This similarity measurement progress can accurately reflect the local matching results between two input images. In contrast, we use adaptive convolution to measure the similarity between local features and global feature in this paper. As shown in Fig. 2(a), we use local features as the kernels, and adopt the convolution on global feature respectively. Moreover, we construct an adaptive local convolution network to generate more discriminative features.

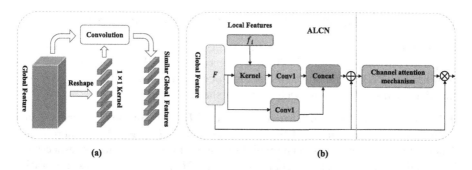

Fig. 2. (a) The process of adaptive convolution. Local features are adopted to the convolution of global feature as the convolution kernels. (b) The architecture of ASFE which consists of the ALCN and channel attention mechanism.

3   Proposed Method

The diagram of proposed similar feature extraction network (SFE-Net) is as shown in Fig. 3. Firstly, global feature is extracted through the CNN and fed into the ASFE. For separating the features of person and occlusions, adaptive convolution is adopted to generate similar global features. Then, we apply the AAP to weight similar global features effectively. Finally, the TSFC is utilized to make the model pay more attention to salient person features. The global branch is involved in the baseline to establish the relationship between global feature and local features. This way can significantly improve the alignment performance of the model.

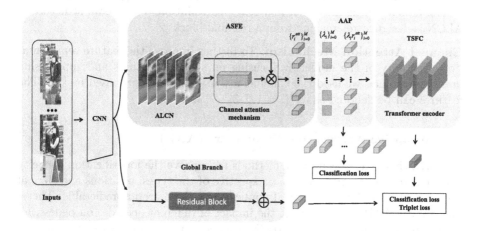

Fig. 3. Illustration of our proposed SFE-Net.

3.1 Similar Adaptive Features Extracting Module (ASFE)

In this module, we separate the features of occluded and non-occluded regions by adaptive convolution method. Then, we construct the adaptive local convolution network (ALCN) to generate robust similar global features. Finally, we use an effective channel attention mechanism, as shown in Fig. 3.

Adaptive Local Convolution Network (ALCN). Suppose a person image x, the feature map generated by CNN backbone is $F \in \mathbb{R}^{h \times w \times c}$, where h, w, c denote the height, width and channel dimension of the feature maps respectively. We equally partitioned the global feature into M horizontal stripes, generating the local features $f_i(i = 1, 2, 3, \ldots, M)$ which have the size of $1 \times 1 \times c$ through max pooling. Then ALCN is adopted to separate occluded and non-occluded features combined with their respective classification supervision, as shown in Fig. 2(b). Adaptive convolution is the process of measuring similarities between local most salient features and global features, so we convolute F with f_i. And after that, all similar global features will be aggregated in their respective localities. Inspired by [10], we construct a relation network to build the relations between the stripes and the global rather than simply concatenation. Through the relation network, more discriminative features are generating to over the interference of the similar local features in different pedestrians. The relation network finally generate more discriminative features $r_i(i = 1, 2, 3, \ldots, M)$ which contain information of the convoluted features and global feature. Thus the feature maps through the ALCN can be denoted as:

$$r_i = F + C\left(R\left(F\right), R\left(A\left(f_i, F\right)\right)\right), \tag{1}$$

where A denotes an adaptive local convolution process and f_i is the convolution kernel. C is a concatenation, R is a 1×1 convolution. The overall structure of ALCN is designed referencing from a residual block [2].

Channel Attention Mechanism. To further improve the feature representation capability of the model by focusing on salience features and suppressing unnecessary ones, we used the Channel attention module proposed by [17]. The features can be denoted as $r_i^{att}(i = 1, 2, 3, \ldots, M)$.

3.2 Adaptive Aggregating Parameters (AAP)

Recently, there have been many methods to improve the identification capacity of occluded pedestrians from the perspective of the types, positions and sizes of occluded objects, but in practical applications, these are unpredictable. Therefore, we use AAP to auto adjust the impact of different occludes on pedestrian detection in images. After the ALCN, we get similar global features measured by various stripes, but inevitably there are stripes that are mostly occluded. So, the similar global features generated by these stripes are likely to contain full of occlusions, which will greatly impact the reference stage. AAP is used to adjust the classification score automatically. After that, higher weights are assigned to features that are helpful for pedestrian classification and lower weights to helpless or noisy features. Our method can flexibly distinguish different occludes without relying on manual labels. The features for classification can be denoted as:

$$Q_i = \lambda_i r_i^{att}, \tag{2}$$

where $\lambda_i (i = 1, 2, 3, \ldots, M)$ is the adaptive aggregating parameters which are learnable.

Loss Function. Cross entropy loss is adopted for learning supervision, and label smoothing [13] is also used to prevent the model from overfitting. The classification loss is defined as:

$$L_{ALFE} = \frac{1}{M} \sum_{k=1}^{M} \sum_{j=1}^{N} -q_j \log Q_i \quad \begin{cases} q_j = \varepsilon/N & y \neq j \\ q_j = 1 - \frac{N-1}{N}\varepsilon & y = j \end{cases}, \tag{3}$$

where N represents the number of pedestrian identities, q_j is the smoothing label and ε is set to be 0.1, true label of the image is y.

3.3 Transformer-Based Salience Feature Capture Module (TSFC)

To satisfy the input requirements of the transformer encoder, we first concatenate all M similar global features r_i^{att} and reduce the channel dimension to d with a 1×1 convolution, then flatten the spatial dimension to create a 1D feature map $\hat{Q}_i \in \mathbb{R}^{hw \times d} (i = 1, 2, 3, \ldots, M)$ without embedding spatial position. We set multi-head to M, so a sequence $[\hat{Q}_1, \hat{Q}_2, \hat{Q}_3, \ldots, \hat{Q}_M]$ is fed into the transformer encoder, and each element in the sequence corresponds to the similar global feature for each stripe. Finally, we get a salient feature $\left[\hat{Q}_1^{sal}, \hat{Q}_2^{sal}, \hat{Q}_3^{sal}, \ldots, \hat{Q}_M^{sal}\right]$.

Training Loss. We use cross entropy loss which can be found in Eq. 1 and triplet loss [3] as our targets.

$$\begin{aligned} L_{MSF} &= L_{CE}\left(\hat{Q}_M^{sal}\right) + L_{tri}\left(\hat{Q}_M^{sal}\right) \\ &= -q_i \log\left(\hat{Q}_M^{sal}\right) + \left|\alpha + d_{\hat{Q}_{AM}^{sal}, \hat{Q}_{PM}^{sal}} - d_{\hat{Q}_{AM}^{sal}, \hat{Q}_{NM}^{sal}}\right|_+, \end{aligned} \tag{4}$$

where α is a margin, $d_{\hat{Q}_{AM}^{sal}, \hat{Q}_{PM}^{sal}}$ denote the distance between a same identity pair and $d_{\hat{Q}_{AM}^{sal}, \hat{Q}_{NM}^{sal}}$ is from different identities.

3.4 Global Branch

A residual module [2] behind the ResNet50 for the extraction of deep features, trying to extract more detailed and discriminative features from pedestrians. In addition, the position of pedestrians in the image can be well located to compensate for the alignment ability of local features. The features generated by the residual module is defined as F^{res}. The global training loss also combines cross entropy loss with triplet loss.

4 Experiments

In this section, we first evaluate the SFE-Net on Occluded-DukeMTMC [9], two widely generic person Re-ID benchmarks: Market1501 [19] and DukeMTMC [20]. Then we conduct ablation studies to validate the effectiveness of each component.

4.1 Datasets and Evaluation Metric

Market1501 contains 1501 identities from 6 cameras. All 32668 images are split into two parts, 12936 images of 751 identities for training and 19732 images for testing. DukeMTMC contains 1812 identities from 8 cameras. It consists of 36411 images which has 16522 images of 702 persons for training, 2228 images of another 702 persons for query and the remaining 17661 images for gallery. Occluded-DukeMTMC is selected from DukeMTMC, which consist of 15618 training images, 2210 query images and 17661 gallery images. It is more challenging due to containing large scale of obstacles and overlapping camera views. We adopt the widely used evaluation metrics, the cumulative matching characteristic (CMC) curve and the mean average precision (mAP). All experiments are conducted in a single query setting.

4.2 Implementation Details

All input images are resized to 384×128 and augmented with random horizontal flipping, random cropping and random erasing [21] with a probability of 0.5. We adopt ResNet50 [2] as our convolutional neural backbone, and set the last spatial down-sampling operation to 1 for higher spatial resolution. M is set to 6. The embedded dim for the transformer encoder is set to 768, and the depth is 4. SFE-Net is trained on 2 GeForce GTX 1080 Ti GPUs with a batch size of 64. We adopt Adam optimizer with 170 epochs. The initial learning rate is set to 4×10^{-6} and we adopt warm-up strategy to linearly grow the learning rate to 4×10^{-4} at 10 epochs, decaying every 20 epochs after 40 epochs with the factor 0.5.

4.3 Comparison with State-of-the-art Methods

Results on Occluded-DukeMTMC. We evaluate the performance of our model on Occluded-DukeMTMC. As shown in Table 1, two types of occluded Re-ID methods are compared with our method: holistic Re-ID methods (HA-CNN [6], Adver Occluded [5], PCB [12]) and occluded ReID methods (PGFA [9], HOReID [16], PAT [7]. Our proposed SFE-Net achieves 54.6% mAP, 65.1% Rank-1 accuracy, 79.8% Rank-5 accuracy, 85.0% Rank-10 accuracy, which set a new SOTA performance. Compared to the holistic method PCB [12], our SFE-Net surpasses by 20.9% in mAP and 22.5% in Rank-1. Compared with the SOTA occluded ReID method, our model surpasses them by at least 1% mAP and 0.6% Rank-1 accuracy without using any external clues such as person masks, human parsing or pose estimation.

Table 1. Comparison with state-of-the-arts on Occluded-DukeMTMC.

Method	Occluded-DukeMTMC			
	mAP (%)	Rank-1 (%)	Rank-5 (%)	Rank-10 (%)
HA-CNN [6] (CVPR 18)	26.0	34.4	51.9	59.4
Adver occluded [5] (CVPR 18)	32.3	44.5	–	–
PCB [12] (ECCV 18)	33.7	42.6	57.1	62.9
PGFA [9] (ICCV 19)	37.3	51.4	68.6	74.9
HOReID [16] (CVPR 20)	43.8	55.1	–	–
PAT [7] (CVPR 21)	53.6	64.5	–	–
SFE-Net (Ours)	**54.6**	**65.1**	**79.8**	**85.0**

Results on Holistic Re-ID Datasets. We also conduct an experiment on two holistic Re-ID datasets: Market1501 and DukeMTMC, and reported in Table 2. All the methods are divided into three groups, the first group are the part-based Re-ID methods (PCB [12], VPM [11]), the second group are the attention based methods (IANet [4], MHN-6 [1]) and the third group using the external cues (AANet [14], HOReID [16]). Our proposed SFE-Net achieves 95.6%/88.4% Rank-1/mAP and 89.7%/78.6% Rank-1/mAP on Market1501 and DukeMTMC datasets, which saturate competitive performance with the SOTA methods on both datasets.

Table 2. Comparison with state-of-the-arts on Market1501 and DukeMTMC.

Method	Market1501		DukeMTMC	
	mAP (%)	Rank-1 (%)	mAP (%)	Rank-1 (%)
PCB [12] (ECCV 18)	77.4	92.3	66.1	81.8
VPM [11] (CVPR 19)	80.8	93.0	72.6	83.6
IANet [4] (CVPR 19)	83.1	94.4	73.4	87.1
MHN-6 [1] (ICCV 19)	85.0	95.1	77.2	89.1
AANet [14] (CVPR 19)	82.5	93.9	72.6	86.4
HOReID [16] (CVPR 20)	84.9	94.2	75.6	86.9
SFE-Net(Ours)	**88.4**	**95.6**	**78.6**	**89.7**

4.4 Ablation Studies

In this section, we evaluate the effectiveness of each component on Occluded-DukeMTMC, as shown in Table 3. In Index-1, our baseline remove all the modules and only use ResNet50 to extract M stripes local features $f_i, (i = 1, 2, 3, \ldots, M)$ combined with an additional residual global feature F^{res}. It only achieves 46.2% Rank-1 and 38.6% mAP.

Table 3. Effectiveness of our proposed SFE-Net on Occluded-Duke.

Index	ASFE	AAP	TSFC	mAP (%)	Rank-1 (%)
1	×	×	×	38.6	46.2
2	√	×	×	53.0	64.2
3	√	√	×	53.5	64.6
4	√	√	√	**54.6**	**65.1**

At first, we add the ASFE to the baseline. As shown in Index-2, the ASFE can significantly improve the performance by 14.4%/18% in mAP/Rank-1 over the baseline. It proves that different from [8], our proposed adaptive convolution of global features with local features can play an important role in the distinction between pedestrians and occlusions.

Secondly, we evaluate the effectiveness of the AAP based on the incorporation of baseline, the ASFE and the AAP. In Index-3, the performance can be further improved by 0.5%/0.4% in mAP/Rank-1. It indicates that optimization of weight distribution for different similar global features can generate more discriminative features, which will further improve the performance of adaptive convolution.

Finally, we use all the modules to verify the effectiveness of the TSFC in Index-4. The mAP and Rank-1 can be improved by 1.1% and 0.5% respectively, and achieves SOTA performance. In this experiment, we set the number of multi-head to M for preventing features from interfering with each other. The experiment proves that the features extracted after transformer architecture become more robust.

5   Conclusion

In conclusion, to solve the problem of data bias and the dependency on the accuracy of other detectors in occluded person Re-ID, we propose a novel similar feature exaction network (SFE-Net). The ASFE is designed to distinguish the pedestrian and occlusion features by adaptive convolution. The AAP is proposed to weight various local features effectively. The TSFC is utilized to generate discriminative features. Through the combination of these three components, salient person features are extracted effectively, and occlusions, background noise are filtered. Experimental results show our method achieves the state-of-the-art performance and demonstrates the effectiveness of each module.

Acknowledgements. This paper is supported by Shandong Province Key Innovation Project (Grant No. 2020CXGC010903 and Grant No. 2021SFGC0701).

References

1. Chen, B., Deng, W., Hu, J.: Mixed high-order attention network for person re-identification. In: 2019 IEEE/CVF International Conference on Computer Vision (ICCV), pp. 371–381 (2019). https://doi.org/10.1109/ICCV.2019.00046
2. He, K., Zhang, X., Ren, S., Sun, J.: Deep residual learning for image recognition. In: 2016 IEEE Conference on Computer Vision and Pattern Recognition (CVPR), pp. 770–778 (2016). https://doi.org/10.1109/CVPR.2016.90
3. Hermans, A., Beyer, L., Leibe, B.: In defense of the triplet loss for person re-identification. CoRR abs/1703.07737 (2017). http://arxiv.org/abs/1703.07737
4. Hou, R., Ma, B., Chang, H., Gu, X., Shan, S., Chen, X.: Interaction-and-aggregation network for person re-identification. In: 2019 IEEE/CVF Conference on Computer Vision and Pattern Recognition (CVPR), pp. 9309–9318 (2019). https://doi.org/10.1109/CVPR.2019.00954
5. Huang, H., Li, D., Zhang, Z., Chen, X., Huang, K.: Adversarially occluded samples for person re-identification. In: 2018 IEEE/CVF Conference on Computer Vision and Pattern Recognition, pp. 5098–5107 (2018). https://doi.org/10.1109/CVPR.2018.00535
6. Li, W., Zhu, X., Gong, S.: Harmonious attention network for person re-identification. In: 2018 IEEE/CVF Conference on Computer Vision and Pattern Recognition, pp. 2285–2294 (2018). https://doi.org/10.1109/CVPR.2018.00243
7. Li, Y., He, J., Zhang, T., Liu, X., Zhang, Y., Wu, F.: Diverse part discovery: occluded person re-identification with part-aware transformer. In: 2021 IEEE/CVF Conference on Computer Vision and Pattern Recognition (CVPR), pp. 2897–2906 (2021). https://doi.org/10.1109/CVPR46437.2021.00292
8. Liao, S., Shao, L.: Interpretable and generalizable person re-identification with query-adaptive convolution and temporal lifting. In: Vedaldi, A., Bischof, H., Brox, T., Frahm, J.-M. (eds.) ECCV 2020. LNCS, vol. 12356, pp. 456–474. Springer, Cham (2020). https://doi.org/10.1007/978-3-030-58621-8_27
9. Miao, J., Wu, Y., Liu, P., Ding, Y., Yang, Y.: Pose-guided feature alignment for occluded person re-identification. In: 2019 IEEE/CVF International Conference on Computer Vision (ICCV), pp. 542–551 (2019). https://doi.org/10.1109/ICCV.2019.00063
10. Park, H., Ham, B.: Relation network for person re-identification. In: Proceedings of the AAAI Conference on Artificial Intelligence, vol. 34, pp. 11839–11847 (2020)
11. Sun, Y., et al.: Perceive where to focus: learning visibility-aware part-level features for partial person re-identification. In: 2019 IEEE/CVF Conference on Computer Vision and Pattern Recognition (CVPR), pp. 393–402 (2019). https://doi.org/10.1109/CVPR.2019.00048
12. Sun, Y., Zheng, L., Yang, Y., Tian, Q., Wang, S.: Beyond part models: person retrieval with refined part pooling (and a strong convolutional baseline). In: Ferrari, V., Hebert, M., Sminchisescu, C., Weiss, Y. (eds.) ECCV 2018. LNCS, vol. 11208, pp. 501–518. Springer, Cham (2018). https://doi.org/10.1007/978-3-030-01225-0_30
13. Szegedy, C., Vanhoucke, V., Ioffe, S., Shlens, J., Wojna, Z.: Rethinking the inception architecture for computer vision. In: 2016 IEEE Conference on Computer Vision and Pattern Recognition (CVPR), pp. 2818–2826 (2016). https://doi.org/10.1109/CVPR.2016.308

14. Tay, C.P., Roy, S., Yap, K.H.: AANet: attribute attention network for person re-identifications. In: 2019 IEEE/CVF Conference on Computer Vision and Pattern Recognition (CVPR), pp. 7127–7136 (2019). https://doi.org/10.1109/CVPR.2019.00730

15. Vaswani, A., et al.: Attention is all you need. In: Advances in Neural Information Processing Systems, vol. 30 (2017)

16. Wang, G., et al.: High-order information matters: learning relation and topology for occluded person re-identification. In: 2020 IEEE/CVF Conference on Computer Vision and Pattern Recognition (CVPR), pp. 6448–6457 (2020). https://doi.org/10.1109/CVPR42600.2020.00648

17. Woo, S., Park, J., Lee, J.-Y., Kweon, I.S.: CBAM: convolutional block attention module. In: Ferrari, V., Hebert, M., Sminchisescu, C., Weiss, Y. (eds.) ECCV 2018. LNCS, vol. 11211, pp. 3–19. Springer, Cham (2018). https://doi.org/10.1007/978-3-030-01234-2_1

18. Zhao, L., Li, X., Zhuang, Y., Wang, J.: Deeply-learned part-aligned representations for person re-identification. In: 2017 IEEE International Conference on Computer Vision (ICCV), pp. 3239–3248 (2017). https://doi.org/10.1109/ICCV.2017.349

19. Zheng, L., Shen, L., Tian, L., Wang, S., Wang, J., Tian, Q.: Scalable person re-identification: a benchmark. In: 2015 IEEE International Conference on Computer Vision (ICCV), pp. 1116–1124 (2015). https://doi.org/10.1109/ICCV.2015.133

20. Zheng, Z., Zheng, L., Yang, Y.: Unlabeled samples generated by GAN improve the person re-identification baseline in vitro. In: 2017 IEEE International Conference on Computer Vision (ICCV), pp. 3774–3782 (2017). https://doi.org/10.1109/ICCV.2017.405

21. Zhong, Z., Zheng, L., Kang, G., Li, S., Yang, Y.: Random erasing data augmentation. In: Proceedings of the AAAI Conference on Artificial Intelligence, vol. 34, August 2017. https://doi.org/10.1609/aaai.v34i07.7000

T-Distribution Based BFO for Life Classification Using DNA Codon Usage Frequencies

Shuang Yang, Zhipeng Xu, Chen Zou, and Gemin Liang[(✉)]

College of Management, Shenzhen University, Shenzhen 518060, China
lianggemin2021@email.szu.edu.cn

Abstract. Biological classification based on gene codon sequence is critical in life science research. This paper aims to improve the classification performance of conventional algorithms by integrating bacterial foraging optimization (BFO) into the classification process. To enhance the searching capability of conventional BFO, we leverage adaptive T-distribution variation to optimize the swimming step size of BFO, which is named TBFO. Different degree of freedom for t-distribution was used according to the iteration process thus to accelerate converging speed of BFO. The parameters of Artificial Neural Network and Random Forest are then optimized through the TBFO thus to enhance the classification accuracy. Comparative experiment is conducted on six standard data set of DNA codon usage frequencies. Results show that, TBFO performs better in terms of accuracy and convergence speed than PSO, WOA, GA, and BFO.

Keywords: Bacterial foraging optimization · Artificial neural networks · Random forest · Life classification · T-distribution based BFO

1 Introduction

According to genetic central dogma [1], genetic information begins with DNA, and the genome's coding DNA describes the proteins that make up an organism in 64 different codons and a stop signal. These codons map to 21 different amino acids, and then go through Transcription and translation processes ultimately produce proteins that perform biological functions. In the process of protein translation, there are different frequencies of use among synonymous codons, and the phenomenon of preferring to use certain codons is called codon bias [2], which may point to class of life. Life classification helps to understand biodiversity and is a basic method for studying organisms [3]. Biological classification divides organisms into diverse levels such as species and genera, and scientifically describes the morphological structure and physiological functions of each level to clarify the affinities and evolutionary relationships between different taxa. Classifying organisms helps to gain deeper understanding of biological diversity and to protect biological diversity. Since the newly discovered organisms continues posting challenges for classification and there are currently 9 million unnamed species in insects and invertebrates alone, and scientists typically spend long hours in the lab

© Springer Nature Switzerland AG 2022
Y. Tan et al. (Eds.): ICSI 2022, LNCS 13345, pp. 331–342, 2022.
https://doi.org/10.1007/978-3-031-09726-3_30

sorting through collected specimens [4], proposing effective and efficient classification algorithms can yield important practical significance for researchers.

This paper attempts to classify properties of nucleic acids from the use and frequency of various synonymous codons by comparing the performance of a variety of machine learning methods including k-Nearest Neighbors [5] (KNN), Random Forest [6] (RF), Naïve Bayesian [7] (NB), Decision Tree [8] (DT), Artificial Neural Network [9] (ANN), Support Vector Machine [10] (SVM), Logistic Regression [11] (LR). Moreover, we leverage swarm intelligence optimization algorithms including Bacterial Foraging Optimization [12] (BFO), Particle Swarm Optimization [13] (PSO), and Whale Optimization Algorithm [14] (WOA) to optimize the classification process thus to enhance the classification accuracy due to their efficiency and simplicity.

The combination of heuristic algorithms and specific classifiers is typically realized in the feature selection step [15]. Traditional feature selection method generates feature subsets by traversing the entire data set and evaluates the features through an evaluation function, thus to sort and select the features. Heuristic algorithms have excellent performance in optimization and can be combined with the evaluation procedure of feature selection [16]. Among the heuristic optimization algorithms, BFO has the advantage of simple structure and global optimization capability and has been widely used in many fields. Wu [17] proposed a power optimization control strategy based on an improved bacterial foraging optimization (IBFO) algorithm. Xing [18] used Bayesian network structure learning combined with improved BFO to solve the problem of track circuit fault diagnosis. Zhang [12] proposed a prediction control model designed with ANN optimized by the hybrid algorithm with quantum particle swarm optimization and improved bacterial community foraging. The BFO can be applied to train the classifier of feature selection. However, on account of the complexity of some datasets, training a basic BFO often takes long time. Besides, the conventional BFO algorithm adopts fixed step size, which makes it difficult to balance exploration and utilization capabilities.

This paper proposes a BFO based on the adaptive t-distribution mutation (TBFO). In the TBFO algorithm, the adaptive t distribution [19] variation is used to optimize the moving step, so that the individual bacteria have higher chance to leap out of the local optimum and converge to the global extreme point, which enhances the convergence speed as well.

The primary contributions of this paper are as follows:

1. Improve BFO by the adaptive t-distribution transmutation to converge the moving step.
2. Integrate the improved BFO with ANN and Random Forest method to enhance the accuracy of the organism classification.

Rest of the paper is organized as follows: Sect. 2 describes two main machine learning methods applied in the experiment, Sect. 3 introduces t-distribution variation based BFO combined with ANN and RF, Sect. 4 deals with the simulation experiment and Sect. 5 discusses the results obtained. Finally, Sect. 6 makes conclusions and looks to the future.

2   Related Work

2.1   Random Forests

Random forest [6] refers to a predictor that uses multiple trees to train samples and make regression predictions. Bootstrap is used for random sampling of data sets to extract N sub-data sets, and M features are selected from all features of the sub-data sets for segmentation of each node. M is a predefined number. Random forest will find the optimal segmentation point of each tree, and the rest of the tree is similar to the decision tree. Then, the predicted value of all branches is averaged to get the final predicted value. RF has the advantages of fast training speed, excellent prediction effect, strong generalization ability and strong robustness.

2.2   Artificial Neural Networks

Neural networks [13] originate from the practice of mimicking human intelligence through computer models. Neural network is composed of many artificial neurons through certain interconnection. Although the structure of individual neurons is relatively simple, their complex connections (topology) will form powerful networks. A neuron generally has multiple inputs, which are weighted and summed by combinatorial functions, and then output is generated by the activation function of the neuron. The strength of connections between neurons is represented by weights. Activation functions commonly used are Sigmoid function, Tanh (hyperbolic tangent function) and so on.

2.3   Random Forests and Artificial Neural Networks with Swarm Intelligence

Due to the fact that optimal or near-optimal values of machine learning 's parameters play an essential role in solving a concrete case, [20] one of the biggest challenges as well as problems is to confirm them. However, there are no common rules. Aiming to address each concrete case, determining a distinct set of parameter's values is indispensable. Acquiring optimal or near optimal values of the parameters is an NP-hard task [21], besides, heuristics algorithms like swarm intelligence methods can be applied to solve it.

   In general, meta-heuristic algorithms, especially swarm intelligence algorithms, have many successful applications in different fields such as image segmentation, classification, clustering [22]. Swarm intelligence is commonly inspired by some social behaviors conducted by a group of universal and simple individuals, like geese, ants, bees, fireflies, moths, bats, dragonflies, etc. In the same group, these sample units coordinated efficiently through intelligent behaviors. Especially, no specialized center will organize or command the units of all individuals. This property of swarm is generally applied as inspiration for swarm intelligence algorithms.

   Stochastic or deterministic methods are exploited to train ANN and RF. Gradient-based training and backpropagation are most commonly used for neural networks optimization. These are deterministic methods, but they often suffer from local optima stagnation, disappearing gradients, and slow speed convergence. Two optimization strategies for random forests [23]: feature selection and parameter optimization, which can lead to overfitting problems. Swarm intelligence has strong self-learning, self-adapting, self-organizing and other intelligent characteristics, the algorithm structure is simple, and the convergence speed is fast. This feature effectively establishes optimal or near-optimal parameter values in machine learning algorithms, which can avoid being stuck in local optimization and speed up convergence.

2.4 Bacterial Foraging Optimization (BFO)

BFO [13] is a swarm intelligence optimization algorithm suggested by Passino Kevin (2002) [12] based on the behavior of Ecoli Escherichia coli. It has been widely used in many fields and has become another hot spot of heuristic algorithm. According to the bacterial foraging theory, bacterial populations have a strong preference for nutrients. But under the constraints of their own physiology and environment, the goal of their movement is to move towards the position of maximum energy. The foraging behavior is primarily comprised of three basic steps: chemotaxis, replication, and migration. Introduce symbols for the convenience of description: S is the population size, N_C is the number of trending operations, N_S is the maximum number of steps that the trending operation moves in any direction, Nre is the number of breeding behaviors, Ne is the number of migration behaviors, and Ped is the migration probability.

Chemotaxis: Bacteria swim in any direction. If the fitness increases, they will continue to swim in that direction until the maximum number of steps Ns is reached, otherwise they will switch directions and swim until the number of trending times Nc is reached. The trend operation of bacteria i is expressed as Eq. (1):

$$\theta^i(j+1,k,l) = \theta^i(j,k,l) + C(i) \times \frac{\Delta(i)}{\sqrt{\Delta^T(i)\Delta(i)}} \qquad (1)$$

where the $\theta_i(j,k,l)$ indicates the concentration of nutrients in j chemotaxis,k reproduction and l elimination-dispersal. $C(i)$ is the chemotaxis step and $\Delta(i)$ is a random vector limited in $[-1,1]$.

Reproduction: According to Eq. (2), the fitness of bacteria after chemotaxis is evaluated and bacteria are sorted accordingly. The superior half population will replace the other larger half. By no means, the population of bacteria keeps the same. This algorithm executes Nre times reproduction programs.

$$f_{i,health} = \sum_{j=1}^{N_c} J(i,j,k,l) \qquad (2)$$

where the $f_{i,health}$ is the health value of bacteria i and J represents fitness value given by the objective function. The algorithm generally retains and replicates the larger health value once, and discards half of the smaller health value to ensure that the foraging ability of the bacteria community is improved concurrently.

Elimination and Dispersal: Bacteria may gradually migrate to other environments in the living environment to cope with sudden adverse environmental changes (such as natural disasters). In order to simulate this behavior, the algorithm gives a predetermined migration probability, and after several replication operations, some bacteria migrate. For the algorithm, this operation shows massive assistance in jumping out of the local optimal solution, so as to confirm the global optimal result.

3   The Proposed Method

3.1   BFO Based on T-distribution Variation (TBFO)

In the conventional BFO algorithm, the most critical step in the BFO algorithm is the chemotaxis process of bacteria, which ensures the local searching ability of bacteria. However, the swimming step size $C(i)$ of the original algorithm is a fixed value, which makes it difficult to determine the step size and cannot reflect the difference between bacteria with different fitness values. This reduces the accuracy and speed of the search process. To overcome this problem, an improved method for the walking step $C(i)'$ is proposed by using adaptive t-distribution variation to optimize the walking step [24].

The t-distribution is characterized by the parameter of degree of freedom n, and its curve shape is related to the size of the degree of freedom n. when $t(n \to \infty) \to N(0, 1)$, $t(n = 1) = C(0, 1)$, where $N(0, 1)$ is Gaussian distribution, $C(0, 1)$ is Cauchy distribution. The standard Gaussian distribution and the Cauchy distribution are two boundary special distributions of the t distribution, and the function distributions of the three are as follows (Fig. 1):

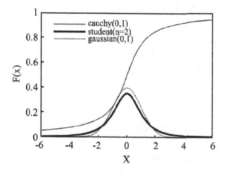

Fig. 1. T-distribution mutation (Han Feifei 2018).

We introduce the adaptive t distribution to improve step size of bacteria as shown in Eq. (3):

$$C(i)' = C(i) + C(i) * t(iter) \tag{3}$$

where $C(i)$ is the i_{th} walking step; $t(iter)$ is the t distribution that use the BFO iteration number as the degree of freedom. The random interference term of t distribution type $C(i) \cdot t(iter)$ is added, so that the individual bacteria can leap out of the local optimum and converge to the global extreme point. Additionally, this helps to improve the converging speed [24].

We used the iteration of BFO algorithm as the freedom degree to restraint curve shape of t distribution. In the early period of the algorithm when the value of the iteration number is small, the mutation of t distribution substantially resembled the Cauchy distribution with a wide and global exploration ability; the variation of t distribution is close to the variation of Gaussian distribution, which has a good local development capability; in the middle stage, the variation of t distribution is between the variation of Cauchy distribution and Gaussian variation. It combines the advantages of Gaussian operator and Cauchy operator, which can simultaneously improve the global exploration and local exploitation capability of algorithm. This interference also enhances the diversity of the population. The pseudocodes of TBFO are as follows:

Initialize the parameters of the TBFO algorithm:
$p, S, Nc, Ns, Nre, Ned, Ped, C(i), \theta i$
Initialize bacterial population
Evaluate the fitness value $J(i, j, k, l)$
For l in 1: Ned
  For k in 1:Nre
    For j in 1:Nc
      For i in 1:S
          Initialize t-distribution with the iteration
          Update the swim length of the chemotaxis with the equation (3)
          Calculate new fitness value $J(i, j + 1, k, l)$
          Let t=0 (initialize a counter)
          While t>Ns
            t=t+1
            If new fitness < original fitness
               Replace the original fitness
            End
          End
      End
      Do the Reproduction with equation (2)
    End
    Do the Elimination as basic BFO
    Do the Dispersal as basic BFO
  End
End

3.2   TBFO Combined with ANN and RF

The parameters that need to be adjusted in the ANN and RF models can be optimized by TBFO algorithm. As for ANN, a biggest challenge need to be solved is how to select the best numbers of nodes of hidden layer and the optimal numbers of hidden layers. Historical experiment shows that if the number is too small, ANN would be weak to own superefficient learning ability as well as the great competence to process information, like Fig. 2 (left). On the contrary, when it grows big exceedingly, the complexity of ANN's structure would increase out of control which continues to be a hard-run model. Furthermore, falling into a local minimum would become another deadly reason to low accuracy during the process of running. What's worse, the lower learning speed should be taken into considerations, like Fig. 2 (right). Anyway, the right network structure will be of severe help. Therefore, we use TBFO to optimize the value of nodes in the hidden layer and the number to decide the hidden layers.

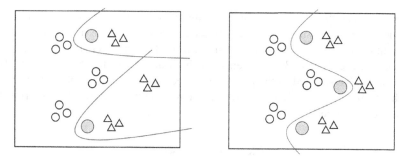

Fig. 2. Influence of hidden layer in ANN.

As for RF, n_estimators is the number of trees in the forest. The effect of this parameter to the random forest model is substantially monotonic. The larger n_estimators is, the better the model tends to be. Correspondingly, there is a decision boundary in any training model. After n_estimators reaches a certain value, the training result will never rise or start to fluctuate. Moreover, when n_estimators grows exceedingly, the amount of calculation and memory need to be required, and the time of training takes longer. For this reason, we are eager to strike this parameter into a balance between lower training difficulty and greater performance. Hence, TBFO was used and the specific implementation of the code can be seen at https://github.com/unfold333/Random-Forests-and-Artificial-Neural-Networks-with-TBFO.git.

The iterative steps of model parameter optimization are as follows (Fig. 3):

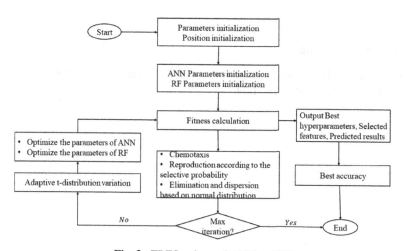

Fig. 3. TBFO enhanced ANN and RF

4 Experiment

4.1 Data Description

We used the data set (http://archive.ics.uci.edu/ml/datasets/Codon+usage) from the UCI machine learning repository, and the data set covers DNA codon usage frequencies of large sample of diverse biological organisms from different taxa. The data set used in the experiment contains 13028 instances with 69 attributes. The Kingdom and DNA type columns are selected as the prediction objects of the model. And to reduce some complexity, both of two columns are di-vided into several categories. The description of it is showed in Table 1.

Table 1. Data set description.

Column	Introduction
Kingdom	Code that classifies the genome, corresponding to 'xxx' in the Codon Usage Tabulated from Genbank (CUTG) database [22]; 1-'arc'(archaea), 2-'bct'(bacteria), 3-'euk'(eukaryotes), 4-'png'(phage),5-'vrl'(virus)
DNA type	An integer for the genomic composition; 0-genomic, 1-mitochondrial, 2-cyanelle
Species ID	The integer that indicates the entries of an organism;
Ncodons	The sum of the numbers listed for the different codons in an entry of CUTG;
Species name	Descriptive label of the name of the species
Codon	The codon frequencies, which are recorded as floats

4.2 Parameter Setting

The parameter setting of ANN and optimization algorithms, namely, PSO, WOA, GA, and BFO, TBFO are reported in Table 2.

Table 2. Parameter setting.

Method	Parameters
ANN	Solver = 'adam' Activation = 'tanh' max_iter = 10000 random_state = 1
PSO	w (inertia weight) = 0.8 c1 (learning speed) = 6 c1 (learning speed) = 6 Iteration = 30 r1 (random constant) = 0.8 r2 (random constant) = 0.5 Pn (umber of particles) = 50
WOA	Dim(dimension of parameter) = 5 Whale (number of whales) = 20 Iteration = 30
GA	Ps (population_size) = 50 Cl (chromosome_length) = 20 Pc (probability threshold) = 0.6 Pm (probability threshold) = 0.01 Generation = 30
BFO TBFO	Ned (the migration of algebraic) = 2 Nre (reproduction algebra) = 4 Nc (steps in a chemotactic step) = 50 Ns (the length of the swimming) = 4 C (running length unit) = 50 Ped (eliminate − scatter probability) = 0.25 Iteration = 30

5 Results and Discussion

The experiment results are shown in Table 3. The accuracy of classification ranged from 53.09% to 99.34%. ANN achieved the highest accuracy of 99.34%, followed by RF, KNN, DT, SVM and LR, with the accuracy of 98.46%, 99.23%, 97.72%,95.80% and 92.75% respectively. Bayes has the lowest accuracy, only 71.54%. In a word, Ann and RF classifiers perform well, which shows that these two machine learning algorithms can effectively predict biological species. Therefore, a variety of optimization algorithms are used to optimize these two models.

Table 3. Results.

	Kingdom		DNA type	
	Train	Test	Train	Test
KNN	0.9724	0.9533	0.9953	0.9923
Bayes	0.5309	0.5356	0.7129	0.7154
Decision tree	0.9425	0.8535	0.9943	0.9772
RF	0.9980	0.9140	0.9991	0.9846
SVM	0.7712	0.7644	0.9572	0.9580
LR	0.6902	0.6907	0.9234	0.9275
ANN	0.9189	0.9190	0.9974	0.9934

Table 4 provides the average accuracy of five heuristic algorithms. It shows that TBFO-ANN has the highest classification accuracy of 96.18% and 99.36%. Specifically, in the two models of Kingdom and DNA type, the accuracy of the optimized ANN and RF models is significantly improved, indicating that TBFO can effectively optimize their own parameters.

Table 4. The optimized results.

		PSO	WOA	GA	BFO	TBFO
Kingdom	RF	0.9572	0.9518	0.9533	0.9520	**0.9580**
	ANN	0.9580	0.9614	0.9515	0.9492	**0.9618**
DNA type	RF	0.9923	0.9904	0.9911	0.9901	**0.9940**
	ANN	0.9942	0.9942	0.9939	0.9931	**0.9936**

The optimization iteration diagram of the two machine learning algorithms is shown in Fig. 4. As the iterative process progresses, the classification accuracy gradually increases It can be observed that in limited iteration TBFO shows fast convergence speed. When applied TBFO to RFs, it has higher accuracy in the early stage of iteration compared to other algorithms. As for ANNs, in the early stage of iteration, it performs commonly, but increases quickly in short term and gain the best result.

Furthermore, TBFO also performs splendidly in terms of the model stability. It greatly shows that in the most cases the convergence speed of TBFO is more stable than the other corresponding algorithms, which is the symbol to declare the smaller volatility TBFO owns. Except that the accuracy of TBFO is exceedingly better than BFO as well as other algorithms, TBFO displays a fantastic competence of stability in the early iterations.

Comprehensively, in limited iterations, the ability of global delivering and sufficient convergence speed is what TBFO maintains to be the more competitive and fantastic

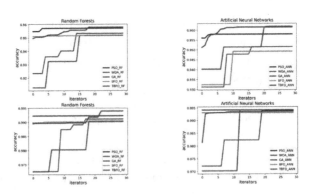

Fig. 4. The accuracy value in iteration process

model compared with other models. Additionally, TBFO gains the highest accuracy of the results which reflects its splendid capability of searching stable precision.

6 Conclusions and Future Work

In this paper, a biodiversity prediction model based on TBFO enhanced ANN and RF was proposed to predict the species according to the frequency of bio-codon use, which has certain theoretical significance to predict as well as identify the taxonomic and genetic character of organisms [12]. It demonstrated that BFO has advantages to jump out of the local optimum and improve the convergence speed. The effectiveness of TBFO in enhancing the classification performance is demonstrated.

Due to the fact that TBFO has shortcomings in the different effect of different machine learning methods, more study should be further searched. More heuristic algorithms may combine with T-distribution to prove the better characteristic. Concurrently, we will take the improvement direction into consideration to increase the global and local optimization capability during the stagy of replication of TBFO. In the future, more extensive experiments shall be conducted to further strengthen the TBFO for classification task parameter optimization.

Acknowledgement. This study is supported by Natural Science Foundation of Guang-dong (2022A1515012077), Shenzhen Higher Education Support Plan (20200826144104001).

References

1. Diao, S.F.: The central dogma and the development of modern biology. Dialectics Nat. 51–5 (2000)
2. Chen, Y.Z., Li, X.Y., Wu, J.L., Hu, J., Liu, Y.Y., Ye, X.Y.: Codon preference and characteristics of the genes highly expressed in flammulina filiformis. Mycosystema, 1–14 (2022)
3. Song, W.Y.: Microorganisms and their biological classification. Hubei Educ. 72–5 (2019)
4. Gao, L.Y.: Artificial intelligence helps biologists classify tiny organisms. Knowl. Mod. Phys. **33**(04), 66 (2022)
5. Bian Z.K.: Class-aware based KNN classification method. **34**(10), 873–84 (2021)
6. Dong, H.Y.: A review of random forest optimization algorithms. **33**(17), 34–7 (2021)
7. Wang, J.M.: Research on improvement of Naive Bayes classification algorithm: Harbin Univ. of Sci. Technol. (2021)
8. Wang, X., Shen, B.G., Guo, R.Z.: Improved ID3 decision tree algorithm induced by neighborhood equivalence. **21**(09), 17–20 (2021)
9. Aliakbari Sani, S., Khorram, A., Jaffari, A., Ebrahimi, G.: Development of processing map for InX-750 superalloy using hyperbolic sinus equation and ANN model. Rare Met. **40**(12), 3598–3607 (2018). https://doi.org/10.1007/s12598-018-1043-9
10. Li, C., Huang, W., Sun, W., Hao, H.: An association method for mulisource date based on SVM classification telecommunications technology, 1–4
11. Hu, X.M., Xie, Y., Jiang, H.F.: Prediction of breast cancer based on Penaltized logistic regression. Data Acquisition Process. **36**(06), 1237–1249 (2021)
12. Zhang, C.S., Mao, H.: Prediction model for tap water coagulation dosing based on BPNN optimized with BFO. Chin. Environ. Sci. **41**(10), 4616–4623 (2021)

13. Gong, R.K.: WSN node coverage optimization based on improved PSO-BFO algorithm. J. North China Univ. Sci. Technol. (Nat. Sci. Ed.) **43**(04), 52–59 (2021)

14. Wan, J.J.: Reliability evaluation of distribution network based on WOA optimized LSTM neural network. Intell. Comput. Appl. **11**(10), 107–112 (2021)

15. Xu, Z.G.: Research on Feature Selection Based on Heuristic Algorithm: Zhejiang University (2020)

16. Huang, Z.Z.: Optimization of outbound contrainer space assignment in automated container terminals based on hyper-heuristic. Comput. Integr. Manuf. Syst.1–26 (2021)

17. Wu, L.Q., Li, Y.Y., Chen, S.T.: Power optimization of EPS system based on improved BFO algorithm. Sens. Microsyst. **40**(08), 61–64 (2021)

18. Xing, D.F., Xue, S.R., Chen, G.W., Shi, J.Q.: Outdoor equipment fault detection based on improved bacterial algorithm. Comput. Simul. **38**(08), 157–161 (2021)

19. Zhang, W.K., Liu, S.: Improved sparrow search algorithm based on adaptive t -distribution and golden sine and its application. Microelectron. Comput. 1–8 (2021)

20. Miodrag, Z.: COVID-19 cases prediction by using hybrid machine learning and beetle antennae search approach. ScienceDirect **6707** (2021). https://doi.org/10.1016/j.scs.2020.102669

21. Fatemeh, H., et al: Computing optimal discrete readout weights in reservoir computing is NP-hard. ScienceDirect, 233–236 (2019). https://doi.org/10.1016/j.neucom.2019.02.009, ISSN 0925-2312

22. Bezdan, T., et al.: Hybrid fruit-fly optimization algorithm with K-means for text document clustering. Mathematics **9**, 1929 (2021). https://doi.org/10.3390/math9161929

23. Bacanin, N., et al.: Artificial neural networks hidden unit and weight connection optimization by quasi-refection-based learning artificial bee colony algorithm. IEEE Access **9**, 169135–169155 (2021). https://doi.org/10.1109/ACCESS.2021.3135201

24. Han, F.F., Liu, S.: Adaptivesatin bower bird optimization algorithm based on adaptive Tdistribution mutation. Microelectron. Comput. **35**(08), 117–121 (2018)

25. Nakamura, Y.: Codon usage tabulated from international DNA sequence databases: status for the year 2000. Nucleic Acids Res. **28**, 292 (2000)

Simpler is Sometimes Better: A Dynamic Aero-Engine Calibration Study

Hao Tong[1,2], Qingquan Zhang[1,2], Chengpeng Hu[1,2], Xudong Feng[3], Feng Wu[4],
and Jialin Liu[1,2(✉)]

[1] Research Institute of Trustworthy Autonomous System (RITAS),
Southern University of Science and Technology, Shenzhen, China
{11930582,12132333}@mail.sustech.edu.cn, liujl@sustech.edu.cn
[2] Department of Computer Science and Engineering,
Southern University of Science and Technology, Shenzhen, China
[3] Xidian University, Xi'an, China
fengxd@stu.xidian.edu.cn
[4] AECC, Chengdu, China
wufeng_my@mail.nwpu.edu.cn

Abstract. Many real-world optimisation problems are in dynamic environments such that the search space and the optimum usually change over time. Various algorithms have been proposed in the literature to deal with such dynamic optimisation problems. In this paper, we focus on the dynamic aero-engine calibration, which is the process of optimising a group of parameters to ensure the performance of an aero-engine under an increasing number of different operation conditions. A real aero-engine is considered in this work. Three different types of strategies for tackling dynamic optimisation problems are compared in our empirical studies. The simplest strategy shows the superior performance which provide an interesting conclusion: Given a new dynamic optimisation problem, the algorithm with complex strategies and having excellent performance on benchmark problems is likely to be applied due to the lack of prior knowledge, however, the simplest restart strategy is sometimes well enough to solve real-world complex dynamic optimisation problems.

Keywords: Dynamic optimisation · Restart strategy · Multi-swarm · Particle swarm optimisation · Real-world application

This work was supported by the AECC, the Research Institute of Trustworthy Autonomous Systems (RITAS), the Guangdong Basic and Applied Basic Research Foundation (Grant No. 2021A1515011830), the Guangdong Provincial Key Laboratory (Grant No. 2020B121201001), the Program for Guangdong Introducing Innovative and Enterpreneurial Teams (Grant No. 2017ZT07X386), the Shenzhen Science and Technology Program (Grant No. KQTD2016112514355531) and the National Natural Science Foundation of China (Grant No. 61906083).

© Springer Nature Switzerland AG 2022
Y. Tan et al. (Eds.): ICSI 2022, LNCS 13345, pp. 343–352, 2022.
https://doi.org/10.1007/978-3-031-09726-3_31

1   Introduction

Many real-world optimisation problems are usually subject to dynamic environments, which probably make the search space, objective functions, or problem constraints change over time. The dynamic changes of a problem's landscape, as well as its optimum solution, make such dynamic optimisation problems (DOPs) much more challenging than stationary optimisation problems [6]. In the past decades, dynamic optimisation has been investigated in different types of problems [16], including continuous problems [2], combinatorial problems [13] and multi-objective problems [5].

To deal with these challenging DOPs, many excellent algorithms have been proposed in the literature. For example, evolutionary algorithms and swarm intelligence algorithms have been widely used for solving DOPs, and many effective strategies have been proposed to help algorithms efficiently track the changing optimum in dynamic environments [11,15]. They have been successfully applied to a series of real applications [9,14].

In the existing literature, the essential part of an efficient dynamic optimisation algorithm is the strategy that helps the algorithm quickly locate a new optimum once the dynamic environment changes, usually called the dynamic strategy [15]. The dynamic strategies could be categorised into two classes, keeping the diversity of the population and benefiting from the past search history. The former, such as increasing the population diversity after a change [4], maintaining diversity during the optimisation [3] and multi-population [1], aims to keep the diversity of the population so that the algorithm is capable of quickly locating the position of the moving optimum. On the other hand, dynamic optimisation based on the past search history is also possible to speed up the process of tracking the new optimum in dynamic environments if a change exhibits some patterns [11]. For example, in the work of [8], the memory schemes saved the potentially better solutions found in the past environments and added them to the new environment.

Although many different effective dynamic strategies have been proposed in the literature to tackle the DOPs, we probably have no knowledge of which one may be the best for a new, unknown real-world DOP. As in our case, we consider a dynamic aero-engine calibration problem (detailed in Sect. 2.1), a black-box optimisation problem without any knowledge about its fitness landscape. Three different dynamic strategies, including a simple restart strategy, a multi-swarm strategy, and a hybrid strategy which combines maintaining diversity during the optimisation and past search history. The experimental studies conclude that the three strategies have no significant difference in multiple dynamic settings. An interesting conclusion is further inferred that the simplest strategy, i.e., the restart strategy which just re-optimises the new problem instance without any adaptions, is sometimes well enough for a new unknown DOP, according to our studies.

The remainder of this paper is organised as follows. Section 2 introduces the dynamic aero-engine calibration problem and an effective corresponding algorithm, i.e., saPSO. Section 3 introduces the three strategies compared in our experiments. Section 4 presents the empirical studies and the experimental results. The conclusion and the future work are provided in Sect. 5.

2   Background

2.1   Dynamic Aero-Engine Calibration

Engine optimisation is a typical kind of problem in the industry, including a wide range of applications, such as engine calibration and optimising engine control systems [17]. In this paper, an aero-engine calibration problem is considered, which aims at adjusting a set of parameters to ensure the performance of the given engine model under different operation conditions [7]. The calibration problem is modelled as a single-objective optimisation problem in this paper because we have found that the calibration results of modelling it as a multi-objective optimisation problem are similar with the single-objective modelling in [7].

In the static calibration process, assuming the d parameters to be adjusted are $x = \{x_1, x_2, ..., x_d\}$ and the m measurements of the aero-engine at an operation point s are $y = \{y_{1,s}, y_{2,s}, ..., y_{m,s}\}$, the calibration process is targeted to ensure that the measurement error, measured with root mean square error (RSME), of the engine is within a certain threshold. The $RMSE$ is calculated by Eq. (1):

$$RMSE = \sqrt{\frac{1}{k*m}\sum_{j=1}^{k}\sum_{i=1}^{m}\left(\frac{y_{i,j}-y_{i,j}^*}{y_{i,j}^*}\right)^2}, \tag{1}$$

where $y_{i,j}^*$ denotes the desired value of measurement i in operation point s_j and k denotes the total number of operation points in the engine calibration. A demonstration of the engine calibration process considered in this paper is presented in Fig. 1.

In our dynamic calibration process, the actual measurements, i.e., $y_{i,j}$ in Eq. (1), are obtained from the aero-engine performance simulation based on various operation conditions $\{s_1, s_2, ..., s_k\}$. However, in the real world, the data of operation conditions is obtained from the real computationally expensive experiments, so that the number of operation conditions will increase progressively during the calibration process. Therefore, the k in Eq. (1) would be $k(t)$ which is related to time t. Thus, the objective function of the dynamic engine calibration problem can be formulated by Eq. (2):

$$f(x,t) = \sqrt{\frac{1}{k(t)*m}\sum_{j=1}^{k(t)}\sum_{i=1}^{m}\left(\frac{y_{i,j}(x)-y_{i,j}^*}{y_{i,j}^*}\right)^2}. \tag{2}$$

2.2   Self-adaptive Particle Swarm Optimisation

For solving the stationary aero-engine calibration problem, a self-adaptive particle swarm optimisation (saPSO) algorithm was proposed, in which a new

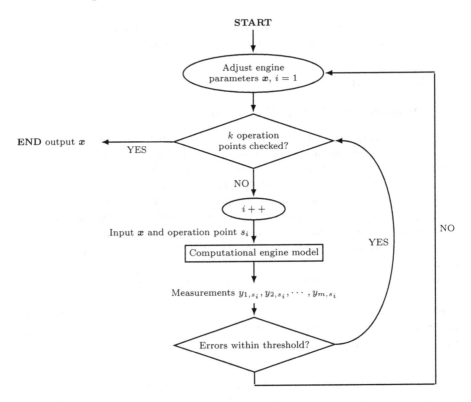

Fig. 1. Image edited from Fig. 1 of [7] with authors' permission. A demonstration of engine calibration process. An acceptable parameter setting x should ensure measurement errors within certain threshold at several given operation points $\{s_1, s_2, \ldots, s_k\}$.

re-sampling strategy was introduced to handle the out-of-range or infeasible solutions [7], and has shown competitive performance in calibrating the real aero-engine considered in this work. In a conventional PSO [12], the position x and velocity v of each particle is updated by following equations:

$$v = \omega * v + C_1 * r_1 * (x_{gbest} - x) + C_2 * r_2 * (x_{pbest} - x)$$
$$x = x + v, \tag{3}$$

where the x_{gbest} is the global best solution and the x_{pbest} is the personal best solution. C_1, C_2 are two constants and r_1, r_2 are two random vectors $\in (0, 1)^d$. In saPSO, the ω is assigned with ω^2 in each re-sampling process when the position of a particle is out-of-range. With decreasing inertia, the feasibility of new sampled particles is increasing. Then, the personal best positions and global best positions will be updated after evaluating the new feasible solutions. More details about saPSO can be found in [7].

3   Dynamic Strategies

To calibrate a real aero-engine considered in this paper, we apply three classic dynamic strategies, including a restart strategy, a multi-swarm strategy, and a hybrid strategy. This section describes how the considered dynamic strategies are integrated to saPSO.

3.1   Restart Strategy

The simplest strategy for dynamic optimisation is the restart strategy. Once a change occurs, the algorithm restarts to optimise the new problem instance without any adaptive operation. For saPSO, we keep the previous particle positions and re-initialise the personal best solutions and the global best solution when the new operation conditions are added into the calibration. All parameters in the algorithm are kept the same.

3.2   Multi-swarm Strategy

The second strategy is a multi-swarm strategy, in which multiple swarms are used during the optimisation to maintain the search diversity. In saPSO, the whole population is first divided into K sub-swarms. Each sub-swarm is optimised independently. If two different sub-swarms are considered to be close enough, i.e., the Euclidean distance between the global best solutions of two sub-swarms is smaller than a threshold, ϵ, the sub-swarm with the better global best solution will remain, and the other one will be replaced by a swarm initialised with randomly generated solutions.

3.3   Hybrid Strategy

The third strategy applied in our studies is a hybrid strategy, in which the population's diversity is controlled during the optimisation, and the solutions found for the past operation conditions are also used in the new conditions. Two strategies used in our hybrid strategies are described as follows.

1. To maintain the diversity of the saPSO, we add a distance criterion to determine the survival of a personal best solution x_{pbest}. If an updated particle is better than its personal best solution, the personal best solution is replaced by the updated particle. On the contrary, if the updated particle is worse than its personal best solution in terms of objective value, we will accept the updated particle with a pr probability when the distance between the global best solution and this particle, $dis(x_{gbest}, x)$, is greater than that of its personal best solution, $dis(x_{gbest}, x_{pbest})$.
2. To make use of the solutions from the past environment, we add all personal best solutions into the new environment. Besides, to increase the diversity of the new population, a set of uniformly randomly generated solutions for the new environment is also added. Thus, the initial population for the new environment includes the previous particles, personal best solutions, and the newly generated solutions.

4   Experimental Study

4.1   Experimental Settings

For the aero-engine we calibrate in this paper, the number of calibrated parameters, i.e., the dimensionality of our DOP, d, is 28. There are $m = 15$ measurements for each operation condition. The number of operation conditions increases during the optimisation process. There are always five initial operation conditions because they are the most basic conditions. To simulate the different situations in the real world, we construct six scenarios by adding one new operation condition in different frequencies, i.e., a new operation condition is added every T_{ef} function evaluations in our experiment. Thus, the number of operation conditions is

$$k(ef) = 5 + \left\lfloor \frac{ef}{T_{ef}} \right\rfloor, \tag{4}$$

where the ef is the number of generations and T_{ef} is set as 200, 400, 600, 800, 1200, 1600 in our experiment. Once the number of operation conditions changes, an optimisation instance is generated in the corresponding scenario. In our experiment, the operation conditions will not be added anymore if $ef > 4800$ because the real-world application usually costs some computational resources to optimise the final instance to guarantee the convergence.

The parameter settings for the saPSO and three dynamic strategies are presented in Table 1. The parameter setting of saPSO follows the setting in the work of [7]. The number of swarms is set as $K = 4$, and the whole population size is 40 so that the sub-population size for each swarm is 10. It is worthy to mention that the parameters related to dynamic strategies in our experiments are all determined by primary configurations with experimental tests.

Table 1. The parameter settings for algorithms in our experiments.

	Restart strategy	Multi-swarm strategy	Hybrid strategy
Specific parameters	–	$K = 4, \epsilon = 4$	$pr = 0.2$
saPSO parameters	$pop\_size = 40, MaxFE = 6000, C1 = C2 = 0.5, \omega = 0.9$		

4.2   Experimental Results

In the experiments, each algorithm is repeated with 20 independent runs in each dynamic scenario, and the best objective values before each dynamic change are recorded. Then, a normalised score performance measurement is used to compare the three algorithms [10] in each dynamic scenario as calculated by Eq. (5):

$$S_{norm}(i) = \frac{1}{p} \sum_{j=1}^{p} |\frac{f_j^{max} - f_{i,j}^*}{f_j^{max} - f_j^{min}}|, \forall i = 1, ..., 3, \tag{5}$$

where p is the number of instances generated in a scenario, and f_j^{max}, f_j^{min} denote the maximum and minimum objective values obtained by all algorithms for instance j, respectively. $f_{i,j}^*$ is the best objective value obtained by the algorithm i in solving instance j in a scenario.

Therefore, the averaged normalised scores and the standard deviation of three algorithms over 20 independent runs are presented in Table 2. The Friedman test with a 0.05 significant level is applied to each dynamic scenario. The p-value for each test in every dynamic scenario is presented in the last column of Table 2. All p-values are significantly smaller than 0.05, indicating that the three algorithms are not equal in all dynamic scenarios. Therefore, we applied the Nemenyi post-hoc test to each dynamic scenario. The ranking for each algorithm according to the test results is presented in the bracket of each cell. Furthermore, the best algorithm(s) with a ranking of '1' is/are highlighted in Table 2 for each dynamic scenario.

Table 2. The mean and standard deviation of normalized scores for saPSO with restart, hybrid, and multi-swarm strategy over 20 independent runs. The last column presents the p-values of statistical tests for three algorithms using the Friedman test with a 0.05 significant level. The ranking of each algorithm in each scenario according to the Nemenyi post-hoc test is included in the bracket.

T_{ef}	Restart strategy	Hybrid strategy	Multi-swarm strategy	p-values
200	**0.182 ± 0.103(1)**	0.439 ± 0.171(3)	0.562 ± 0.236(2)	2.16e−07
400	**0.189 ± 0.140(1)**	**0.585 ± 0.150(1)**	0.318 ± 0.231(3)	4.33e−06
600	**0.142 ± 0.074(1)**	**0.454 ± 0.192(1)**	0.203 ± 0.127(3)	1.06e−05
800	**0.132 ± 0.072(1)**	**0.435 ± 0.197(1)**	0.194 ± 0.149(3)	7.16e−07
1200	**0.179 ± 0.196(1)**	**0.464 ± 0.155(1)**	0.136 ± 0.069(3)	5.03e−06
1600	**0.131 ± 0.092(1)**	**0.385 ± 0.187(1)**	0.099 ± 0.050(3)	8.74e−07

Besides, we also plot the convergence curves of three algorithms in six dynamic scenarios, which are presented in Fig. 2. It is very clear that the restart strategy and hybrid strategy outperform the multi-swarm strategy. In particular, for the scenario with higher frequency of adding new operation conditions, i.e., T_{ef} with a smaller value, the advantage of the restart strategy is even more prominent. Especially when $T_{ef} = 200$, the restart strategy is much better than other two strategies.

Therefore, it is clear that the restart strategy performs the best in all dynamic scenarios according to our experimental results. The hybrid strategy has similar performance to the restart strategy, and the multi-swarm strategy has the worst performance compared with the other two algorithms. Thus, we can conclude that the restart strategy is well enough for our dynamic aero-engine calibration problem. The dynamic strategies added to the static algorithms never improve the algorithm's performance, and sometimes even deteriorate the algorithm's performance, such as the multi-swarm strategy in our experiments.

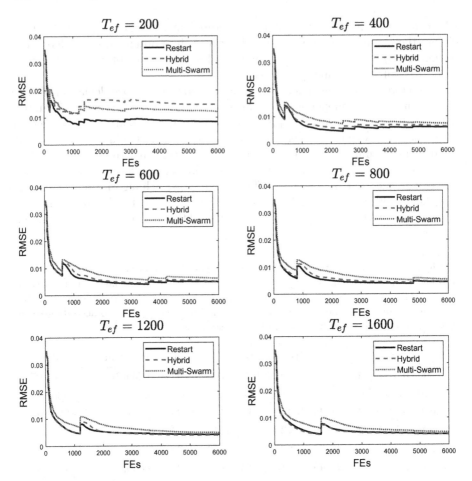

Fig. 2. Averaged RMSE values along the optimisation process obtained by saPSO with restart, hybrid, and multi-swarm strategy over 20 independent runs in various dynamic scenarios with different T_{ef} values (200, 400, 600, 800, 1200, and 1600). T_{ef} denotes the interval function evaluations of adding one new operation condition. *x-axis*: FEs (function evaluations), *y-axis*: RMSE.

4.3   Discussions

From Fig. 2, the algorithm converges to a high-quality solution very quickly. Even for the scenario with $T_{ef} = 200$, the algorithm also converges fast after adding new operation conditions. After the algorithm converges to a high-quality solution, the quality of the best solution does not change much, even after adding new operation conditions. It indicates that the saPSO is very efficient for our dynamic calibration problem, and the diversity control may deteriorate the performance of the original saPSO. As in our restart strategy, we also keep the particle positions and re-initialise the personal best and global best solutions.

The convergence curves is not corrupted by any big jump, indicating the global optimum might not move far from the current particle positions. Therefore, the simplest strategy is the most suitable one in our dynamic aero-engine calibration problem. If we would like to obtain further improvements, it would be better to investigate more in improving the original saPSO.

5   Conclusion

In this paper, we focus on a dynamic aero-engine calibration problem that requires to find an optimal group of parameters to ensure the performance of an aero-engine under an increasing number of different operation conditions. Three different dynamic strategies including a restart strategy, a multi-swarm strategy and a hybrid strategy are used to assist an algorithm shown to be effective in static aero-engine calibration, saPSO, to calibrate a real aero-engine in a dynamic setting. Six different dynamic scenarios with different frequency of adding new operation conditions are considered in our experiments assuming that the desired performance of all operation conditions is not always available at once and are provided periodically. Empirical results indicate that the simple restart strategy performs the best among the three strategies because saPSO is efficient enough for our dynamic aero-engine calibration problem even the number of operation conditions increases during the optimisation process. The location of the new global optimum after adding new operation conditions does not change a lot is also an essential reason.

Our study on dynamic aero-engine calibration provide an interesting insight, thus when facing some new and unknown real-world DOPs, the simple restart-strategy is worth trying before using more sophisticated strategies and it may have surprisingly excellent performance. In the future, we would like to investigate more DOPs including benchmark problems and other real applications to find out in which types of DOPs that the restart strategy can obtain excellent performance.

References

1. Blackwell, T., Branke, J.: Multi-swarm optimization in dynamic environments. In: Raidl, G.R., et al. (eds.) EvoWorkshops 2004. LNCS, vol. 3005, pp. 489–500. Springer, Heidelberg (2004). https://doi.org/10.1007/978-3-540-24653-4_50
2. Blackwell, T., Branke, J.: Multiswarms, exclusion, and anti-convergence in dynamic environments. IEEE Trans. Evol. Comput. **10**(4), 459–472 (2006)
3. Daneshyari, M., Yen, G.G.: Dynamic optimization using cultural based PSO. In: 2011 IEEE Congress of Evolutionary Computation, pp. 509–516. IEEE (2011)
4. Eberhart, R.C., Shi, Y.: Tracking and optimizing dynamic systems with particle swarms. In: Proceedings of the 2001 Congress on Evolutionary Computation, vol. 1, pp. 94–100. IEEE (2001)
5. Jiang, M., Wang, Z., Hong, H., Yen, G.G.: Knee point-based imbalanced transfer learning for dynamic multiobjective optimization. IEEE Trans. Evol. Comput. **25**(1), 117–129 (2021)

6. Jin, Y., Branke, J.: Evolutionary optimization in uncertain environments-a survey. IEEE Trans. Evol. Comput. **9**(3), 303–317 (2005)

7. Liu, J., Zhang, Q., Pei, J., Tong, H., Feng, X., Wu, F.: fSDE: efficient evolutionary optimisation for many-objective aero-engine calibration. Complex Intell. Syst. (2021)

8. Luo, W., Sun, J., Bu, C., Liang, H.: Species-based particle swarm optimizer enhanced by memory for dynamic optimization. Appl. Soft Comput. **47**, 130–140 (2016)

9. Mavrovouniotis, M., Li, C., Yang, S.: A survey of swarm intelligence for dynamic optimization: algorithms and applications. Swarm Evol. Comput. **33**, 1–17 (2017)

10. Nguyen, T.T.: Continuous dynamic optimisation using evolutionary algorithms. Ph.D. thesis, University of Birmingham (2011)

11. Nguyen, T.T., Yang, S., Branke, J.: Evolutionary dynamic optimization: a survey of the state of the art. Swarm Evol. Comput. **6**, 1–24 (2012)

12. Shi, Y., Eberhart, R.C.: A modified particle swarm optimizer. In: 1998 IEEE International Conference on Evolutionary Computation Proceedings. IEEE World Congress on Computational Intelligence. IEEE (1998)

13. Tong, H., Minku, L.L., Menzel, S., Sendhoff, B., Yao, X.: A novel generalised metaheuristic framework for dynamic capacitated arc routing problems. IEEE Trans. Evol. Comput. 1–15 (2022). https://doi.org/10.1109/TEVC.2022.3147509

14. Yang, Z., Jin, Y., Hao, K.: A bio-inspired self-learning coevolutionary dynamic multiobjective optimization algorithm for internet of things services. IEEE Trans. Evol. Comput. **23**(4), 675–688 (2019)

15. Yazdani, D., Cheng, R., Yazdani, D., Branke, J., Jin, Y., Yao, X.: A survey of evolutionary continuous dynamic optimization over two decades—part A. IEEE Trans. Evol. Comput. **25**(4), 609–629 (2021)

16. Yazdani, D., Cheng, R., Yazdani, D., Branke, J., Jin, Y., Yao, X.: A survey of evolutionary continuous dynamic optimization over two decades—part B. IEEE Trans. Evol. Comput. **25**(4), 630–650 (2021)

17. Yu, X., Zhu, L., Wang, Y., Filev, D., Yao, X.: Internal combustion engine calibration using optimization algorithms. Appl. Energy **305**, 117894 (2022)

Other Optimization Applications

Research on the Optimal Management Plan of Forests

Yifan Jiang[1]([⊠]), Yuxuan Liu[2]([⊠]), and Jiaze Li[1]([⊠])

[1] College of Internet of Things Engineering, Hohai University, Changzhou 213022, People's Republic of China
2858478788@qq.com, 2549912122@qq.com
[2] Business School, Hohai University, Changzhou 213022, People's Republic of China
liuyuxuan1105@126.com

Abstract. This paper studied a method for selecting the optimal management plan for forests based on AHP. First, the paper used the control variable method to determine that the scope of application of the forest management plan is North America, Eurasia (mainly temperate continental climate and subtropical, tropical monsoon climate) and South America (mainly tropical climate, plain and plateau terrain). Within this range, we selected the Xing'an larch forest in the northern part of the Greater Xing'an Mountains as a special sample, and calculated the importance and weight scores of the ecological benefit plan (PEB), economic benefit plan (PES) and social benefit plan (PSP) of the forest, and the scheme with the highest score was the plan preferred by the operator. Combined with 9 indicators representing the comprehensive value of forest, this paper used the analytic hierarchy process to design a decision-making model, and determined the optimal forest management plan through calculation, which balances the various values of the forest.

Keywords: AHP · Control variable method · Decision-making model · Optimal forest management plan

1 Introduction

As we know, climate change has posed a threat to the survival and development of global life. In addition to reducing greenhouse gas emissions at the source, we can also reduce emissions from the transmission. Generally, we choose forests as carbon sequestration sites.

Therefore, we need to protect the forest and reduce forest accidents. Many factors need to be considered in this process, such as the age and type of trees, the geology and topography of the forest, and the benefits and longevity of forest products.

Early scholars paid more attention to excessive deforestation and forest crisis in forest management. The solution is to establish forest dynamic resource detection and cutting quota management [1, 2]. During this period, many scholars paid attention to the planned management of forest resources [3–5] and the total value of forests. It is agreed that forest management should combine ecological value with social and economic value

© Springer Nature Switzerland AG 2022
Y. Tan et al. (Eds.): ICSI 2022, LNCS 13345, pp. 355–366, 2022.
https://doi.org/10.1007/978-3-031-09726-3_32

to maximize the total value of forests [6, 7]. Finally, some scholars are concerned about natural disturbance factors such as fire on the forest ecosystem [8].

2 Problem Restatement

We comprehensively consider the various values of forest, use the analytic hierarchy process to design the decision-making model, and use the control variable method to divide the application scope of the decision-making model management plan. At the same time, an optimal forest management plan is determined through the model, which can balance the various values of forests. Finally, we selected the most representative Xing'an larch forest in the northern part of the Greater Xing'an Mountains to verify the model and then obtained the availability and adaptability of our forest management plan.

3 Models and Methods

3.1 Decision-Making Model Based on AHP

We established a hierarchical structure; the target layer was to select the optimal plan for forest management, the criterion layer was the impact indicators of different programs, and the measure layer was the three programs of the forest management plan, which are the priority ecological benefit program (PES), the priority economic benefit program (PEB) and priority social benefit program (PSP).

(1) Hierarchical Division

We decided on three schemes on the basis of scheme scores. They are optimal scheme of ecological benefits (PES), economic benefits (PEB) and social benefits (PSP) (Table 1).

Table 1. Hierarchy and index of Decision-making model

Target layer	Criterion layer		Plan layer
Optimal forest management plan decision system	Carbon Sequestration	CSS	PES
	Biodiversity	BDS	
	Average annual logging	AAL	
	Average annual income from forest products	AAP	PEB
	Number of development projects such as tourism	NDT	
	Forest protection fee	FPF	
	Employment situation	EPS	PSP
	Monitoring the quality of air and water resources in human settlements	MWH	
	Surrounding population density	SPD	

(2) Construct Judgment Matrix

Based on the above process analysis, the index data obtained are normalized. The judgment matrix is constructed by the pairwise comparison method, and the relative importance of each index in each level relative to its upper index is expressed in the form of a matrix (Table 2).

Table 2. Scale of AHP

Scale	Importance
1	i and j are equally important
3	i is slightly more important than j
5	i is significantly more important than j
7	i is more important than j
9	i is extremely important than j

Note: 2, 4, 6, and 8 are the intermediate values, and the reciprocal is used for the opposite importance.

We used aij to represent the ratio of the influence degree of xi and xj on the target Z, where xi represents the row index and xj represents the column index. When aij > 1, xi is more important than xj for the target Z, and its numerical value indicates the degree of importance. When aij = 1, xi and xj are equally important for the target Z; when aij ≤ 1, xi is less important than xj for the target Z, and its numerical value indicates the magnitude of the importance.

(3) Hierarchical Single Sort

The so-called hierarchical single sorting refers to the evaluation order of all elements of this layer relative to the previous layer. In short, it is the problem of calculating the eigenroots and eigenvectors of the judgment matrix.

(4) Consistency Test

To ensure the accuracy of the index, a consistency test is carried out on the judgment matrix after the index weight is calculated.

$$CI = \frac{\lambda_{max} - n}{n - 1}, CR = \frac{CI}{RI} \tag{1}$$

Table 3. Consistency testing standard

n	1	2	3	4	5	6	7	8	9	10	11
RI	0	0	0.58	0.90	1.12	1.24	1.32	1.41	1.45	1.49	1.51

CI is the consistency index, λmax is the largest eigenvalue of the comparison matrix, and n is the order of the comparison matrix; the smaller the value is of CI, the closer the judgment matrix is to complete consistency;

RI is the random consistency index, and CR is the consistency ratio; the smaller the value of CR is, the higher the consistency ratio is, which means the higher the consistency accuracy of the judgment matrix.

3.2 Regional Division Based on Control Variable Method

We adopt the idea of controlling variables, that is, keeping other variables constant during each discussion, and only study the change of one variable to the problem. We take three factors of the natural belt, topography, and climate as the basis for judging and classifying, and divide the world into 5 groups (Table 4 and Fig. 1).

Table 4. Division of global forest groups

[30.7°N,39.2°S]	[13.2°N,55.6°S]	[80°N,35°N]	[34.2°N,10.2°N]	[10.3°S,45.1°N]
[25.3°W,35°E]	[82.3°W,30°W]	[30°E,60°W]	[30°E,60°W]	[114.23°E,150°E]
A1	A2	A3	A4	A5

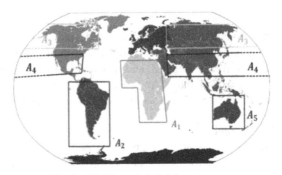

Fig. 1. Division of global forest groups

Among them, the average altitude of A1 was above 500, and the average altitude of A2 was about 200; the probability of natural disasters of A3 was low, and the average annual probability of natural disasters of A4 exceeded 80%.

A2: the main grassland climate, A4: the main monsoon climate, A5: the main desert climate.

We chose three indicators: carbon sequestration, forest product income, and the annual rate of change in employment, which corresponded to ecological benefits, economic benefits, and social benefits, respectively. We established 4 groups of comparison objects. For each of the two categories in each group, two of the variables were held constant and only one of the factors was changed, as described below:

We collected relevant data on three indicators of major forest resources in five regions in the past ten years. The indicator was $X = \{X_1, X_2, X_3\}$, and the indicator value was $X^{(0)} = \left\{X_1^{(0)}, X_2^{(0)}, X_3^{(0)}\right\}$. Each indicator had 10 sample data, namely

$$X_i^{(0)} = \left\{X_i^{(0)}(1), X_i^{(0)}(2), X_i^{(0)}(3)\right\}, i = 1, 2, 3 \tag{2}$$

In the number of j area we selected, its indicator value was: $X^{(j)} = \left\{ X_1^{(j)}, X_2^{(j)}, X_3^{(j)} \right\}$. Each indicator has 10 sample data, namely

$$X_i^{(j)} = \left\{ X_i^{(j)}(1), X_i^{(j)}(2), X_i^{(j)}(3) \right\}, i = 1, 2, 3 \tag{3}$$

In order to analyze the degree of response of one structure to the shock of another structure, we defined the structural change synergy coefficient. Its calculation formula was,

$$C_{jk} = \frac{\sum_{i=1}^{p} X_i^{(0)}(k) \cdot X_i^{(j)}(k)}{\sqrt{\sum_{i=1}^{p} \left[X_i^{(0)}(k) \right]^2 \cdot \sum_{i=1}^{p} \left[X_i^{(j)}(k) \right]^2}}, k = 1, 2, \cdots, 10, j = 1, 2, \cdots, 5 \tag{4}$$

$$C_j = \frac{1}{10} \sum_{k=1}^{10} C_{jk} \tag{5}$$

In the formula, k referred to the k group of data in the sample, j represened the j region, and C_j was the adaptability coefficient of the j region. The value of the coefficient was between 0–1, the closer it was to 1, the stronger was the adaptability of the decision-making model to regional changes. The more sensitive the model was to changes in regional structure, and the faster it can respond to changes in regional structure.

To better solve the problem, we performed hierarchical quantization on the fitness coefficient (Table 5):

Table 5. Comparison table of grade division

Fitness coefficient interval	Adaptation level
(0, 0.375]	Dangerous level
(0.375, 0.575]	Early warning level
(0.575, 0.805]	Balanced level
(0.805, 1]	Excellent level

4 Result Analysis

4.1 Scope of Application

We have included areas with adaptation levels at balance level and above within the scope of the forest management plan. From this we got the following results:

$$C_{A1} = 0.5642, C_{A2} = 0.6012, C_{A3} = 0.8932, C_{A4} = 0.8451, C_{A5} = 0.4911$$

According to this analysis, we can conclude that the scope of application of the forest management plan is North America, Eurasia (mainly temperate continental climate and subtropical, tropical monsoon climate) and South America (mainly tropical climate, plain and plateau terrain). That is, we believed that our decision-making model are applicable to all forests in this area.

Based on the above analysis, we decided to choose Xing'an larch forest in the northern part of the Greater Xing'an Mountains.

4.2 Best Management Plan

According to AHP, we get the final scores of the three schemes, as shown in the table below,

Table 6. The impact of the 9 indicators

	CSS	BDS	AAL	AAP	NDT	FPF	EPS	MWH	SPD
CSS	1	1/4	5	1/2	1/3	5	2	3	7
BDS	4	1	8	3	2	6	3	5	7
AAL	1/5	1/8	1	1/5	1/7	1/2	1/5	1/3	2
AAP	2	1/3	5	1	1/2	3	2	3	3
NDT	3	1/2	7	2	1	3	3	3	2
FPF	1/5	1/6	2	1/3	1/3	1	1/5	5	2
EPS	1/2	1/3	5	1/2	1/3	5	1	3	2
MWH	1/3	1/5	3	1/3	1/3	1/5	1/3	1	2
SPD	1/7	1/7	1/2	1/3	1/2	1/2	1/2	1/2	1

To compare the impact of the 9 indicators in the table on the target "Optimal Forest Program", we used pairwise comparisons, as shown in Table 6.

In the first row of the table: the carbon sequestration in the first column is as important as the carbon sequestration in the first row, so the row indicator (carbon sequestration)/column indicator (carbon sequestration) = 1; the second column, biodiversity (column index) is more important than carbon sequestration (row index), so row index (carbon sequestration)/column index (biodiversity) = 1/4 < 1; in the third column, the average annual logging amount is less important than the carbon sequestration amount, So row index (carbon sequestration)/column index (average annual logging volume) = 5 > 1; since the first row and the second column are 1/4, the second row and the first column are 4; the first row and the third column is 5, so the first column of the third row is 1/5, and so on. Relative scales are used in the importance analysis to minimize the difficulty of comparing indicators with different properties and improve the accuracy.

Finally, the judgment matrix A is obtained as follows:

$$
A = \begin{bmatrix}
1 & \frac{1}{4} & 5 & \frac{1}{2} & \frac{1}{3} & 5 & 2 & 3 & 7 \\
4 & 1 & 8 & 3 & 2 & 6 & 3 & 5 & 7 \\
\frac{1}{5} & \frac{1}{8} & 1 & \frac{1}{5} & \frac{1}{7} & \frac{1}{2} & \frac{1}{5} & \frac{1}{3} & 2 \\
2 & \frac{1}{3} & 5 & 1 & \frac{1}{2} & 3 & 2 & 3 & 3 \\
3 & \frac{1}{2} & 7 & 2 & 1 & 3 & 3 & 3 & 2 \\
\frac{1}{5} & \frac{1}{6} & 2 & \frac{1}{3} & \frac{1}{3} & 1 & \frac{1}{5} & 5 & 2 \\
\frac{1}{2} & \frac{1}{3} & 5 & \frac{1}{2} & \frac{1}{3} & 5 & 1 & 3 & 2 \\
\frac{1}{3} & \frac{1}{5} & 3 & \frac{1}{3} & \frac{1}{3} & \frac{1}{5} & \frac{1}{3} & 1 & 2 \\
\frac{1}{7} & \frac{1}{7} & \frac{1}{2} & \frac{1}{3} & \frac{1}{2} & \frac{1}{2} & \frac{1}{2} & \frac{1}{2} & 1
\end{bmatrix}
$$

On this basis, we carry out hierarchical single sorting. The single-level ranking is to put all the indicators of this layer in the order of a certain indicator of the previous layer, and this order is expressed by relative numerical values, as shown in Tables 7, 8, 9, 10, 11, 12, 13, 14 and 15.

Table 7. .

CSS	PES	PEB	PSP
PES	1	5	3
PEB	1/5	1	1/3
PSP	1/3	3	1

Table 8. .

BDS	PES	PEB	PSP
PES	1	4	3
PEB	1/4	1	1/2
PSP	1/3	2	1

Table 9. .

AAL	PES	PEB	PSP
PES	1	3	2
PEB	1/3	1	1/3
PSP	1/2	3	1

Table 10. .

AAP	PES	PEB	PSP
PES	1	1/7	1/3
PEB	7	1	2
PSP	3	1/2	1

Table 11. .

NDT	PES	PEB	PSP
PES	1	1/2	4
PEB	2	1	6
PSP	1/4	1/6	1

Table 12. .

FPF	PES	PEB	PSP
PES	1	1/2	5
PEB	2	1	7
PSP	1/5	1/7	1

Table 13. .

EPS	PES	PEB	PSP
PES	1	1/3	1/5
PEB	3	1	1/2
PSP	5	2	1

Table 14. .

MWH	PES	PEB	PSP
PES	1	5	1/2
PEB	1/5	1	1/7
PSP	2	7	1

Table 15. .

SPD	PES	PEB	PSP
PES	1	1/3	1/5
PEB	3	1	1/4
PSP	5	4	1

As shown in Table 7, for the single consideration of carbon sequestration, the first row and first column are equally important, and the ratio is 1; for the first row and second column, in order to ensure carbon sequestration, PES is more important than PEB, so the row Index (PES)/Column Index (PEB) = 5 > 1; For the first row and third column, in order to ensure carbon sequestration, PES is more important than PSP, so row index (PES)/column index (PSP) = 3 > 1. All of the following matrices are analyzed in turn:

$$B_1 = \begin{bmatrix} 1 & 5 & 3 \\ \frac{1}{5} & 1 & \frac{1}{3} \\ \frac{1}{3} & 3 & 1 \end{bmatrix} \quad B_2 = \begin{bmatrix} 1 & 4 & 3 \\ \frac{1}{4} & 1 & \frac{1}{2} \\ \frac{1}{3} & 2 & 1 \end{bmatrix} \quad B_3 = \begin{bmatrix} 1 & 2 & 2 \\ \frac{1}{3} & 1 & \frac{1}{3} \\ \frac{1}{2} & 3 & 1 \end{bmatrix}$$

$$B_4 = \begin{bmatrix} 1 & \frac{1}{7} & \frac{1}{3} \\ 7 & 1 & 2 \\ 3 & \frac{1}{2} & 1 \end{bmatrix} \quad B_5 = \begin{bmatrix} 1 & \frac{1}{2} & 4 \\ 2 & 1 & 6 \\ \frac{1}{4} & \frac{1}{6} & 1 \end{bmatrix} \quad B_6 = \begin{bmatrix} 1 & \frac{1}{2} & 5 \\ 2 & 1 & 7 \\ \frac{1}{5} & \frac{1}{7} & 1 \end{bmatrix}$$

$$B_7 = \begin{bmatrix} 1 & \frac{1}{3} & \frac{1}{5} \\ 3 & 1 & \frac{1}{2} \\ 5 & 2 & 1 \end{bmatrix} \quad B_8 = \begin{bmatrix} 1 & 5 & \frac{1}{2} \\ \frac{1}{5} & 1 & \frac{1}{7} \\ 2 & 7 & 1 \end{bmatrix} \quad B_9 = \begin{bmatrix} 1 & \frac{1}{3} & \frac{1}{5} \\ 3 & 1 & \frac{1}{4} \\ 5 & 4 & 1 \end{bmatrix}$$

First, we normalized each column of matrix A, and then summed the rows of the column-normalized matrix to get the eigenvectors, and then normalized the resulting eigenvectors to calculate the largest eigenvalue.

The column normalized matrix, the eigenvectors, the eigenvector normalization are:

$$W^{(0)} = \begin{bmatrix} 0.1295 \\ 0.2929 \\ 0.0278 \\ 0.1322 \\ 0.1812 \\ 0.0581 \\ 0.0993 \\ 0.0450 \\ 0.0341 \end{bmatrix} \quad W^{(0)} = \begin{pmatrix} 0.1295 \\ 0.2929 \\ 0.0278 \\ 0.1322 \\ 0.1812 \\ 0.0581 \\ 0.0993 \\ 0.0450 \\ 0.0341 \end{pmatrix} \quad Q = AW^{(0)} = \begin{bmatrix} 1.3310 \\ 2.9025 \\ 0.2747 \\ 1.3286 \\ 1.8506 \\ 0.6059 \\ 1.0209 \\ 0.4475 \\ 0.3442 \end{bmatrix}$$

The largest eigenroot of matrix A:

$$\lambda_{max}^{(0)} = \frac{1}{9} \sum_{i=1}^{9} \left(Q_i \div W_i^{(0)} \right) = 10.1203$$

In the formula, i represents the number of columns of the column matrix, and Qi and $W_i^{(0)}$ represent the data of the i-th column.

Similarly, the eigenvectors of B_1–B_9 can be found:

$w^{(1)} = (0.6358\ 0.1052\ 0.2590)^T$, $w^{(2)} = (0.6472\ 0.0725\ 0.2802)^T$
$w^{(3)} = (0.4972\ 0.1481\ 0.3547)^T$, $w^{(4)} = (0.0926\ 0.6152\ 0.2923)^T$
$w^{(5)} = (0.3235\ 0.5874\ 0.0891)^T$, $w^{(6)} = (0.3334\ 0.5914\ 0.0752)^T$
$w^{(7)} = (0.1095\ 0.3090\ 0.5815)^T$, $w^{(8)} = (0.3334\ 0.0752\ 0.5914)^T$
$w^{(9)} = (0.1017\ 0.2274\ 0.6709)^T$

The largest eigenroots of B_1–B_9 are:

$\lambda_{max}^{(1)} = 3.0385$, $\lambda_{max}^{(2)} = 3.0122$, $\lambda_{max}^{(3)} = 3.0049$, $\lambda_{max}^{(4)} = 3.0183$, $\lambda_{max}^{(5)} = 3.0291$, $\lambda_{max}^{(6)} = 3.0084$, $\lambda_{max}^{(7)} = 3.0233$, $\lambda_{max}^{(8)} = 3.0301$, $\lambda_{max}^{(9)} = 3.0217$.

We used the consistency indicators CI and CR and the consistency test standard table to test.

The conformance testing standard table is shown in Table 3:

For the judgment matrix A, n = 9, $\lambda_{max}^{(0)}$ =10.1203, RI = 1.45, $CI = \frac{10.1203-9}{9-1} = 0.140038$, $CR = \frac{CI}{RI} = \frac{0.140038}{1.45} = 0.0965779$.

From this we got the final eigenvector.

CR < 0.1 indicates that the degree of inconsistency of A is within the allowable range, at this time, the eigenvector of A can be used to replace the weight; similarly, the judgment matrix B_1-B_9 has passed the consistency test using the above principles.

Assuming: $w^1 = \left(w^{(1)}w^{(2)}w^{(3)}w^{(4)}w^{(5)}w^{(6)}w^{(7)}w^{(8)}w^{(9)}\right)$,

So the final eigenvector: $w = w^l w^{(0)} = \left(0.4173\ \ 0.2999\ \ 0.2829\right)^T$.

Finally, we got the importance and weight scores of the three schemes, as shown in Table 16 and Table 17 below:

Table 16. Index weight value

	Metric weight	PES	PEB	PSP
CCS	0.1295	0.7283	0.0817	0.19
BDS	0.2929	0.6472	0.0725	0.2802
AAL	0.0278	0.4972	0.1481	0.3547
AAP	0.1322	0.0926	0.6152	0.2923
NDT	0.1812	0.3235	0.5874	0.0891
FPF	0.0581	0.3334	0.5914	0.0752
EPS	0.0993	0.1095	0.309	0.5815
MWH	0.045	0.3334	0.0752	0.5914
SPD	0.0341	0.1017	0.2274	0.6709

According to the final scheme scores, forest managers should choose the priority ecological benefit scheme.

Table 17. Scheme score

	PES	PEB	PSP
Plan score	0.4173	0.2999	0.2829

The result of the priority ecological benefit plan is reasonable. The reason is that in the three criteria for measuring ecological benefits, carbon sequestration and biodiversity have a larger weight than other benefit criteria. The weight of biodiversity reaches is 0.2929, which has the largest weight among all indicators. At the same time, carbon sequestration and biodiversity are also more important than other indicators, reaching 0.7283 and 0.6472, respectively. The average annual logging volume, another criterion indicator of ecological benefits, has a relatively small weight, but its relative importance is also relatively high, reaching 0.4972. Combining above, the priority ecological benefit plan will have a higher score than the other two plans.

5 Conclusion

Aiming at the characteristics of forest management planning, this paper established a comprehensive evaluation system for forest management plans. This paper uses AHP to determine the judgment matrix, the weight of each index, and the final score of each program, to determine the optimal program for forest management. On this basis, the range of regions suitable for our management plan was determined according to the control variable method, and a special sample was selected to analyze the adaptability of our model. Compared with the previous papers, which only studied one type of forest, this paper has a breakthrough in the research scope. And adding more indicators helps to get better results. The model is suitable for forests with different characteristics in different regions.

References

1. Lin, L., Xue, Y.: Establishing a logging quota plan management system to comprehensively manage the excessive consumption of forest resources: how to solve the problem of unplanned logging in forest industry enterprises in Northeast and Inner Mongolia. For. Resour. Manage. (04), 10–17 (1989). https://doi.org/10.13466/j.cnki.lyzygl.1989.04.001
2. Zhang, C.: Establishing forest resource archives to strengthen forest resource management. Sichuan Arch. (06), 13 (1989)
3. Epp, H.T.: Application of science to environmental impact assessment in boreal forest management: the Saskatchewan example. Water Air Soil Pollut. **82**(1–2), 179–188 (1995)
4. Huang, R., Huang, D.: Reflections on how to strengthen the management of forest resources in Jiangxi under the market economy. For. Resour. Manage. (04), 5–8 (2002). https://doi.org/10.13466/j.cnki.lyzygl.2002.04.002
5. Fredrik, I., Lars, H., Bo, D.: Nature conservation in forest management plans for small-scale forestry in Sweden. Small-Scale For. Econ. Manage. Policy **3**(1), 17–34 (2004)
6. Xie, Z., Liu, A., Xu, Z., Chen, X., Chen, A.: Research on forest park tourism products. J. Beijing For. Univ. (03), 72–75 (2000). https://doi.org/10.13332/j.1000-1522.2000.03.018

7. Luo, J.: Development method of forest tourism products based on green marketing theory. For. Econ. Issues (04), 344–348 (2008). https://doi.org/10.16832/j.cnki.1005-9709.2008.04.01

8. Kessell, S.R.: Gradient modeling: a new approach to fire modeling and wilderness resource management. Environ. Manage. 1(1), 39–48 (1976)

Forest Management Based on Carbon Sequestration

Wenqi Jiang[✉], Xiaoxue Wang[✉], Jiafeng Ding[✉], and Chenkai Fang[✉]

College of Internet of Things Engineering, Hohai University,
Changzhou 213022, The People's Republic of China
2427585123@qq.com, 1542387739@qq.com, 1756622574@qq.com,
1253478745@qq.com

Abstract. The greenhouse gases have exacerbated global warming. We modeled carbon sequestration, annual growth, and expected cutting profits to mitigate climate change to determine appropriate and sustainable forest management. Initially, we established a Carbon Sequestration Model based on a hierarchical model to estimate the carbon storage and its products. The calculation shows that the carbon stored in aboveground biomass accounts for considerable weight. So in the process of forest management, we should increase wood density, forest area, and biomass per unit area to increase the amount of carbon sequestration. Then, considering environmental, economic, and social issues, we build a Single Objective model to realize the balance of maximizing annual growth, the expected cutting profits, and carbon sequestration.

Keywords: Carbon sequestration · Hierarchical model · Single objective model

1 Introduction

Forests are indispensable to mitigate climate change. Researchers have developed many methods for estimating forest carbon domestically and overseas in recent years. Zhao Miaomiao analyzed the main characteristics of the forest carbon measurement methods based on sample plot inventory, the forest carbon measurement method based on a model, and the flux observation method based on ecosystem positioning observatory [1]. S. Etemad used goal programming techniques to estimate the optimum stock level of different tree species considering economics, environmental and social issues [3]. L. Demidova and M. Ivkina developed forecasting models based on multidimensional time series, which are features used in the formation of the datasets, dividing further into the training and test sets [5].

Nowadays, sustainable forest management involves forestry activities to achieve sustainability. Furthermore, forest management is an intricate problem, so new tools and approaches will be needed to reach the multiple goals in forest management. This study presents a Single Objective model to attain optimization and study optimal stock or forest harvest with economic, environmental, and social objectives towards sustainable forest harvesting.

© Springer Nature Switzerland AG 2022
Y. Tan et al. (Eds.): ICSI 2022, LNCS 13345, pp. 367–378, 2022.
https://doi.org/10.1007/978-3-031-09726-3_33

2 Problem Description

Forest managers and policy-makers are being encouraged to incorporate carbon sequestration as a criterion for decision-making. Besides, the ways their forest is valued may
include potential carbon sequestration, conservation, biodiversity aspects, recreational
uses, cultural considerations, etc. The description of the problem can be expressed as
follows:

To increase the carbon storage, we are required to develop a carbon sequestration
model to determine how much carbon dioxide a forest and its products can be expected
to sequester over time. Our model should decide which indicator is the most effective.

The forest management plan that is best for carbon sequestration is not necessarily
the best for society given the other ways that forests are valued. So we need to develop a
model which should determine a forest management plan that balances the various ways
that forests are valued.

3 Calculation and Simplifying the Model

Firstly, we use formulas of forest carbon measurement to determine how much carbon
dioxide a forest and its products can be expected to sequester as time goes by. In addition,
we construct a Hierarchical model to evaluate the most effective impact indicators on
carbon sequestration. Secondly, we develop a single objective model to inform forest
managers of the best forest management. We need to consider the balance of interests
and ecological value.

3.1 Model Assumption

For Carbon Sequestration Model, economic forests are ignored, and only the public
welfare forest is considered. The economic forest is unstable, so its carbon sequestration
is unstable. Furthermore, we miss the interactions between factors affecting carbon
sequestration to simplify the model.

Emergencies such as forest fires are not considered. Special patrols will be carried
out in some areas with frequent wildfires, and fire barriers will be set up in advance in
the high season. Therefore, the frequency of forest fires has dramatically decreased, and
we will not consider it here.

We classify the four age groups of young forest, age near a forest, medium forest,
mature forest into the immature forest, and over-ripe forest into the mature forest in
ecology. The mature forest in the ecological sense means that the forest biomass reaches
a stable stage, and the mature forest's age often exceeds that of the over-mature forest
in the inventory.

3.2 A Carbon Sequestration Model

We thoroughly considered land cover, precipitation, and carbon pools based on forestry
for carbon sequestration. Carbon pools include carbon stored in aboveground biomass,

carbon stored in belowground biomass, carbon is stored in the soil and carbon is stored in dead organic matter.

In terms of the carbon pools, we calculated the carbon storage of forest ecosystems, and carbon sequestration is calculated from changes in carbon storage over time.

Model Formulation

The estimation of forest carbon storage is generally based on the calculation method of forest biomass. That is, the biomass of forest vegetation is measured directly or indirectly. We found a series of forest survey data (such as tree species, vertical structure, stand height and stand density, etc.), which are known to be obtained by forest resources inventory. Finally, we get the calculation equation of biomass, and then we multiply it by the percentage of carbon in the biomass. The carbon storage of the forest system is the sum of the calculated values of the four-carbon pools.

Total carbon storage mainly includes carbon stored in aboveground biomass, belowground biomass, soil, and dead organic matter [2].

$$C = C_{above} + C_{below} + C_{soil} + C_{dead} \tag{1}$$

where, C_{above} is the carbon stored in aboveground biomass, C_{below} is the carbon stored in belowground biomass, C_{soil} is the carbon that is stored in soil, C_{dead} is the carbon stored in dead organic matter.

The carbon stored in aboveground biomass is calculated by the biomass expansion factor method:

$$C_{above} = CF \cdot W_{ij} = CF \cdot A_i \cdot W_{ij} = CF \cdot A_i \cdot V_{ij} \cdot BEF_{ij} \cdot SVD_{ij} \tag{2}$$

where, CF is the carbon content, A_i is the area of Forest Type i. W_{ij} is the biomass per unit area of Forest Type i, Tree species j. V_{ij} is accumulation per hectare of Forest Type i, Tree species j. BEF_{ij} is the biomass expansion coefficient of Forest Type i, Tree species j. SVD_{ij} is the wood density of Forest Type i, Tree species j.

The carbon stored in belowground biomass is determined as below:

$$C_{below} = CF \cdot W_{ij} \cdot RhizomeRatio \tag{3}$$

where, $RhizomeRatio$ is the parameters of rhizome ratio selected according to forest age and forest species.

The biomass is shown in Appendix B. The carbon stored in dead organic matter is follow as:

$$C_{dead} = CF \cdot N_{ij} \tag{4}$$

where, N_{ij} is the biomass of dead organic matter.

In most research that estimates carbon storage and sequestration rates in a forest, soil pool measures only include soil organic carbon (SOC) in mineral soils. According to the observation and measurement, the calculation formula is as follows:

$$C_{soil} = A_{ij} \cdot 0.58 \cdot C \cdot D \cdot E \cdot \frac{(1-G)}{100} \tag{5}$$

where, C is the soil organic matter content. D is the soil bulk density. E is the soil thickness. G is the percentage of volume of gravel with diameter ≥ 2 mm.

Then, we construct a hierarchical model to evaluate the most effective impact indicators on carbon sequestration. The decision-making problem is decomposed into three levels. The top layer is the target layer M, that is, the selection of key indicators affecting carbon sequestration. The lowest layer is the scheme layer, namely nine influencing factors Pl, P2, P3, P4, P5, P6, P7, P8, P9. The middle layer is the criterion layer, which includes four indicators of carbon stored in aboveground biomass, carbon stored in belowground biomass, carbon stored in soil and carbon stored in dead organic matter (as shown in Fig. 1).

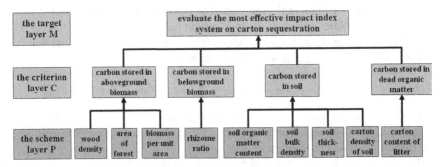

Fig. 1. Analytic hierarchy process diagram

3.3 Decision Making on Sustainable Forest Management

Model Formulation

We use single objective model to estimate the optimum stock level of different tree species considering economics, environmental and social issues. We take a forest in Beijing as an example and select the stocks of different tree species as decision variables. Furthermore, we consider constraints in the process of decision making to realize the balance of maximizing annual growth, the expected cutting profits, carbon sequestration.

We take the stock of different tree species type as the decision variable. All of the included goals in the single objective model are handled in a similar way: indicated by goal restriction. The included objective constraint consists of objective variables that evaluate the quantity by which the augmentation of all actions to the target in question have a shortage and a surplus with respect to the goal level [3].

Undesirable deviations should be minimized in an achievement function. Consequently, constraints and objective functions of single objective model are follow as determined as below:

$$\text{Min } Z = \sum_{j=1}^{10} W_j \left(d_j^- \right) \tag{6}$$

$$\sum_{i=1}^{5} X_i + d_T^- = g_T \tag{7}$$

$$X_i + d_{Vi}^- = g_{Vi} \tag{8}$$

$$\sum_{i=1}^{5} a_i X_i + d_C^- = g_C \tag{9}$$

$$\sum_{i=1}^{5} b_i X_i + d_G^- = g_G \tag{10}$$

$$\sum_{i=1}^{5} m_i X_i + d_L^- = g_L \tag{11}$$

$$\sum_{i=1}^{5} n_i X_i + d_{NPV}^- = g_{NPV} \tag{12}$$

$$d_T^-, d_{Vi}^-, d_C^-, d_G^-, d_L^-, d_{NPV}^-, d_j^-, W_j \geq 0 \tag{13}$$

where, j ranges from 1 to 10, which refer to total stock, beech stock, hornbeam stock, oak stock, alder stock, other species stock, sequestrated carbon, growth, labor and NPV, respectively. Besides, i ranges from 1 to 5, which indicates decision variables such as beech, hornbeam, oak, alder and other species. The key mathematical notations used in this model formulation are listed in Table 1.

Furthermore, we design the questionnaire based on the Likert scale. The most important goals and weights are determined by questionnaires. The most important goals are minimum total feasible stock, minimum feasible stock of each species, minimum feasible carbon sequestration, minimum feasible growth per hectare, minimum feasible personnel or labor and minimum suitable Net Present Value (NPV) which are given by experts. The questionnaires were used for weighting the goals for the aim of multipurpose forest management.

Eventually, we solve the multi-objective model using the LINGO software from LINDO systems.

Table 1. Notations used in this paper

Symbol	Description
W_j	The weight allocated to each deviation
X_i	decision variables, the stocks of different tree species
d^-	Negative deviation from goal value
g_T	Minimum total feasible stock
g_{Vi}	Minimum feasible stock of each species
g_C	Minimum feasible carbon sequestration
g_G	Minimum feasible growth per hectare
g_L	Minimum feasible personnel or labor
g_{NPV}	Minimum suitable Net Present Value (NPV)
a	Coefficient of sequestrated carbon
b	Coefficient of growth
m	Coefficient of labor
n	Coefficient of NPV

Table 2. The values of coefficients in GP model

Species name	a	b	m	n
Beech	269.533	8.371	61	7931.098
Hornbeam	313.889	13.721	61	4012.206
Oak	729.607	8.517	61	5271.125
Alder	238.329	22.786	61	6824.013
Other species	360.011	31.262	61	5577.614

Table 3. The values of goals and weights based on questionnaires

j	g	w	j	g	w
total stock	408	24.51	other species stock	20.4	490.20
beech stock	256.2	39.03	sequestrated carbon	128783.16	0.0776
hornbeam stock	61.2	163.4	growth	4509.834	2.2174
oak stock	40.8	245.1	labor	25000	0.4
alder stock	20.4	490.2	NPV	2816929.37	0.0035

4 Experimental Results and Analysis

4.1 A Carbon Sequestration Model

Carbon Sequestration Over Time

We applied the carbon sequestration model to a forest in Beijing. The actual data can be found for all the unknowns in the formula in the model. Table 4 and Table 5 below show the parameters of this forest in 1992. The reason we only classify immature forest and over-ripe forest is the convenience of plugging in unknown parameters.

Carbon sequestration refers to the change in carbon storage over a period of time. We put the collected data into the formula to calculate the carbon storage in different years, and the carbon storage that can be sealed in the forest over time can be determined by linear fitting. We find the data from 1992 to 2008. As can be seen from the diagram, the slope of the carbon storage curve is basically unchanged, so the change of carbon sequestration is stable (Fig. 2).

In addition, we use the InVEST model [6] to estimate the carbon sequestration of the forest over a period of time. The model uses maps of land use and land cover types, as well as wood harvesting, degradation rates of logging products, and carbon storage of the four carbon pools. The carbon sequestration distribution of this forest from 2013 to 2015 is shown below (Fig. 3).

The Optimal Evaluation Index

To evaluate the most effective indicators on carbon sequestration, we construct the judgment Matrix M-C [7]. Firstly, we compare the four elements in the criterion layer C in pairs. We can get pairs of comparison matrices, as shown in Table 6.

The weight vector can be obtained by solving the eigenvalues of the matrix M-C. We define ω as the weight vector, therefore we have:

$$\omega = \begin{pmatrix} 0.4717 & 0.1644 & 0.2562 & 0.1078 \end{pmatrix} \tag{14}$$

We have known:

$$CI = \frac{\chi_{max} - n}{n - 1} \tag{15}$$

$$CR = \frac{CI}{RI} \tag{16}$$

We can calculate the value of CR, which is 0.0172. The value of CR is less than 1. So the matrix passed the consistency test. Our model determine that the carbon stored in aboveground biomass is the most effective factor at sequestering carbon dioxide.

4.2 Decision Making on Sustainable Forest Management

The parameters values of constraints and objective is shown in Tables 2 and 3. We solve the single objective model using the LINGO software from LINDO systems. The result shows the negative deviations of beech, hornbeam, alder and other species in the forest of Beijing are zero. It means that they quite achieved the goal and also their optimal

Table 4. The actual data can be found for all the unknowns in the formula

Species	BEF					RSR					SVD	CF
	young	me-dium	age near	ripe	over-ripe	young	me-dium	age near	ripe	over-ripe		
Beech	1.526	1.395	1.252	1.109	1.18	0.229	0.279	0.235	0.19	0.212	0.541	0.491
Horn-beam	1.446	1.376	1.411	1.393	1.402	0.241	0.232	0.237	0.235	0.236	0.414	0.515
Oak	1.297	1.178	1.165	1.138	1.151	0.219	0.221	0.181	0.27	0.226	0.578	0.525
alder	1.847	1.497	1.233	1.245	1.535	0.218	0.233	0.329	0.384	0.365	0.478	0.51
Other species	1.38	1.327	1.36	1.474	1.587	0.26	0.275	0.41	0.281	0.153	0.676	0.5

Table 5. The actual data can be found for all the unknowns in the formula

Forest types	Age group	Rhizome ratio	Litter biomass
Coniferous forest	Young, medium, age near, ripe	0.2	15.24
	Overripe	0.22	16.17
Broadleaf forest	Young, medium, age near, ripe	0.3	8.87
	Overripe	0.28	7.84
Coniferous mixed forest	Young, medium, age near, ripe	0.25	6.76
	Overripe	0.21	5.86
Coniferous mixed forest	Young, medium, age near, ripe	0.16	0.53
	Overripe	0.2	0.53
Broadleaf mixed forest	Young, medium, age near, ripe	0.2	11.7
	Overripe	0.25	11.02

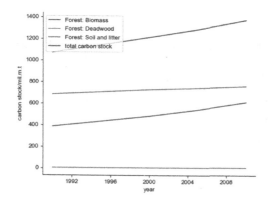

Fig. 2. Increasing trend of carbon storage

stock is 256.2, 61.2, 20.4 and 20.4 $m^3 \cdot ha^{-1}$, respectively. In conclusion, we accomplish to determine the optimal stock with a single objective model based on different factors.

Because carbon storage decreases with age after reaching the transition point, as is shown in Fig. 4. Appropriate harvesting strategies need to be developed in the forest management. Forests sequester carbon dioxide in living plants and in the products created from their trees including furniture, lumber, plywood, paper, and other wood products. These forest products sequester carbon dioxide for their lifespan. Some products have a short lifespan, while others have a lifespan that may exceed that of the trees from which they are produced. The carbon sequestered in some forest products combined with the carbon sequestered because of the regrowth of younger forests has the potential to allow for more carbon sequestration over time when compared to the carbon sequestration benefits of not cutting forests at all.

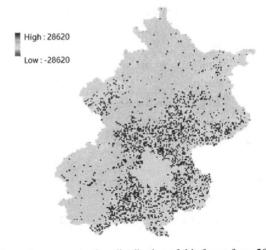

Fig. 3. The carbon sequestration distribution of this forest from 2013 to 2015

Table 6. Comparison matrix of analytic hierarchy process

M	C1	C2	C3	C4
C1	1	3	2	4
C2	1/3	1	1/2	2
C3	1/2	2	1	2
C4	1/4	1/2	1/2	1

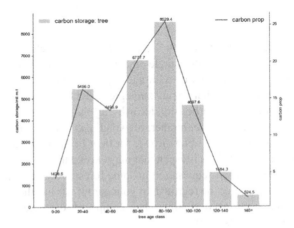

Fig. 4. Analytic hierarchy process diagram

4.3 Sensitivity Analysis

Confronted with complex forest system, we need to adjust the evaluation model of forest carbon storage and management plan, which includes the modification of constraint conditions and index system.

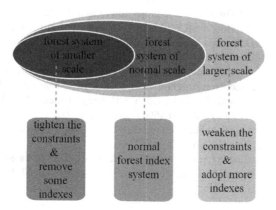

Fig. 5. Expansion method of goal programming model

We carry out scalability analysis under different forest systems as is shown in Fig. 5. Firstly, If we want to migrate to a smaller scale, we need to tighten the constraints and remove some indexes, which is hard or impossible to measure. On the contrary, if a model used to evaluate a forest wants to be migrated to a larger scale, we should adopt more indexes and weaken the constraints to consider the evaluation.

5 Conclusion

In this study, we establish a Carbon Sequestration Model to evaluate the most effective factor on carbon sequestration. The calculation shows that the carbon stored in above-ground biomass accounts for the largest weight. So in the process of forest management, we should increase wood density, forest area and the biomass per unit area to increase the amount of carbon sequestration.

Furthermore, we accomplish to determine the optimal stock with Single Objective model based on different factors, which include sequestrated carbon, minimum acceptable labor, growth, NPV of stand. However, there can be no optimal management suitable for all forests because of the various composition, climate, population, interests and values of forests around the world. Therefore, we conclude that this type of study is rather data sensitive and needs to be performed on a case-to-case basis. This is the main contribution by this study.

References

1. Zhao, M., Zhao, N., Liu, Y., Yang, J., Liu, Y., Yue, T.-X.: An overview of forest carbon measurement methods. Acta Ecol. Sin. **39**(11), 3797–3807 (2019)

2. National Forestry and Steppe Bureau: Guideline on Carbon Stock Accounting in Forest Ecosy, China (2018)
3. Etemad, S., Limaei, S.M., Olsson, L., Yousefpour, R.: Decision making on sustainable forest harvest production using goal programming approach (case study: Iranian hyrcanian forest). In: 2018 IEEE International Conference on Industrial Engineering and Engineering Management (IEEM), pp. 736–740 (2018). https://doi.org/10.1109/IEEM.2018.8607503
4. Schelhaas, M.-J., et al.: Forest resource projection tools at the European level. In: Barreiro, S., Schelhaas, M.-J., McRoberts, R.E., Kändler, G. (eds.) Forest Inventory-based Projection Systems for Wood and Biomass Availability. MFE, vol. 29, pp. 49–68. Springer, Cham (2017). https://doi.org/10.1007/978-3-319-56201-8_4
5. Demidova, L., Ivkina, M.: Development and research of the forecasting models based on the time series using the random forest algorithm. In: 2020 2nd International Conference on Control Systems, Mathematical Modeling, Automation and Energy Efficiency (SUMMA), pp. 359–264 (2020). https://doi.org/10.1109/SUMMA50634.2020.9280771
6. Tao, H., Yujun, S.: Dynamic monitoring of forest carbon storage based on InVEST model. J. Zhejiang Agric. For. Univ. 33(03), 377–383 (2016)
7. Bayram, B.Ç.: A sustainable forest management criteria and indicators assessment using fuzzy analytic hierarchy process. Environ. Monit. Assess. 193(7), 1–15 (2021). https://doi.org/10.1007/s10661-021-09176-x

Effects of Reservation Profit Level on the Efficiency of Supply Chain Contracts

Shoufu Wan[1], Haiyang Xia[2], and Lijing Tan[1(✉)]

[1] School of Management, Shenzhen Institute of Information Technology, Shenzhen 518172,
China
mstlj@163.com
[2] School of Business, East China University of Science and Technology, Shanghai 200237,
China

Abstract. The level of reservation profit is one of the important factors affecting the efficiency of supply chain. Consider a two-tier supply chain system consisting of a manufacturer and a retailer to study the effects of reservation profit levels of different supply chain members on the feasibility and efficiency of wholesale price contract and agency selling contract. The results show that if the unit production cost is sufficiently low, the agency selling contract is more efficient (less efficient) than the wholesale price contract when the reservation profit level of the retailer is below (above) a certain threshold. However, if the unit production cost is sufficiently high, the wholesale price contract is always superior to the agency selling contract in efficiency. For the feasibility of the two types of contracts, we find that when the unit production cost is sufficiently high, the feasibility range of the wholesale price contract is wider than that of the agency selling contract.

Keywords: Supply chain · Wholesale contract · Agency selling contract · Contract efficiency · Reservation profit

1 Introduction

Supply chain, regarded as an integrated network of providing goods or services to the end customers, consists of multiple members such as suppliers, manufacturers, logistics service companies, and retailers. Often, these parties in the supply chain are endowed with their own economic interests, which will lead to the well-known phenomenon called double marginalization, resulting in the Price of Anarchy. It has been a hot topic in both industry and academia which considers how to design various supply chain contracts to boost supply chain system efficiency including quantity discount contracts, two-part tariff contracts, buyback contracts, quantity flexibility contracts, and revenue sharing contracts [1–5]. [6] conducts an excellent and comprehensive review regarding supply chain coordination with contracts.

The existing literature has examined how to align the incentive of supply chain members through complex management costs in the implementation process. In practice, it is commonly observed that the members of a supply chain always trade with each other

Y. Tan et al. (Eds.): ICSI 2022, LNCS 13345, pp. 379–388, 2022.
https://doi.org/10.1007/978-3-031-09726-3_34

merely using some simple contracts, such as the wholesale price contract, the agency selling contract and so on. Knowing this, it is of great importance and significance to explore how these simple contracts affect the efficiency of the supply chain (i.e., the ratio of decentralized supply chain profit with the contract to the centralized supply chain's optimal profit).

There exists a vast of research on this topic, among which [7] coined the concept of Price of Anarchy (PoA) to quantify the efficiency of the wholesale price contract in a decentralized supply chain system with different configuration structures. [8] utilized Price of Anarchy to quantify the efficiency of the quality-dependent wholesale price contract in competing decentralized reverse supply chain systems. [9] focused on how to improve contract's efficiency by increasing contract parameters in a decentralized supply chain system. Specifically, the author analyzed the contract's efficiency improvement of two-part tariff contract, revenue-sharing contract, and effort cost sharing contract (two parameters contract) compared with the wholesale price contract (one parameter contract).

Different from the aforementioned studies assuming that the reservation profit level of supply chain members are zero. Our paper aims to investigate the impact of the reservation profit level of supply chain members on the supply chain coordination. The main research questions studied are as follows. (1) To what extent will the wholesale price contract or the agency selling contract cause a loss regarding the supply chain system performance, i.e., how efficient are these two types of contracts? And which one is more efficient? (2) Under what conditions could the supply chain members agree to establish the wholesale contract or the agency selling contract between them? (3) What are the preferences of different supply chain members for the two types of contract forms?

2 Model Description

Consider a two-echelon supply chain system (S) consisting of an upstream manufacturer (He) and a retailer (She), where the manufacturer produces a product at a constant unit cost c and sells it to an independent retailer, who then sells the product to the end consumer at a retail price p. We assume that consumer demand is a linear function of the retail price of this product, i.e., $D(p) = a - bp$, where a is the market size and parameter b is the sensitivity coefficient of the retail price. Distinct from the previous literature, different levels of reservation profit for the manufacturer and the retailer are allowed in our model, which reflects the capability they can earn from external alternatives. The two reservation profit levels for the manufacturer and the retailer are represented respectively by V_M and V_R, and we assume that $V_M \geq 0$, $V_R \geq 0$ to avoid trivial cases.

In this paper, we consider two types of widely used contracts between the manufacturer and the retailer: the wholesale price contract (WPC) and the agency selling contract (ASC). Under WPC, the manufacturer first announces the unit wholesale price w, and then the retailer decides whether to accept the offer after observing w. Under ASC, the manufacturer sells his product to the end consumer through the retailer by paying a commission fee α, different from WPC, the retailer does not have ownership of the product throughout the whole sales period.

To provide a benchmark for comparison, we first consider the centralized supply chain scenario under which the manufacturer and the retailer can be considered as a whole to determine the retail price p maximizing the total supply chain profit:

$$\Pi_S^C(p) = (p - c)(a - bp) \tag{1}$$

where the superscript C represents the centralized supply chain. It is easy to find that the optimal price is $p^{C^*} = \frac{a+bc}{2b}$ and the optimal profit is $\Pi_S^{C^*} = \frac{(a-bc)^2}{4b}\pi$. To ensure the two parties have incentive to participate, the condition $0 \le V_M + V_R \le \frac{(a-bc)^2}{4b}$ should be satisfied. In the following, we introduce the definition of contract's efficiency.

Definition 1. *For a supply chain contract K, its contract efficiency is the ratio of the equilibrium profit of the decentralized supply chain system to the optimal profit of the centralized supply chain system, i.e., $E^K = \frac{\Pi_S^{K^*}}{\Pi_S^{C^*}}$.*

3 Feasible Range of Establishing the Wholesale Price Contract and Efficiency Analysis

In this section, we consider the case where the manufacturer and the retailer are governed by a wholesale price contract. As such, a Stackelberg game is played between them with the manufacturer as a leader and the retailer being a follower. Given that a wholesale price w is charged by the manufacturer, the retailer needs to decide whether to accept this offer or not. If she takes it, then she should determine the retail price p of the product. First, we analyze the pricing problem if the retailer accepts the contract by maximizing $\pi_R^W(p) = (p - w)D(p)$. It is readily to obtain that the retailer's optimal retail price is $p^{W^*}(w) = \frac{a+bw}{2b}$ and the corresponding profit is $\pi_R^{W^*}(w) = \frac{(a-bw^2)}{4b}$. Therefore, it is obvious that if the retailer's reservation profit level V_R satisfies $V_R \le \frac{(a-bw^2)}{4b}$, she will accept WPC proposed by the manufacturer, otherwise the retailer rejects.

In the first stage of the game, the manufacturer needs to decide whether to propose a wholesale price contract with wholesale price w to the retailer, which ensures the retailer's equilibrium profit is no less than V_R. Thus, the model of manufacturer is given by

$$max \ \pi_M^W(w) = (w - c)D\left(p^{W^*}(w)\right)$$

$$s.t. \ \frac{\left(a - bw^2\right)}{4b} \ge V_R \tag{2}$$

Solving the above problem yields the manufacturer's optimal wholesale price as

$$w^* = \begin{cases} \frac{a+bc}{2b}, & \text{if } V_R \le \frac{(a-bc^2)}{16b}, \\ \frac{a-2\sqrt{(bV_R)}}{b}, & \text{if } \frac{(a-bc^2)}{16b} \le V_R \le \frac{(a-bc^2)}{4b} \end{cases} \tag{3}$$

The corresponding equilibrium profit of manufacturer is

$$\pi_M^{W*} = \begin{cases} \frac{\left(a-bc^2\right)}{8b}, & \text{if } V_R \leq \frac{\left(a-bc^2\right)}{16b}, \\ (a-bc)\sqrt{\left(\frac{V_R}{b}\right)} - 2V_R, & \text{if } \frac{\left(a-bc^2\right)}{16b} \leq V_R \leq \frac{\left(a-bc^2\right)}{4b} \end{cases} \tag{4}$$

Proposition 1. *In the decentralized supply chain system described above, WPC can be established between the manufacturer and the retailer when one of the following two conditions is satisfied:*

$$(i) \; V_M \leq \frac{\left(a-bc^2\right)}{8b} \; and \; V_R \leq \frac{\left(a-bc^2\right)}{16b};$$

$$(ii) \; V_M \leq (a-bc)\sqrt{\left(\frac{V_R}{b}\right)} - 2V_R \; and \; \frac{\left(a-bc^2\right)}{16b} \leq V_R \leq \frac{\left(a-bc^2\right)}{4b}. \tag{5}$$

Proposition 1 gives the feasible range of establishing WPC when both the manufacturer and retailer have non-zero reservation profits. As shown in Fig. 1, the regions I, II and III characterize the feasible ranges for all possible supply chain contracts, i.e., $0 \leq V_M + V_R \leq \frac{(a-bc^2)}{4b}$, $V_M \geq 0$, $V_R \geq 0$. According to Proposition 1, WPC between the manufacturer and the retailer can be reached if the combination of their reservation profits lies in region I ((i) is satisfied) or region II ((ii) is satisfied).

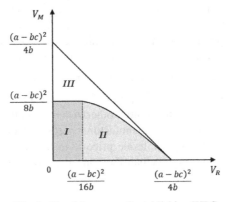

Fig. 1. Feasible range of establishing WPC

Corollary 1.

(i) *Denoted by w^{i*}, p^{i*} the equilibrium wholesale price and retail price when (V_R, V_M) is located in region $i (i \in \{I, II\})$, then $w^{I*} \geq w^{II*}, p^{I*} \geq p^{II*}$;*

(ii) *Denoted by π_M^{i*}, π_R^{i*}, the equilibrium wholesale price and retail price when (V_R, V_M) is located in region $i (i \in \{I, II\})$, then $\pi_M^{I*} \geq V_M, \pi_R^{I*} \geq V_R, \pi_R^{II*} \geq V_R$;*

According to Corollary 1, compared to the reservation profit levels combination (V_R, V_M) in region *II*, when the reservation profit levels of the manufacturer and the retailer fall in region *I* (i.e., the retailer's reservation profit level is low and the manufacturer's reservation profit level is moderate), the manufacturer will impose a higher wholesale price and the retailer will set a higher retail price. As a result, both will obtain equilibrium profits higher than their respective reservation profit levels. In region *II*, the retailer's equilibrium profit is equal to her reservation profit level, and the manufacturer's equilibrium profit is no less than his reservation profit level.

Proposition 2. *With respect to the efficiency of WPC, the following holds.*

(i) *In region I, the efficiency of WPC established between the manufacturer and the retailer satisfies $E_I^W = \frac{3}{4}$, and that in region II satisfies $E_{II}^W = \frac{(a-bc)\sqrt{(bV_R)}-4bV_R}{(a-bc^2)}$;*

(ii) *$E_{II}^W \geq E_I^W$, and E_{II}^W is monotonically increasing in the retailer's level of reservation profit V_R. Furthermore, when V_R reaches the upper bound $\frac{(a-bc^2)}{4b}, E_{II}^{W*} = 1$.*

In Proposition 2, we reveal that the efficiency of WPC in feasible regions *I*, *II*, and the effect of reservation profit level on the contract's efficiency. We find that the efficiency of wholesale price contract does not contingent on the manufacturer's reservation profit V_M (although the feasible establishment range of WPC is related to V_M). The impact of the retailer's reservation profit level V_R on the efficiency of WPC can be divided into two subcases. As shown in Fig. 2, when $V_R \in [0, \frac{(a-bc^2)}{16b}]$, the efficiency of WPC does not vary with the V_R and remains at $\frac{3}{4}$; when $V_R \in [\frac{(a-bc^2)}{16b}, \frac{(a-bc^2)}{4b}]$, the efficiency of WPC increases in V_R until it reaches 1.

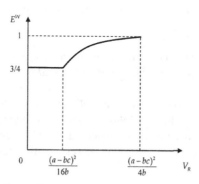

Fig. 2. The impact of the retailer's reservation profit level V_R on the efficiency of WPC

4 Feasible Range of Establishing the Agency Selling Contract and Efficiency Analysis

In this section, we consider the case where an agency selling contract is considered between the manufacturer and the retailer, which is also a two-stage Stackelberg game between the two parties. In the second stage, the retailer needs to decide: first, whether to accept ASC provided by the manufacturer; second, if so, how to determine the retail price p of the product. If the retailer accepts this contract, say, she agrees on the commission percentage α, then she needs to determine the retail price p by maximizing $\pi_R^{AS}(p) = \alpha p D(p)$. Using the first-order condition, it is easy to obtain that the retailer's optimal retail price for a given commission percentage α is $p^{AS^*}(\alpha) = \frac{a}{2b}$, which is independent of α, and the equilibrium profit is $\pi_R^{AS^*}(\alpha) = \alpha \frac{a^2}{4b}$. Given the above results, it is safely to say that if the retailer's reservation profit level meets $V_R \leq \alpha \frac{a^2}{4b} (\alpha \geq \frac{4bV_R}{a^2})$, the retailer will accept ASC provided by the manufacturer, otherwise the retailer rejects the contract.

In the first stage, the manufacturer needs to determine whether to initiate an agency selling contract and how to pay unit commission fee α to the retailer such that the retailer's equilibrium profit at least reaches V_R. Thus, the objective of manufacturer can be expressed as follows:

$$max \; \pi_M^{AS}(\alpha) = (1-\alpha)p^{AS^*}(\alpha)D\left(p^{AS^*}(\alpha)\right) - cD\left(p^{AS^*}(\alpha)\right)$$

$$s.t. \; \frac{a^2\alpha}{4b} \geq V_R \qquad (6)$$

Since the objective function of the manufacturer $\pi_M^{AS}(\alpha) = \frac{a}{4b}(a - 2bc - a\alpha)$ is strictly decreasing in α, to induce the retailer accept the contract, the commission percentage must satisfy $\alpha \geq \frac{4bV_R}{a^2}$, hence, the optimal sale commission percentage set by the manufacturer is $\alpha^{AS^*} = \frac{4bV_R}{a^2}$. In this case, the retailer's equilibrium profit is equal to her reservation profit, i.e., $\pi_R^{AS^*} = V_R$ and the equilibrium profit of the manufacturer is $\pi_M^{AS^*} = \frac{a}{4b}(a - 2bc) - V_R$.

Proposition 3. *In the above decentralized supply chain system, ASC can be established if the level of reservation profit of the manufacturer and the retailer (V_M, V_R) satisfies $V_M + V_R \leq \frac{a}{4b}(a - 2bc)$.*

Proposition 3 presents conditions for establishing ASC between the manufacturer and the retailer. As shown in Fig. 3, if the combination of the reservation profits of the manufacturer and the retailer lies in the shaded area, ASC can be established. As the production cost c increases, $\frac{a}{4b}(a - 2bc)$ decreases, i.e., the feasible establishment area for ASC keeps dwindling.

Corollary 2. *If the unit production cost satisfies $c > \frac{a}{2b}$, the manufacturer and the retailer will never choose ASC between them.*

Next, we analyze the efficiency of ASC within the feasible establishment ranges.

Proposition 4. *The efficiency of ASC holds for the following statements:*

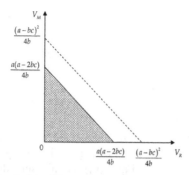

Fig. 3. Feasible establishment range of ASC

(i)   *The efficiency of ASC is* $E^{AS} = \frac{a(a-2bc)}{(a-bc^2)}$;

(ii) *The efficiency of ASC E^{AS} is irrelevant to each member's reservation profit level V_M and V_M;*

(iii) *The efficiency of ASC E^{AS} is strictly decreasing in the unit production cost c.*

In Proposition 4, we characterize the efficiency of ASC and analyze the effect of the manufacturer's reservation profit level V_M, the retailer's reservation profit level V_R, and the unit production cost c on the contract's efficiency.

5   Comparative Analysis of Wpc and the Agency Selling Contract

In this section, we compare WPC and ASC in term of the feasible establishment range of contracts, contract's efficiency and the equilibrium profits of the manufacturer and the retailer.

Proposition 5. *If the unit production cost $c \geq \frac{(\sqrt{2}-1)a}{b}$, the feasible establishment range under WPC is broader than that of ASC.*

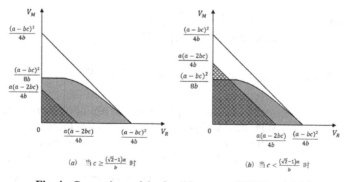

Fig. 4. Comparison of the feasible range of WPC and ASC

Figure 4 illustrates the feasible establishment ranges between WPC and ASC. It can be seen that when the unit production cost satisfies $c \geq \frac{(\sqrt{2}-1)a}{b}(\frac{a(a-2bc)}{4b} \leq \frac{(a-bc)^2}{8b})$, the feasible establishment region for ASC (grid shaded area in Fig. 4(a)) lies within the feasible establishment region for WPC (gray shaded area in Fig. 4(a)). It means that WPC has a broader feasible region than ASC, i.e., given any combination of reservation profit level set (V_M, V_R) in the region, it is always possible to choose WPC if ASC is in consideration, however, the opposite is not true.

When the unit production cost satisfies $c < \frac{(\sqrt{2}-1)a}{b}(\frac{a(a-2bc)}{4b} > \frac{(a-bc)^2}{8b})$, a part of the feasible establishment region of ASC (grid shaded area in Fig. 4(b)) is outside that of WPC (gray shaded area in Fig. 4(b)). As a result, there exists a region, i.e., $\{(V_M, V_R)| \frac{(a-bc)^2}{8b} < V_M \leq \frac{a(a-2bc)}{4b}, V_M + V_R \leq \frac{a(a-2bc)}{4b}\}$, in which ASC can be established, but WPC is out of consideration.

Proposition 6. *Under ASC, the retailer's profit is always equal to her reservation profit level* V_R; *while under WPC, the retailer's equilibrium profit is* $\max\{V_R, \frac{(a-bc)^2}{16b}\}$.

According to the Proposition 6, when the retailer's reservation profit level is lower than $\frac{(a-bc)^2}{16b}$, she will better off if the manufacturer offers WPC. However, if the retailer's reservation profit level satisfies $V_R \geq \frac{(a-bc)^2}{16b}$, it makes no difference which kind of contract form is provided by the manufacturer, because the retailer can only gain an equilibrium profit equal to her reservation profit level under either contract.

Proposition 7. *Comparing the manufacturer's equilibrium profit between WPC and ASC, we obtain:*

1) *If* $0 < c \leq \frac{a}{3b}$, *then*

   (i)   *when* $V_R \in [0, \frac{(a-2bc)^2}{4b}]$, $\pi_M^{AS*} \geq \pi_M^{W*}$;
   (ii) *when* $V_R \in [\frac{(a-2bc)^2}{4b}, \frac{a(a-2bc)}{4b}]$, $\pi_M^{W*} \geq \pi_M^{AS*}$;

2) *If* $\frac{a}{3b} < c \leq \frac{(\sqrt{2}-1)a}{b}$, *then*

   (i)   *when* $V_R \in [0, \frac{a^2-2abc-b^2c^2}{8b}]$, $\pi_M^{AS*} \geq \pi_M^{W*}$;
   (ii) *when* $V_R \in [\frac{a^2-2abc-b^2c^2}{8b}, \frac{a(a-2bc)}{4b}]$, $\pi_M^{W*} \geq \pi_M^{AS*}$;

3) *if* $\frac{(\sqrt{2}-1)a}{b} < c \leq \frac{a}{2b}$ *and* $V_R \in [0, \frac{a(a-2bc)}{4b}]$, $\pi_M^{W*} \geq \pi_M^{AS*}$.

Proposition 7 shows that the relationship between the equilibrium profits received by the manufacturer under WPC and ASC is related to the value of the unit production cost c. On the one hand, when the product cost c is relatively small, there exists a threshold for the retailer's reservation profit level V_R. On the other hand, if the unit product cost c is sufficiently large, i.e., $c > \frac{(\sqrt{2}-1)a}{b}$, the manufacturer can always gain a higher equilibrium profit level by adopting WPC compared with ASC.

Proposition 8. *Comparing the efficiency of WPC with that of ASC, we derive:*

1) *when* $0 < c \leq \frac{a}{3b}$, *then*

   (i) *if* $V_R \in [0, \frac{(a-2bc)^2}{4b}]$, $E^{AS} \geq E^W$;

   (ii) *when* $V_R \in [\frac{(a-2bc)^2}{4b}, \frac{a(a-2bc)}{4b}]$, $E^W \geq E^{AS}$;

2) *when* $\frac{a}{3b} < c \leq \frac{a}{2b}$ *and* $V_R \in [0, \frac{a(a-2bc)}{4b}]$, $E^W \geq E^{AS}$.

Proposition 8 reveals the relationship of the efficiency of WPC and ASC. We show that if the unit production cost c is relatively small, i.e., $c \leq \frac{a}{3b}$, ASC is more efficient than WPC given the retailer reservation profit level V_R is relatively low (i.e., $V_R \leq \frac{(a-2bc)^2}{4b}$). When the retailer is endowed with a high level of reservation profit V_R (i.e., $V_R \geq \frac{(a-2bc)^2}{4b}$), WPC is more efficient than ASC. Likewise, the manufacturer will also choose WPC as claimed by Proposition 6.

When the unit production cost satisfies $c \geq \frac{a}{3b}$, the efficiency of WPC is always higher than that of ASC. It means that WPC can generate higher profit level to the whole supply chain compared to ASC under the same condition. However, from Proposition 7, when $\frac{a}{3b} < c < \frac{(\sqrt{2}-1)a}{b}$ and $V_R \in [0, \frac{a^2 - 2abc - b^2c^2}{8b}]$, we have $\pi_M^{AS*} \geq \pi_M^{W*}$, which means that the manufacturer should choose ASC under this circumstance.

6 Conclusion

This paper examines how the reservation profit levels of the manufacturer and the retailer affect the feasible establishment range of wholesale price contract and the agency selling contract as well as the corresponding contract's efficiency. First, we show that when the unit production cost is relatively low, the agency selling contract is more efficient if the retailer's reservation profit level is low, whereas higher supply chain performance can be achieved with the wholesale price contract when the retailer's reservation profit level is high. Second, we find that the wholesale price contract has a broader feasible establishment range than the agency selling contract when the manufacturer's unit production cost is sufficiently high. Third, when the unit production cost is relatively low, if the retailer's reservation profit level is below (above) a certain threshold, the manufacturer prefers the agency selling contract (the wholesale price contract). When the unit production cost is relatively high, the manufacturer betters off under the wholesales price contract which can also benefit the retailer given her reservation profit is low.

Acknowledgement. The work was supported by The Natural Science Foundation of Guangdong Province (No. 2020A1515010752).

References

1. Jeuland, A.P., Shugan, S.M.: Managing channel profits. Mark. Sci. **2**(3), 239–272 (1983)
2. Weng, Z.K.: Channel coordination and quantity discounts. Manage. Sci. **41**(9), 1509–1522 (1995)
3. Pasternack, B.A.: Optimal pricing and return policies for perishable commodities. Mark. Sci. **4**(2), 166–176 (1985)
4. Tsay, A.A.: The quantity flexibility contract and supplier-customer incentives. Manage. Sci. **45**(10), 1339–1358 (1999)
5. Cachon, G.P., Lariviere, M.A.: Supply chain coordination with revenue-sharing contracts: strengths and limitations. Manage. Sci. **51**(1), 30–44 (2005)
6. Cachon, G.P.: Supply chain coordination with contracts. Handb. Oper. Res. Manage. Sci. **11**, 227–339 (2003)
7. Perakis, G., Roels, G.: The price of anarchy in supply chains: quantifying the efficiency of price-only contracts. Manage. Sci. **53**(8), 1249–1268 (2007)
8. Ye, Y.S., Ma, Z.J., Dai, Y.: The price of anarchy in competitive reverse supply chains with quality-dependent price-only contracts. Transp. Res. Part E: Logist. Transp. Rev. **89**, 86–107 (2016)
9. Yan, X.: Contract efficiency for a decentralized supply chain in the presence of quality improvement. Int. Trans. Oper. Res. **22**(4), 713–734 (2015)

Optimal Synthesis of a Motion Generation Six-Bar Linkage

Pakin Phromphan, Jirachot Suvisuthikasame, Metas Kaewmongkol,
Woravech Chanpichitwanich, and Suwin Sleesongsom[✉] ⓘ

Department of Aeronautical Engineering, International Academy of Aviation Industry, King
Mongkut's Institute of Technology Ladkrabang, Bangkok 10520, Thailand
suwin.se@kmitl.ac.th

Abstract. This research presents a new motion generation technique combined
with the latest updated of teaching-learning-based optimization with a diversity
archive (ATLBO-DA), which has been proved the performance in the previous
study. The six-bar prototype in this study is formulated by creating the second loop
attached to the simple four-bar model, which expects to follow the Watt I model.
Two specific motion generation problems are used to test the performance of this
technique. The objective function is created in the form of weighted sum between
the position error and the angle error to make it can perform in a single objective
optimization problem. The performance of the optimizations with different weight
is investigated in this research. The results show that the purposed technique
performs with a good accuracy in the specific motion without prescribed timing
problems.

Keywords: Motion generation · Six-bar linkage · Optimization · Metaheuristic

1 Introduction

The mechanism is a processing device used to transfer the input motion into the desired
output motion, for example, a mechanism operation of a machine that relies on gears
joined together to make the machine work [1]. The four-bar mechanism has been used
regularly in a lot of applicants throughout the world such as plane flap, bicycle paddling,
pump jack, etc. Unfortunately, there are many specific works that require a unique
mechanism, which is beyond the movement limitation of the four-bar linkage motion. In
the past, steering linkage had been performed with a four-bar linkage, but at present it was
replaced with the six-bar linkage. It has become the reason why many researchers aim
to develop a new mechanism to receive the specific task [2], which expected the task can
be alleviated using a six-bar linkage. The six-bar linkage is separated into two categories
based on their linkage rearrangement such as the Watt chain and Stephenson chain [3,
4]. There are an analysis studies of the metaheuristic optimizer for solving the six bar
linkage [5, 6]. The model in this research is the six-bar watt I mechanism extended from
the traditional four-bar linkages model from the previous work [7, 8]. The method has
an expect to synthesize and demonstrate the behavior of the mechanism can be varied

© Springer Nature Switzerland AG 2022
Y. Tan et al. (Eds.): ICSI 2022, LNCS 13345, pp. 389–398, 2022.
https://doi.org/10.1007/978-3-031-09726-3_35

depending on the objective task i.e., type synthesis, and dimensional synthesis. The major synthesis of the last category is motion generation which is based on dimensional synthesis. The motion generation technique is used to find the optimum form of the linkages and focuses on the discrepancy of the desired point movement and angle on the coupler link. Since full motion rotating constraint techniques relied on the Grashof criterion are included in the dimensional synthesis in many types of research [7–9], this experiment will adopt the same constraint techniques to simplify the mechanism and enhance the optimization performance. Evolutionary optimization has been applied in this analysis for the convenience and the efficiency result. The previous research showed that the important factors leading to the computational synthesis performance were the constraint handling technique and the type of the optimizer [10–12]. Yet, some studies on four-bar synthesis found that the traditional model might not be appropriate for the complicated or specific task [12] due to the limitation of the mechanism itself, which triggers the aim of this research to study the six-bar mechanism.

The optimizer used to synthesize in this study is the adjusted version of the TLBO called teaching-learning-based optimization with a diversity archive which provides a good result in the path generation of the four-bar linkage [13]. The problem used in this study is divided into two cases: rectilinear shape [7] and full circular shape [7]. The objective function in the problem is using the weighted sum technique to find the best solution [7, 12]. The quantity difference of the inserted weight becomes another factor that needs to be investigated in this study as it has an influence on the ability of the optimizer.

The paper starts with a mathematical expression of the position in brief followed by the detail of constraint handling technique. The various design problems and design results are given in order. The design result is reported in statistics to compare the difference of the solution. The overall discussion and conclusion of this study are placed in the last section.

2 Methodology

2.1 The Position and Angle Analysis of the Six-Bar Model

The six-bar model is extended from the simple four-bar model with the additional loop attached to the coupler link and the last link of the traditional four-bar. The connected joint C between the based model and the extended model has been fixed to reduce the output degree of freedom. This technique makes the model more convenient to study. This type of model can be categorized as the Watt I model from the ternary link on link 3(BD) and link 4(DC). Link 1 (AD) is designed as a fixed frame. The input link is link 2 (O_2B), while the output is link 6 (FG). Link 5 (EF) and link 6 (FG) are attached to the extended link 4, 2 (CE) and link 3, 2(CG), respectively. The point of interest is held by link 6 (FG). Normally, the possibility to create another form of the six-bar model is varied to formed open and cross mode which can happen at link 4, 1 (CD) and link 6 (FG). In study, the major focus of the model will be considered only the formed open mechanism. The kinematic diagram shown in Fig. 1 is the targeted model in the global axis which included 7 joints with 6 linkages. The constraint handling technique is used at the diving link to ensure that the machine can operate properly in one revolution without any issue. The position and angle of the point of interest can be evaluated by the known pertinent parameter $r_1, r_2, r_{3,1}, r_{3,2}, r_{4,1}, r_{4,2}, r_5, r_6, r_{px}, r_{py}$ and others to define the motion of the target point. The outcome of the expression is shown as the equation below

$$x_p = x_{O2} + r_2 \cos(\theta_2 + \theta_1) + (r_{3,1} + r_{3,2})\cos(\theta_3 + \theta_1) \tag{1}$$

$$- r_{px} \cos(\theta_6 + \theta_1) + r_{py}\sin(\theta_6 + \theta_1)$$

$$y_p = y_{O2} + r_2 \sin(\theta_2 + \theta_1) + (r_{3,1} + r_{3,2})\sin(\theta_3 + \theta_1)$$

$$+ r_{px} \cos(\theta_6 + \theta_1) + r_{py}\sin(\theta_6 + \theta_1)$$

where x_{O2} and y_{O2} are the starting position of the mechanism at revolute joint O_2 in the global coordinate axis. Noticeably, other parameters have to append with θ_1 to transform their presence to be the same coordinate as the origin pin joint.

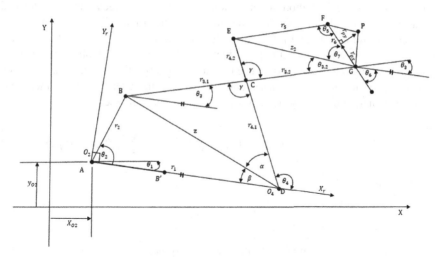

Fig. 1. Six bar linkage in the global coordinate axis

The values θ_3 and γ are defined by the known of the values $r_1, r_2, r_{3,1}, r_{4,1}$ which can be changed respectively with any degree of the crank angles (θ_2).

The angle $\theta_{3,2}$ and θ_7 is required to achieve the angle of the point of interest (θ_6) which is expressed in the equation below:

$$\theta_{3,2} = \cos^{-1}\left[\frac{Z_2^2 + r_{3,2}^2 - r_{4,2}^2}{2z_2 r_{3,2}}\right] \tag{2}$$

$$\theta_7 = \cos^{-1}\left[\frac{Z_2^2 + r_6^2 - r_5^2}{2z_6 r_6}\right] \tag{3}$$

The angle θ_6 can determine by the equation expressed as follow:

$$\theta_6 = \theta_{3,2} + \theta_7 - \theta_3 \tag{4}$$

2.2 Optimization Constraint Handling Techniques

To analyze the motion generation problem, two factors needed to be concerned are the error of the distance and the error of the angle. The error from the distance is calculated by the interval between the desired point (x_d, y_d) and the actual point $P_d(x_p, y_p)$, while the error of the angle can be measured by the difference in radian from the desired angle (θ_{6d}) and the actual angle (θ_{6p}). Both errors will become the objective function for the optimization problem in terms of the sum square. Two constraints are needed in this study. The first one is the mechanism according to the Grashof criterion to assure that the crank-rocker mechanism will move the crank linkage without any issue. Since the first loop of the model is the driving part of the mechanism, the constraint handling technique will only be considered at the first half of the mechanism. The second constraint is a sequence of the crank which should be rotated in a direction (clockwise or counterclockwise). The motion generation problem in this study can be performed with a non-prescribed time synthesis meaning that the driving link angle and other parameters are un-determined [10]. The weighted-sum technique is one of a classical multi-objective optimization technique, which sets up weight of each objective function before performing optimization run. The priority of the objective functions is preferred as it may affect the performance in searching of the optimizer. The weighting factor has to be investigated in this study. The constraint equation is expressed below:

$$\min f(x) = w_1 \sum_{i=1}^{N} [(x_{d,i} - x_{p,i})^2 + (y_{d,i} - y_{p,i})^2] + w_2 \sum_{i=1}^{N} [(\theta_{6d,i} - \theta_{6p,i})^2] \tag{5}$$

Subject to

$$w_1 = 1 - w_2 \tag{6}$$

$$\min(r_1, r_2, r_{3,1}, r_{4,1}) = crank(r_2) \tag{7}$$

$$2\min(r_1, r_2, r_{3,1}, r_{4,1}) + 2\max(r_1, r_2, r_{3,1}, r_{4,1}) < (r_1, r_2, r_{3,1}, r_{4,1}) \tag{8}$$

$$\theta_2^1 < \theta_2^2 \ldots < \theta_2^N \tag{9}$$

$$x_L \leq x \leq x_U \tag{10}$$

where $x = \{r_1, r_2, r_{3,1}, r_{3,2}, r_{4,1}, r_{4,2}, r_5, r_6, r_{px}, r_{py}, \theta_2^i\}^T$ is the optimum design variables, which can obtain from the optimizer. The series of the link length values will be combined to create the mechanism except the crank angle value θ_2^i which has been prepared before the optimization process. The crank angle in this research θ_2^i is divided into 200 intervals equally from $0-360°$ due to the first constraint. The efficiency of the result can be diversified according to the number used in the crank angle sequence (9). The limitation of a design variable (10) is included to scope the working space of the mechanism, which made the huge role in the optimization performance. The boundary of the limitation can affect the possibility of the designed mechanism and the precision of the result. Once the synthesis process is done and mechanism motion is defined, the sum square error will be calculated by the equation below:

$$f(x) = \sum_{i=1}^{N} min\,(w_1 d_{ij}{}^2 + w_2 \theta_{ij}^2) \tag{11}$$

where $d_{ij}{}^2 = (x_{d,i} - x_{p,j})^2 + (y_{d,i} - y_{p,j})^2$ and $\theta_{ij}^2 = (\theta_{6d,i} - \theta_{6p,i})^2$ for $j = 1, \ldots, N$. The origin of this equation was explained in [10–13].

3 Numerical Experiment

The optimization problem in this study is the traditional problem that used to be solved by the four-bar mechanism [7, 12]. The purpose of these models is to compare and indicate the efficiency of this mechanism. Two problems are selected for this experiment. The design variable limitations are assigned as the previous research to make the comparison more precise.

Case 1: Rectilinear Shape Problem
Design variables:

$$\left\{ r_1, r_2, r_{3,1}, r_{3,2}, r_{4,1}, r_{4,2}, r_5, r_6, r_{px}, r_{py}, x_0, y_0, \theta_1, \theta_2^1, \theta_2^2, \theta_2^3, \theta_2^4, \theta_2^5, \theta_2^6 \right\}^T$$

Desired point:

$$r_d^i = \{(20, 20), (20, 25), (20, 30), (20, 35), (20, 40), (20, 45)\}$$

$$\theta_d^i = \{1.9937, 1.9220, 1.8434, 1.7599, 1.6709, 1.5735\}\,rad$$

Limitation of design variables:

$$5 \le r_1, r_2, r_{3,1}, r_{3,2}, r_{4,1}, r_{4,2}, r_5, r_6 \le 60$$

$$-60 \le x_0, y_0, r_{px}, r_{py} \le 60$$

$$0 \le \theta_1, \theta_2^1, \theta_2^2, \theta_2^3, \theta_2^4, \theta_2^5, \theta_2^6 \le 2\pi$$

Case 2: Full Circular Shape Problem
Design variables:

$$\left\{ r_1, r_2, r_{3,1}, r_{3,2}, r_{4,1}, r_{4,2}, r_5, r_6, r_{px}, r_{py}, x_0, y_0, \theta_1, \theta_2^1, \theta_2^2, \theta_2^3, \theta_2^4, \theta_2^5, \theta_2^6, \theta_2^7, \theta_2^8, \theta_2^9, \theta_2^{10} \right\}^T$$

Desired point:

$$r_d^i = \left\{ \begin{array}{l} (20, 10), (17.66, 15.142), (11.736, 17.878), (5, 16.928), (0.60307, 12.736), \\ (0.60307, 7.2638), (5, 3.0718), (11.736, 2.1215), (17.66, 4.8577), (20, 10) \end{array} \right\}$$

$$\theta_d^i = \{0.4208, 0.5117, 0.7433, 0.9910, 1.1394, 1.1296, 0.9599, 0.7322, 0.5257, 0.4208\} \text{ rad}$$

Limitation of design variables:

$$5 \le r_1, r_2, r_{3,1}, r_{3,2}, r_{4,1}, r_{4,2}, r_5, r_6 \le 60$$

$$-60 \le x_0, y_0, r_{px}, r_{py} \le 60$$

$$0 \le \theta_1, \theta_2^1, \theta_2^2, \theta_2^3, \theta_2^4, \theta_2^5, \theta_2^6, \theta_2^7, \theta_2^8, \theta_2^9, \theta_2^{10} \le 2\pi$$

These two design problems will be solved by the new optimizer called the teaching-learning-based optimization with a diversity archive (ATLBO-DA). The algorithm already proved the dimensional synthesis effectiveness in the previous four-bar mechanism research [14, 15]. The variables of the optimizer are determined as IReset = 20, IRange = 5, and $\delta = 1$. The population is set up as $n_p = 100$ while the maximum iteration is about 2000 times. Each problem is solved for 30 times to gain an accurate result. A simple design experimental study for weighted-sum factor value is investigated its effect to the motion generation design results. Each of the weight factors is made up the trend of the objective function. The objective with higher weighting factor tends to have the priority in the objective function than the other one. The weight set varies closed to 0.5, which was consistent with the previous work [12]. The chosen weighted-sum factor values are:

First case:

$$w_1 = 0.35, w_2 = 0.65$$

Second case:

$$w_1 = 0.5, w_2 = 0.5$$

Third case:

$$w_1 = 0.65, w_2 = 0.35$$

4   Design Result

The design result is shown in Table 1. There is mean objective function value (mean), best objective function value (min), standard deviation (std), and the number of successful runs (Success). To find the best minimum error (min) from those 30 algorithms, there are different weighting values of each case. From case1, the best minimum error is equal to 0.0011; the second-best is 0.005; and the worst is 0.00693. Obviously, "0.5 weight" represents the best result for the problem in case 1. However other weight values (0.35, 0.65) are accounted for moderate result. So, when comparing to the previous work, it can be seen that the result of mechanism with six-bar linkages is more accurate than the traditional four bar model.

Table 1. The design result in case 1 with different weight

Parameters	Case-1		
Weight	0.35	0.5	0.65
r_1	5.2267	55.8482	35.5164
r_2	5.1957	5.0002	17.3689
r_3	13.8803	15.4604	22.7408
$r_{3,2}$	43.0049	52.6132	31.0391
r_4	13.8998	57.6218	40.8685
$r_{4,2}$	11.2591	9.7714	5.0218
r_5	47.2722	11.1257	20.7905
r_6	59.9493	46.6557	33.5357
r_{px}	43.3628	53.9578	31.8485
r_{py}	2.9665	-35.5398	1.2379
x_0	36.0942	59.2403	29.2874
y_0	23.8601	-11.5585	22.5163
θ_0	2.3803	2.9193	2.5352
Target Error	0.0044	1.54E-04	0.0033
Angle Error	5.59E-04	9.49E-04	3.60E-03
mean	0.4078	0.099649	0.156968
max	0.21439	0.40267	0.32741
min	5.00E-03	1.10E-03	6.93E-03
std	0.98172	0.106241	0.116965
Success.*	24	23	21

In case 2 the results are reported in Table 2. The best minimum error is equal to 0.15074 at 0.5 weighting factor. The second best of 0.65 weight is equal to 0.36586, and the worst case is 0.36629 of 0.35 weight. It found that the result gives a better performance than the previous work. The statistical values also affirmed the improvement of the result by the decrease in mean (0.099649) and standard deviation (0.106241). As expected, when focusing on the optimum path error by controlling weight at 0.65, the solution in path objective is insignificant. On the other hand, the amount of error obtained from the angle is the highest from all of three cases due to the lack of the weight itself which made the optimizer more careless in finding the minimum angle error, so the final result of the weight factor, either 0.65 or 0.35, is still not the optimum solution that we want to find out. In summary, the best-balanced weight factor ($w = 0.5$) is the optimum solution from this problem since it focuses on finding the solution in path and in angle equally,

Table 2. The design result in case 2 in different weighting factor

Parameter	Case-2		
Weight	0.35	0.5	0.65
r_1	46.1114	35.5721	43.5532
r_2	5	9.8878	5
r_3	13.7932	18.0779	15.6828
$r_{3,2}$	5	6.0493	5.6885
r_4	49.2639	35.9806	48.0339
$r_{4,2}$	7.6663	34.8798	6.4985
r_5	5.5256	24.8591	5.81
r_6	5	54.9828	5
r_{px}	-22.8141	-77.9131	-47.3881
r_{py}	-33.4587	-2.6495	-37.2992
x_0	24.9098	-45.5196	14.215
y_0	57.1212	60.5772	79.4463
θ_0	0.3436	0.1325	0.3573
Target Error	0.245	0.0845	0.0982
Angle Error	0.1213	0.0663	0.2677
mean	1.598197	1.480009	3.874234
max	5.649	5.7393	13.4044
min	0.36629	0.15074	0.36586
std	1.398474	1.721310578	4.942765
Success.*	12	10	7

while the other weighting factors provide the similar result, which is not the best (Figs. 2 and 3).

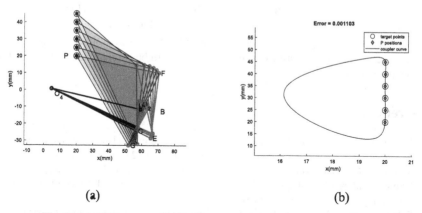

| (a) | (b) |

Fig. 2. (a) Optimum mechanism for case 1 (b) Optimum path for case 1

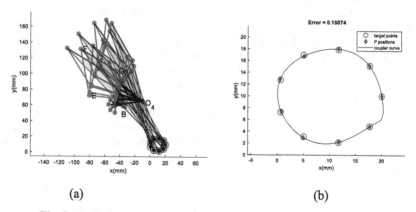

| (a) | (b) |

Fig. 3. (a) Optimum mechanism for case 2 (b) Optimum path for case 2

5 Conclusion

The research first starts with the new model of six-bar linkage by extending the model from the traditional four-bar model with an assumption to improve its performance for a specific task. The task has been studied for a motion generation synthesis with four-bar mechanism. There is not much research conducted by using six-bar linkage with Metaheuristic algorithm. That is our first aim. According to the results of the motion generation six-bar synthesis, the minimum error for case one is 0.0011 and case two is 0.15074 meaning that the new model performs better than the traditional four-bar model in both situations. The proposed six-bar model and the new optimizer also improve the consistency of the solution in motion generation synthesis. Another interesting result is the effect of the weighting factor. It can be seen that the optimum results in each case are varied to the weighting factor. Therefore, it can be concluded that the effect of the weight can be varied depending on the objective of the study. However, the most effective weighting factor for the overall optimum mechanism is 0.5. This result shows that the six-bar motion generation synthesis is effectiveness in problems as same as the path generation did. For future work, this technique will be extended to analyze the high lift mechanism synthesis.

References

1. Han, J., Liang, L., Zhang, H., Zhao, Y.: Dynamics modeling and experimental investigation of gear mechanism with coupled small clearances. Entropy **23**(7), 834 (2021)
2. McCarthy, J.: Mechanism synthesis theory and the design of robots. In: Proceedings 2000 Millennium conference IEEE International Conference on Robotics and Automation. Symposia Proceedings, vol. 1, pp. 55–60 (2000)
3. Soh, G., Robson, N.: Kinematic synthesis of minimally actuated multi-loop planar linkages with second order motion constraints for object grasping. In: ASME 2013 Dynamic Systems and Control Conference, DSCC (2013)
4. Plecnik, M., McCarthy, J.: Kinematic synthesis of Stephenson III six-bar function generators. Mech. Mach. Theory **97**, 112–126 (2016)
5. Jianwei, S., Peng, W., Wenrui, L., Jinkui, C., Luquan, R.: Synthesis of multiple tasks of a planar six-bar mechanism by wavelet series. Inverse Probl. Sci. Eng. **27**(3), 388–406 (2019)
6. Peón-Escalante, R., Jiménez Cuenca, F., Soberanis Escalante, M.A., Peñuñuri, F.: Path generation with dwells in the optimum dimensional synthesis of Stephenson III six-bar mechanisms. Mech. Mach. Theory **144**, 103650 (2020)
7. Phukaokaew, W., Sleesongsom, S., Panagant, N., Bureerat, S.: Synthesis of four-bar linkage motion generation using optimization algorithms. Adv. Comput. Design **4**(3), 197–210 (2019)
8. Sleesongsom, S., Bureerat, S.: Alternative constraint handling technique for four-bar linkage motion generation. In: IOP Conference Series: Materials Science and Engineering, vol. 23, p. 834 (2019)
9. Nafees, K., Mohammad, A.: Optimal dimensional synthesis of six-bar Stephenson I mechanism for path generation. Int. J. Mech. Eng. Technol. **7**(6), 535–546 (2018)
10. Sleesongsom, S., Bureerat, S.: Four-bar linkage path generation through self-adaptive population size teaching-learning based optimization. Knowl.-Based Syst. **135**, 180–191 (2017)
11. Sleesongsom, S., Bureerat, S.: Optimal synthesis of four-bar linkage path generation through evolutionary computation with a novel constraint handling technique. Comput. Intell. Neurosci. Article ID 5462563 (2018)
12. Sleesongsom, S., Bureerat, S.: Optimization of a high-lift mechanism motion generation synthesis using MHS. In: Tan, Y., Shi, Y. (eds.) ICSI 2021. LNCS, vol. 12689, pp. 38–45. Springer, Cham (2021). https://doi.org/10.1007/978-3-030-78743-1_4
13. Bureerat, S., Sleesongsom, S.: Constraint handling technique for four-bar linkage path generation using self-adaptive teaching–learning-based optimization with a diversity archive. Eng. Optim. **53**, 513–530 (2021)
14. Gündoğdu, Ö., Erentürkb, K.: Fuzzy control of a dc motor driven four-bar mechanism. Mechatronics **15**(4), 423–438 (2005)
15. Gogate, G.R.: Explicit input link rotatability analysis of Watt six-link mechanisms. J. Mech. Sci. Technol. **32**(7), 3407–3417 (2018). https://doi.org/10.1007/s12206-018-0644-4

Reliability-Based Design Optimization of a Goland Wing with a Two-Step Approach

Suwapat Chanu, Alfan Wattanathorn, Moses Senpong, and Suwin Sleesongsom[✉][ID]

Department of Aeronautical Engineering, International Academy of Aviation Industry, King Mongkut's Institute of Technology Ladkrabang, Bangkok 10520, Thailand
suwin.se@kmitl.ac.th

Abstract. This research proposes to design a Goland wing structure using a two-step approach that concerns the last step uncertainty. The first step starts with performing the design optimization for multi-purposes followed by the design of wing mass, stress, and buckling factor and the reliability test of all solution set. Due to the aircraft wing being subjected to aerodynamics loads, both the structure failure and the material can deviate from the optimum result. A vortex lattice method is used for aerodynamics analysis, the finite element method is analyzed by the structural failure. These techniques are expected to reduce the complexity of Reliability-Based Design Optimization (RBDO). The Latin hypercube sampling method is used to quantify uncertainties of the aircraft wing structural design so that the experimental results including the solution sets are more acceptable and reliable which the proposed approach can be an alternative way for the RBDO technique.

Keywords: Multi-objective optimization · Reliability test · Latin hypercube sampling · Goland wing

1 Introduction

Inherently, all process of the aircraft wing design is uncertain and always causes the effects that can deviate the actual performance of the aircraft [1]. To decrease the inexact results of the aircraft design process, the Reliability-Based Design Optimization (RBDO), the reliable method used to analyze reliability or failure probability in an optimization problem with uncertainties, has been proposed [2]. For the first time, reliability index can be performed using probabilistic technique. The alternative and non-probabilistic techniques have been accomplished a drawback of the probabilistic group that requires more precise calculation and caused inefficient computation. The traditional technique is also known as Monte Carlo simulation (MCS) has been used to quantify the uncertainties in aircraft design though its drawback is time consumption. Later, various alternative forms have been developed to decrease the drawbacks. The adaptive forms are the first-order and the second moment (FOSM), the first-order reliability method (FORM) and the second-order reliability methods (SORM) [3]. The most successful method has been used to consider the uncertainties in a Goland wing

© Springer Nature Switzerland AG 2022
Y. Tan et al. (Eds.): ICSI 2022, LNCS 13345, pp. 399–410, 2022.
https://doi.org/10.1007/978-3-031-09726-3_36

that it was the polynomial chaos expansion (PCE) [4, 5]. The extension has been used to design a composite plate wing with uncertainty in ply orientations [6]. However, RBDO is a combination of a deterministic design process and the uncertainty quantify technique through an iterative optimization process. The use of various forms of MCS for RBDO still needs a great number of function evaluations, which leads to inefficient computation [7]. Later, the Latin hypercube sampling (LHS) has been used to generate random variable rather than using MCS [8]. The adaptation of LHS was optimum Latin hypercube sampling (OLHS), which is expected to increase the quality of space filling [9]. Recently, this technique is combined with a surrogate model to improve the performance in solving RBDO [10], but this technique still requires a thousand of calculations which the reason for new ideas emerging demands. A second group is non-probabilistic approach that does not need a precise distribution of random variables. The well-known methods are a convex set [3], an interval method [11] and a fuzzy set theory [12]. The general reliability-based design optimization problem is a double-looped nested problem due to the calculation of probability failure and optimization solving. In cases of a non-probabilistic approach, a triple-looped nested problem is needed due to the possible safety index (PSI) calculation. For solving the tasks, the Target Performance-Based Design Approach (TPBDA) [13], the Interval Perturbation Method (IMP) [14], and the Multi-Objective Reliability-Based Design Optimization (MORBDO) using metaheuristics (MHs) [15–17] have been proposed. The first work applied MORBDO using a worst-case scenario technique for solving a classical aircraft aeroelastic [15]. The extension of the worst-case scenario has been studied in a steering linkage design [16]. The last technique [16] is an extension of [13] the Multi-Objective Reliability-Based Topology Optimization (MORBTO) for solving topology optimization problems that can reduce time consumption. As mentioned previously, RBTO has been used to design aeronautical structures, which is expected to be weight-saving [4]. The reliability index significantly affected to final shape layout of the aero-structures [11]. RBTO is necessary to design of aircraft wing structure with considering aeroelastic phenomena.

However, RBTO of aircraft aeroelasticity is a computational burden problem due to the complexity of aircraft structures and the triple-loop nested problem for the non-probabilistic technique. So far, there have been only a few techniques introducing the non-probabilistic reliability index into the aeroelastic topology design of aircraft wings [18]. Except for the work by [19], their work used the PSI technique combining with MH for solving a composite aircraft wing design. The combination is called PSIBDO. The optimization aims to reduce the mass of aircraft wing to meet aeroelastic and strength constraints. The RBDO technique gets rather good in the design of the composite aircraft wing. Unfortunately, the work is still based on the traditional RBDO that operates the problem in form of a single objective with a probability constraint to fail under a defined probability of failure. The difference in the probability of failure causes different reliability solution. The recent work presents different viewing points by the way of multi-objective reliability-based design optimization analysis, the reliability solution set can perform one optimization run [20]. Out of the previous track, to tackle RBDO, a two-step approach for optimization and reliability is needed to tackle RBDO. Thus, this method is the solution to solve the problem by performing the first step with a multi-objective evolutionary optimization together with the reliability analysis. This research reveals a relation between the single objective reliability-based design optimization that can separate into a multi-objective optimization and examine the reliability of solutions as a technique, while the reliability analysis is in the group of probabilistic [21].

As mentioned above, both research groups have a gap in a study RBDO with two-step approaches without performing optimization in the second step to find an optimum reliability solution. The second step can perform only the reliability test of all solution to check the possibility and reliability of each solution. The rest of this paper is divided into four sections. Start with briefly details of aerodynamic, the load, and the finite element analysis. A two-step approach to multi-objective reliability-based design optimization is presented in Sect. 3. Numerical experiment and results are given in Sect. 4 and 5, respectively. The conclusions are detailed in the last section.

2   Aircraft Wing Model

A Goland wing model utilizes in this study has a semi-span wing of 6.096 m, a chord of 1.216 m, and a wing thickness is 0.0508 m. The whole aircraft wing structure is made from aluminum as present in Fig. 1 and the material properties are shown in Table 1. The Goland wing model is selected to be a test case for studying a new two-step approach in reliability-based design optimization of the aircraft wing structure. However, the sizing design thickness of the whole components still be the obstacle in this case, so the aircraft wing is applied with aerodynamic load due to a free stream velocity of air at a speed of 40 m/s and the angle of attack is 5°.

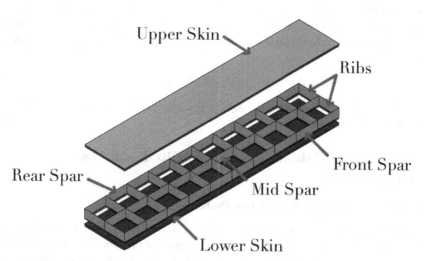

Fig. 1. Geometry model of a Goland wing.

2.1   Aerodynamics and Finite Element Analysis

In this research, the aircraft wing structure is subject to steady aerodynamic load, which is essential to static structural analysis for finding stress (σ) in the whole aircraft wing

Table 1. Material properties of aluminum.

Properties	Value	Unit
Young's modulus (E)	70 × 109	Pa
Poisson's ratio (v)	0.3	–
Density (ρ)	2700	kg/m^3

structure. The static structural analysis accomplishes with finite element analysis (FEA). A vortex ring method is used for aerodynamic analysis in this study. The aerodynamic forces can transfer into aircraft wing structure as shown as follows:

$$[K]u = q[G]^T[S]^T[AIC][G]^T\{\alpha\}. \tag{1}$$

where $[K]$ is a structural stiffness matrix, q is dynamic pressure, $[S]$ is the diagonal matrix of panel areas, $[AIC]$ is the aerodynamic influence coefficient matrix, $[G]$ is the transformation matrix, $\{u\}$ is the vector of structural displacements of the wing finite element model, and $\{\alpha\}$ is the vector of panel angles of attack.

2.2 Buckling Analysis

Aircraft wing structure composes of many thin plate structures, which applies with transverse aerodynamics load that causes structure fail with buckling. In this study, linear buckling analysis is used. When the work done by in-plane stress due to bending displacement on the wing exceeds its elastic potential energy, the buckling phenomena occurs. Therefore, the buckling factor can be solved by considering eigenvalue problem below:

$$[K]\{u\} - \lambda[K_G] = \{0\}. \tag{2}$$

where λ is a buckling factor, $[K]$ depicts the structural stiffness matrix, and $[K_G]$ illustrates a global geometrical matrix.

3 A Two-Step Multi-objective Reliability-Based Design Optimization

3.1 Sizing Design and Optimization Problem

The sizing optimization design in this research is carried out by the way that the aircraft wing structure composes many segments as presented in Fig. 2(a). The design variables are thickness of the wing segments. An example design result of an internal wing structure with a different shade of thickness displays in Fig. 2(b).

A multi-objective deterministic design of the present work can perform into three conflicting objectives which are stress, mass minimization, and buckling factor maximization. The multi-objective optimization design optimization (MODO) problem can be formulated as:

$$Min\{\sigma, M, -\lambda\} \tag{3}$$

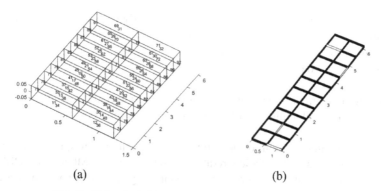

(a)   (b)

Fig. 2. Goland wing segments and internal wing structure.

$$\text{Subject to} \quad \lambda \geq 1$$

$$\sigma \leq 0.5\sigma_y$$

$$0.0001 \leq t_1 \leq 0.01 \text{ m}.$$

where M is the structural mass, σ is the maximum equivalent stress on the whole aircraft wing structure, σ_y is the yield stress, and λ is the buckling factor. The whole thickness is 92 design variables as present in Fig. 2. The MODO, metaheuristics (MHs), and FEA commercial software are combined. All procedures except the FEA are coded in MATLAB. MATLAB language is used to connect between aerodynamic analysis FEA and optimizer as present in Fig. 3. The optimizer is chosen in searching for the optimum solution set is a variant of multi-objective opposite-based population based incremental learning (MOPBIL) [22–24]. The new algorithm is added a self-adaptive learning rate technique, which is important part to improve the performance of MOPBIL in searching.

3.2 Two-Step Approach Reliability-Based Design

The present technique has an idea from the relationship between reliability-based design optimization (RBDO) and multi-objective design optimization (MODO), which has been published [22]. The work reveals the relation of MODO solution set (Pareto solution) is close to the RBDO solution. The work starts with formulating and solving the MODO problem, then, follows with performing the reliability-based design optimization design by using the previous solution from the first step. The technique can reduce the time for finding a reliable solution set by starting searching from the deterministic solution set rather than starting from a random point. Nevertheless, the technique still needs large computation time in both steps (MODO and RBDO). The computation time can be reduced by changing the RBDO testing step to the reliability testing step. The double-loop nested problem in RBDO at the last step can accomplish by performing the reliability test of all solutions from the previous step. The testing will check the reliability index (β), probability of failure (p_f) and reliability value (R). With this concept, the second step can open to choosing various techniques to quantify uncertainty in both groups of probabilistic or non-probabilistic. At the final step, the designer can choose a solution by using p_f value as an additional condition.

A two-step approach:

1. To perform the multi-objective optimization run (3) to find the solution set.
2. To perform the reliability test of each solution in the previous step.

Many techniques in group of probabilistic have been used to quantify uncertainty inherently embedded in the whole aircraft wing design such as MCS, LHS, OLHS etc. The LHS has improved the performance in quantifying uncertainty, which needed less computation time consumption when compared with MCS. In this research, the LHS is chosen to accomplish the task.

Latin Hypercube Sampling

Latin Hypercube Sampling (LHS) is a technique for sampling random variable values, which cuts down on the number of runs required to get a reasonably accurate result by the MCS. The LHS has been proposed for accomplishing a large-scale problem due to the difficulties of the large number of simulations existing. The LHS sprits the range of the random variables into n intervals that accords to the probability values. With the sampling technique can reduce the number of samplings, which is possible to achieve an accurate result. Consequently, the computational time burden can be accomplished.

Reliability Analysis

The problems can quantify the uncertainties by LHS, which is the best way to get a reasonably accurate result when compared with the MCS.

The reliability index (β) of the problem (3) is computed by.

$$\beta_j(x, y) = \frac{mean_{gj(x,y)}}{std_{gj(x,y)}}. \tag{4}$$

where $mean_g$ is the mean, std_g the standard deviation (STD) of values of the constraint $g_j(x)$, x is the design variable vector, and y is the random variable vector.

Then the probability of failure can compute as follow:

$$p_f = 0.5\left(1 + erf\left(\frac{\beta}{\sqrt{2}}\right)\right). \tag{5}$$

where erf is an error function.

The reliability value is

$$R = 1 - p_f \tag{6}$$

More detail of the reliability analysis can see more in [10].

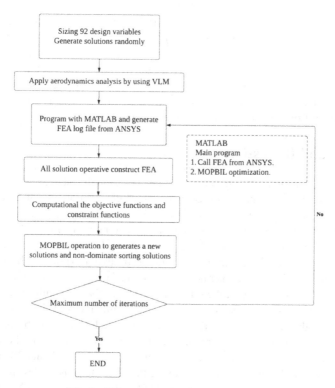

Fig. 3. The process of a simulation model.

4 Numerical Experiment

A numerical simulation is a computer-based simulation that uses software to produce a mathematical model of a physical system. Most non-linear systems require numerical simulations to analyze their behavior because their exact mathematics is too complex to give analytical solutions.

In the first step of performing MODO, the new design concept of the aircraft wing structure is using the combination of MATLAB R2021 and FEA software. The optimization process consists of running the variant MOPBIL algorithm while function evaluations are achieved using finite element analysis (FEA) for stress, mass, and buckling and buckling factor. Aerodynamic loads are computed using the vortex ring method, which is coded in MATLAB. During the flight, the wing is subject to static aerodynamic loads at cruise speed, which means the stress and buckling constraints due to such applied loads are taken into consideration. Aerodynamic force calculated based on the air density ($\rho_{air} = 1.2\,\text{kg/m}^3$), free stream velocity ($v = 40\,\text{m/s}$) and the wing angle of attack (AoA) is $5°$ are applied over the wing surface.

The second step, the reliability test can accomplish by using LHS to quantify uncertainties from the material properties such as Young's modulus (E), Poisson's ratio (v), and yield stress (σ_y). To test the reliability of all solution sets from the first step, the mean value and the variance of the random variable can be assigned as $(\mu_E, \sigma_E) = (70, 19.799)$ GPa, $(\mu_\nu, \sigma_\nu) = (0.3, 0.023)$, and $(\mu_{\sigma y}, \sigma_{\sigma y}) = (100, 24.495)$ MPa.

A personal computer with the following specifications is AMD Ryzen 5 4600 H with Radeon Graphics 3.00 GHz, 8.00 GB, and a 64-bit Windows 10 operating system is used to test the design problem. Initial values setting of the optimizer are set the population size of 30 and number of iterations is 400 and Pareto archive size is 30 in the first step, while the reliability test is used 1000 number of samplings.

5 Result and Discussion

Pareto optimal front and optimal values obtained from the first step are present in Fig. 4, and Table 2 (2nd–4th column), respectively. The figure shows that the Pareto fronts obtained the range of stress as 128360–843130 Pa, while the range of structure mass was 128.231–251.193 kg and the range of buckling factor obtained as 41.291–957.55. All solution meets MODO constraints. The results show the Pareto front No. 1 demonstrates a minimum of stress, the Pareto front No. 6 demonstrates a maximum of buckling factor, and the Pareto front No.7 demonstrates a minimum of wing structural mass.

In Table 2 (5th–7th column), the reliability index, probability failure, and reliability value (RV) of all Pareto solutions are shown. The reliability index of all Pareto solutions in the first row is calculated from Eq. (4) and then applied to Eqs. (5) and (6) for calculation the probability of failure, and the reliability value, which is shown in the second, the third row, respectively. From the Pareto solutions No. 3, 9, 15, 19, and 30 are obtained RV lower than other Pareto solutions, implying that their designs are less reliable. If the designer chooses the solution get lesser $p_f \leq 0.0002$, the Pareto solution No. 3, 9, 15, 19, and 30 are unselected. However, even with some smaller RVs, the overall appearance of the RV is excellent. Some selected aircraft wing structures is presented in Fig. 5(a and b) (sol. No. 5 and 6) and unselected is presented in Fig. 5(b) (sol. No. 15).

Table 2. Objective value and Reliability test of Pareto solution set from Fig. 4.

Pareto front no.	Stress (Pa)	Mass (kg)	Buckling (−)	Reliability Index (β)	Probabilistic of failure (p_f)	Reliability value (R)
1	128360	251.193	527.29	3.5235	2.1295e−04	0.9998
2	135630	234.45	321.17	3.5145	2.2030e−04	0.9998
3	**238110**	**147.817**	**67.361**	**3.4747**	**2.5573e−04**	**0.9997**
4	578160	141.881	749.63	3.5252	2.1155e−04	0.9998
5	169140	218.612	911.93	3.5270	2.1016e−04	0.9998
6	139330	237.211	957.55	3.5252	2.1160e−04	0.9998
7	843130	128.231	307.48	3.5167	2.1847e−04	0.9998
8	395670	141.546	582.45	3.5254	2.1145e−04	0.9998

(continued)

Table 2. (*continued*)

Pareto front no.	Stress (Pa)	Mass (kg)	Buckling (−)	Reliability Index (β)	Probabilistic of failure (p_f)	Reliability value (R)
9	**154430**	**202.753**	**66.186**	**3.4756**	**2.5482e−04**	**0.9997**
10	216610	168.836	365.87	3.5218	2.1432e−04	0.9998
11	371560	140.216	267.77	3.5183	2.1718e−04	0.9998
12	375950	198.471	940.75	3.5251	2.1166e−04	0.9998
13	376020	133.629	282.21	3.5189	2.1667e−04	0.9998
14	191430	180.301	900.3	3.5251	2.1166e−04	0.9998
15	**194340**	**175.442**	**41.291**	**3.4431**	**2.8750e−04**	**0.9997**
16	196310	168.411	107.08	3.4984	2.3399e−04	0.9998
17	306010	168.958	869.45	3.5271	2.1010e−04	0.9998
18	200770	173.26	606.2	3.5253	2.1152e−04	0.9998
19	**173160**	**215.308**	**592.62**	**3.3942**	**2.4418e−04**	**0.9997**
20	328390	158.058	674.69	3.5246	2.1205e−04	0.9998
21	203870	169.844	278.11	3.5188	2.1675e−04	0.9998
22	318080	148.15	80.328	3.4874	2.4386e−04	0.9998
23	408440	146.29	639.29	3.5242	2.1239e−04	0.9998
24	289500	165.216	690.64	3.5247	2.1201e−04	0.9998
25	320320	163.785	411.12	3.5226	2.1367e−04	0.9998
26	151270	214.424	510.83	3.5227	2.1356e−04	0.9998
27	496170	159.717	695.23	3.5245	2.1211e−04	0.9998
28	317650	156.192	323.91	3.5203	2.1557e−04	0.9998
29	159880	213.981	272.22	3.5179	2.1751e−04	0.9998
30	**322260**	**154.635**	**134.73**	**3.3588**	**3.9137e−04**	**0.9996**

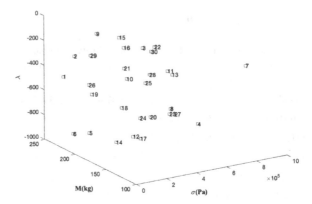

Fig. 4. Pareto frontier from the first step.

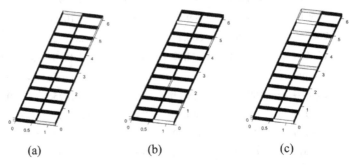

Fig. 5. Some internal aircraft wing structure from Fig. 4. (a) sol. 6 (b) sol. 7 and (c) sol. 15.

6   Conclusions

The purpose of the research is to design an aircraft wing structure using reliability-based design optimization with a two-step approach that have the conflict objective functions such as stress minimization, mass minimization, and buckling maximization. The two-step approach starts with the MODO and follows with reliability test of each solution. Sizing design optimization of a Goland wing was used as a design demonstration aircraft wing structure, and aerodynamic forces were applied to the wing surface. Uncertainties due to material properties can deviate optimum results, which can quantify by LHS. Reliability index, probability failure, and reliability value are accomplished. The experimental results show that the Pareto front solutions have achieved conflict objectives and meet with the design constraints with acceptable reliability. The newly proposed method can achieve acceptable results that can be a choice for the multi-objective reliability-based design optimization technique.

Acknowledgements. The authors are grateful for the program and the financial support provided by King Mongkut's Institute of Technology Ladkrabang and the National Research Council Thailand (N42A650549)

References

1. López, C., Baldomir, A., Hernández, S.: The relevance of reliability-based topology optimization in early deign stages of aircraft structures. Struct. Multidisc. Optim. **55**(3), 1121–1141 (2017)
2. D'Ippolito, R., Donders, S., Hack, M., Van Der Linden, G., Vandepitte, D.: Reliability-based design optimization of composite and steel aerospace structures. In: Proceeding of 47th AIAA/ASME/ASCE/AHS/ASC Structures, Structural Dynamics and Materials Conference, pp. 6621–6633. Newport, RI, USA (2006)
3. Elishakoff, I., Colombi, P.: Combination of probabilistic and convex models of uncertainty when scarce knowledge is present on acoustic excitation parameters. Comput. Methods Appl. Mech. Eng. **104**, 187–209 (1993)
4. Manan, A., Cooper, J.: Design of composite wings including uncertainties: a probabilistic approach. JA Aircr. **46**(2), 601–607 (2009)
5. Scarth, C., Cooper, J.E., Weaver, P.M., Silva, G.H.C.: Uncertainty quantification of aeroelastic stability of composite plate wings using lamination parameters. Compos. Struct. **116**(1), 84–93 (2014)
6. Scarth, C., Cooper, J.E.: Reliability-based aeroelastic design of composite plate wings using a stability margin. Struct. Multidiscip. Optim. **57**(4), 1695–1709 (2017). https://doi.org/10.1007/s00158-017-1838-6
7. Papageorgiou, A., Tarkian, M., Amadori, K., Ölvander, J.: Multidisciplinary design optimization of aerial vehicles: a review of recent advancements. Int. J. Aerosp. Eng. **2018**, 1–21 (2018)
8. Helton, J.C., Davis, F.J.: Latin hypercube sampling and the propagation of uncertainty in analyses of complex systems. Reliab. Eng. Syst. Saf. **81**(1), 23–69 (2003)
9. Pholdee, N., Bureerat, S.: An efficient optimum latin hypercube sampling technique based on sequencing optimisation using simulated annealing. Int. J. Syst. Sci. **46**(10), 1780–1789 (2015)
10. Wansaseub, K., Sleesongsom, S., Panagant, N., Pholdee, N., Bureerat, S.: Surrogate-assisted reliability optimisation of an aircraft wing with static and dynamic aeroelastic constraints. Int. J. Aeronaut. Space Sci. **21**(3), 723–732 (2020). https://doi.org/10.1007/s42405-019-00246-6
11. Fang, J., Smith, S.M., Elishakoff, I.: Combination of anti-optimization and fuzzy-set-based analyses for structural optimization under uncertainty. Math. Probl. Eng. **4**, 187–200 (1998)
12. Moller, B., Graf, W., Beer, M.: Fuzzy structural analysis using α-level optimization. Comput. Mech. **26**, 547–565 (2000)
13. Tang, Z.C., Lu, Z.Z., Hu, J.X.: An efficient approach for design optimization of structures involving fuzzy variables. Fuzzy Sets Syst. **255**, 52–73 (2014)
14. Yin, H., Yu, D., Xia, R.: Reliability-based topology optimization for structures using fuzzy set model. Comput. Methods Appl. Mech. Eng. **333**, 197–217 (2018)
15. Sleesongsom, S., Yooyen, S., Prapamonthon, P., Bureerat, S.: Reliability-based design optimization of classical wing aeroelasticity. In: IOP Conference Series: Materials Science and Engineering, vol. 886, no. 1 (2020)
16. Sleesongsom, S., Bureerat, S.: Multi-objective reliability-based topology optimization of structures using a fuzzy set model. J. Mech. Sci. Technol. **34**(10), 3973–3980 (2020)
17. Sleesongsom, S., Bureerat, S.: Multi-objective and reliability-based design optimization of a steering linkage. Appl. Sci. **10**(17), 5748 (2020)
18. Wang, L., Lui, G., Qiu, Z.: Review: recent developments in the uncertainty-based aerostructural design optimization for aerospace vehicles. J. Harbin Inst. Technol. **25**(3), 1–15 (2018)

19. Winyangkul, S., Sleesongsom, S., Bureerat, S.: Reliability-based design of an aircraft wing using a fuzzy-based metaheuristic. Appl. Sci. **11**(14), 6463 (2021)
20. Sleesongsom, S., Winyangkul, S., Bureerat, S.: Reliability-based design of an aircraft wing using a fuzzy-based metaheuristic. In: Proceedings of the ASME 2021 International Mechanical Engineering Congress and Expositionvol. 13: Safety Engineering, Risk, and Reliability Analysis; Research Posters. Virtual, V013T14A016, pp. 1–5. ASME, November 2021
21. Ho-Huu, V., Duong-Gia, D., Vo-Duy, T., Le-Duc, T., Nguyen-Thoi, T.: An efficient combination of multi-objective evolutionary optimization and reliability analysis for reliability-based design optimization of truss structures. Expert Syst. Appl. **102**, 262–272 (2018)
22. Sleesongsom, S., Bureerat, S.: New conceptual design of aeroelastic wing structures by multi objective optimization. Eng. Optim. **45**(1), 107–122 (2013)
23. Sleesongsom, S., Bureerat, S., Tai, K.: Aircraft morphing wing design by using partial topology optimization. Struct. Multidiscip. Optim. **48**(6), 1109–1128 (2013). https://doi.org/10.1007/s00158-013-0944-3
24. Sleesongsom, S., Bureerat, S.: Morphing wing structural optimization using opposite-based population-based incremental learning and multigrid ground elements. Math. Probl. Eng. **2015**, 730626 (2015)

ICSI-OC'2022: Competition on Single Objective Bounded Optimization Problems

Enhancing Fireworks Algorithm in Local Adaptation and Global Collaboration for Solving ICSI 2022 Benchmark Problems

Yifan Liu, Yifeng Li, and Ying Tan[✉]

Key Laboratory of Machine Perception (MOE), School of Artificial Intelligence, Institute for Artificial Intelligence, Peking University, Beijing, China
{liyifeng,ytan}@pku.edu.cn

Abstract. Fireworks algorithm is a swarm intelligence optimization algorithm with superior performance, which can be used to solve various practical optimization problems. To enhance the performance of fireworks algorithm, we introduce a powerful local search mechanism and add multiple cooperative strategies. These strategies improve the local exploitation capability and global exploration capability of fireworks algorithm. The experimental results on the ICSI'2022 test set demonstrate that the performance of the algorithm is satisfactory.

Keywords: Fireworks algorithm · Swarm intelligence · Optimization · Collaboration optimization

1 Introduction

Optimization problem has always been one of the hottest topics of research in various fields because it is widely used in many real-world applications, especially with the advent of machine learning and deep learning. Due to the complexity of modern optimization problems, the optimization of some functions is relatively difficult. The emergence of stochastic search algorithms such as swarm intelligence optimization algorithms and evolutionary algorithms makes it possible to find the global optimal solutions of some complex functions.

Fireworks algorithm (FWA [7]) is a kind of swarm intelligent optimization algorithm inspired by the phenomenon of firework explosion. During the process of fireworks algorithm, fireworks create sparks around themselves by exploding, which could search the surrounding area. Besides, fireworks could cooperate with each other to improve the efficiency of search. After multiple iterations, the algorithm is likely to find the global optimum of the objective function.

The single-objective bounded optimization problem is one of the basic settings of all optimization problems. Many complex optimization problems can be decomposed into single-objective optimization problems. Therefore, for many current swarm intelligence optimization algorithms and evolutionary algorithms,

© Springer Nature Switzerland AG 2022
Y. Tan et al. (Eds.): ICSI 2022, LNCS 13345, pp. 413–422, 2022.
https://doi.org/10.1007/978-3-031-09726-3_37

how to improve the performance of the algorithm on single-objective optimization problems is a key issue to deal with.

In this paper, we introduce a novel fireworks algorithm. It introduce the local search mechanism in CMA-ES to improve the local search performance of the fireworks algorithm. In addition, a new search space partition strategy has been added to the algorithm to improve the collaborative ability of fireworks, which greatly enhances the global search ability of the fireworks algorithm. The newly proposed algorithm is called Fireworks Algorithm with Search Space Partition (FWASSP). The ICSI'2022 test set is a newly proposed single-objective optimization test set for various intelligent optimization algorithms. We carry out experiment on ICSI'2022 with FWASSP. The experimental results show that the new algorithm performs well in both global search and local search.

The paper is organized as follows. Section 2 shows the background of our research, including the problem definition and related works. Section 3 describes the newly proposed algorithm in detail. The experimental results on the ICSI'2022 test set and the discussion are shown in Sect. 4.

2   Background

2.1   Problem Definition

Without loss of generality, we consider the general bound-constrained optimization problem which targets to find the optimal solution x^*:

$$\mathbf{x}^* = \arg\min_{x \in S} f(\mathbf{x}), \tag{1}$$

where $f : \mathbb{R}^d \rightarrow \mathbb{R}$ is an unknown objective function (also called fitness function). $S = \left\{ x \in \mathbb{R}^d : lb_i < x_i < ub_i \right\}$ is the feasible space of f. lb_i is the lower bound of x_i and ub_i is the upper bound of x_i.

2.2   Related Works

Evolution Strategies (ESs) are a sub-class of nature-inspired optimization methods belonging to the class of Evolutionary Algorithms (EAs). Covariance Matrix Adaptation Evolution Strategy (CMA-ES) [1] is a well-designed evolutionary strategy. It uses the quadratic model to fit local shapes to improve search efficiency. Besides, quite a few mechanisms are employed to control the search direction and the step size. Therefore, CMA-ES has an excellent ability in local search.

Tan and Zhu proposed a firework algorithm by simulating the explosion of fireworks [7]. After fireworks algorithm was proposed, it received extensive attention due to its great performance and excellent optimization efficiency. On the basis of the fireworks algorithm, researchers have proposed many variants of the fireworks algorithm such as EFWA [9], AFWA [3], dynFWA [8], CoFFWA [10] and GFWA [4]. In 2017, Li and Tan proposed the Loser-out Tournament-based Fireworks Algorithm (LoTFWA) [2], which added the loser elimination mechanism to reinitialize the noncompetitive fireworks. This mechanism could restart the fireworks which are trapped in the local optimum to accelerate the process of optimization.

In LoTFWA, each firework optimizes its local area by a uniform explosion within a dynamic amplitude. A guided mutation spark is generated for each firework to accelerate its local exploitation. Then, some unpromising fireworks are detected and restarted to avoid waste of resources. Algorithm 1 outlines the framework of LoTFWA. A detailed explanation and parameter setting of LoTFWA can be found in [2].

Algorithm 1. Loser-out Tournament-based Fireworks Algorithm

1: Randomly initialize μ fireworks in the search space.
2: Evaluate the fireworks' fitness
3: **repeat**
4:     **for** $i = 1$ to μ **do**
5:         Calculate dynamic explodes parameters λ_i and A_i.
6:         Generate explosion sparks.
7:         Generate guiding sparks.
8:         Evaluate all the fitness of the sparks.
9:         Select the best individual (including firework, its explosion sparks and guiding sparks) as the next generation of fireworks.
10:    Perform the loser-out tournament.
11: **until** Termination criterion is met.
12: **return** The position and the fitness of the best solution.

---

3   Proposed Strategies

LoTFWA is an outstanding global optimization algorithm with extremely simple mechanisms. However, there are still two major weaknesses in LoTFWA which need to be improved.

1. The local search efficiency of the explosion operator and mutation operator is limited by a basic uniform trust region scheme. This results in the searched solution being less refined.
2. The collaboration method is too weak because the restart mechanism is rarely triggered and it can only save limited resources rather than guide fireworks to cooperate.

In response to the above problems, we propose a series of strategies to improve it.

3.1   Gaussian Explosion with Adaptation

For the first weakness, the local search capability of the fireworks algorithm is enhanced by introducing the local search strategy in the CMA-ES [1]. CMA-ES uses Gaussian explosion instead of uniform explosion. The advantage of Gaussian explosion is that it has more parameters to control the shape of the explosion, which means that the search efficiency will be higher. By introducing Gaussian explosion, FWASSP is able to estimate the local fitness landscape and generate more effective sparks.

In the g-th generation, the k-th explosion spark $\mathbf{x}_k^{(g+1)}$ is generated from a Gaussian distribution:

$$\mathbf{x}_k^{(g+1)} \sim \mathbf{m}^{(g)} + \sigma^{(g)} \times \mathcal{N}\left(\mathbf{0}, C^{(g)}\right), \tag{2}$$

where $\mathbf{m}$ and C is the mean and covariance matrix. $\sigma^{(g)}$ is the overall step size. In the g-th generation generation, each firework generates the same number of λ sparks.

After evaluation of all sparks $\mathbf{x}^{(g+1)}$, the explosion distribution is adapted according to the strategies in CMA-ES. The detailed explanation and parameter setting of CMA-ES can be found in [1]. The complete adaptation process is provided in [5].

3.2   Restart Mechanism

Since Gaussian explosion and adaptation mechanism accelerate local optimization process significantly, the algorithm requires more detection mechanisms to ensure timely restart of fireworks that are not promising to improve the global optimal. Four extra restart conditions are proposed in our algorithm. These condition are determined by the search status of the firework individual and the relationship between fireworks:

1. Low Value Variance: $\mathrm{var}\left[f\left(x_{1:\lambda}^{(g+1)}\right)\right] \leq \epsilon_v$
2. Low Position Variance: $\sigma^{(g+1)} \times \left\|\mathbf{C}^{(g+1)}\right\| \leq \epsilon_p$
3. Not improving: Not improved for $I_{max\_not\_improve}$ iterations.
4. Covered by Better: More than 85% of the firework's sparks are covered by a better firework's explosion range.

3.3   Collaboration

Since we use Gaussian explosion, the explosion boundary of a firework X with parameters $(\mathbf{m}, C, \sigma)$ is defined as:

$$\left\{\mathbf{x}\Big| \|\mathbf{C}^{-\frac{1}{2}}\left(\frac{\mathbf{x} - \mathbf{m}}{\sigma}\right)\| = E\|\mathcal{N}(\mathbf{0}, \mathbf{I})\|\right\}. \tag{3}$$

There are two principles in collaboration strategies. First, the explosion scope tends to form a segmentation within the global optimization area, which can help fireworks avoid overlapping or omission of search scope. Second, the better fireworks tend to search independently, and the worse fireworks tend to search collaboratively. It guarantees the local optimization of leading fireworks will not be severely affected by collaboration.

Based on these ideas, the proposed algorithm conducts collaboration by the following steps:

Compare Fireworks. We need to compare the search progress of the fireworks for collaboration strategies. A fuzzy comparison between each pair of fireworks is introduced to estimate their relative optimization progress, which is described in Algorithm 2.

Algorithm 2. Fuzzy Comparison of Fireworks

Require: Fireworks $\mathbf{X}_i$ and $\mathbf{X}_j$ with sparks $\mathbf{x}_{i,1:\lambda}^{(g+1)}$ and $\mathbf{x}_{j,1:\lambda}^{(g+1)}$ (if not restarted)
1: **if** Both $\mathbf{X}_i$ and $\mathbf{X}_j$ are just restarted **then**
2:     **return** $\mathbf{X}_i$ and $\mathbf{X}_j$ are similar
3: **if** $\mathbf{X}_i$ is restarted **then**
4:     **return** $\mathbf{X}_j$ is ahead of $\mathbf{X}_i$
5: **else**
6:     **return** $\mathbf{X}_i$ is ahead of $\mathbf{X}_j$
7: **if** $\min \mathbf{x}_{i,1:\lambda}^{(g+1)} > \max \mathbf{x}_{j,1:\lambda}^{(g+1)}$ **then**
8:     **return** $\mathbf{X}_j$ is ahead of $\mathbf{X}_i$
9: **else**
10:     **return** $\mathbf{X}_i$ is ahead of $\mathbf{X}_j$
11: **return** X_i and X_j are similar

The fuzzy comparison method saves the time of the algorithm. At the same time, it can provide enough accurate information.

Compute Dividing Points. Different fireworks cooperate to search different areas, so it is necessary to calculate the dividing points to specify where the search range of both fireworks are divided. Figure 1 shows 4 possible situations of the collaboration method. We use the following steps to calculate the dividing point, which is described in Algorithm 3.

Fit Dividing Points. The boundary of firework $X(\mathbf{m}, C, \sigma)$ is adapted to fit its dividing points. For each dividing point P_k, a new covariance matrix C_k is calculated. On the direction of XP_k, P_k lies right on the boundary. On the conjugate directions, the radii of boundary do not changed. The mean of all adapted covariance matrix $\frac{1}{K} \sum_{k=1}^{K} C_k$ is taken as the overall collaborated result of X.

The mathematical calculation for fitting a single split point can be found in [5].

Algorithm 4 outlines the framework of the proposed collaboration strategy:

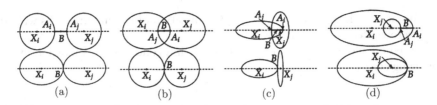

Fig. 1. Four cases of collaboration between two fireworks. A_i and A_j are the closer intersections of line X_iX_j with their boundaries. The actual dividing point could be any point on A_iA_j. The second row shows the collaboration results when taking the midpoint B of A_iA_j as dividing point.

Algorithm 3. Compute Dividing Points

Require: Fireworks $\mathbf{X}_i$, $\mathbf{X}_j$ and their intersections A_i, A_j
1: Calculate the radius $r_{ij} = \|X_iA_i\|$ and $r_{ji} = \|X_jA_j\|$ on line X_iX_j
2: Determine the situation (See Fig. 1) according to r_{ij}, r_{ji} and d_{ij}
3: Calculate the position of A_i and A_j
4: **if** X_i is ahead of X_j **then**
5:     A_i is the dividing point
6: **else if** X_j is ahead of X_i **then**
7:     A_j is the dividing point
8: **else**
9:     the midpoint B of A_iA_j is the dividing point.
10: **return** X_i and X_j are similar

Algorithm 4. Framework of Fireworks Collaboration

Require: n fireworks X_i and parameters $(\mathbf{m}_i, \mathbf{C}_i, \sigma_i)$ in N dimensional feasible space
Ensure: Collaborated parameters of fireworks
1: **for** each pair of fireworks X_i and X_j **do**
2:     Compare the progress of X_i and X_j
3:     Calculate $d_{ij} = |X_iX_j|$, expected sample distance r_{ij} and r_{ji} on X_iX_j
4:     Calculate the dividing point $P_{ij} (= P_{ji})$
5: **for** each firework X_i **do**
6:     Gather $K = \min(N, n-1)$ closest dividing points $P_{i,j_{1:K}}$
7:     Clip the length of $X_iP_{ij_k}$ within $[0.5r_{ij_k}, 2r_{ij_k}]$
8:     **for** $k \leftarrow 1 : K$ **do**
9:         Fit P_{ij_k} on the boundary of X_i and obtain $\mathbf{C}_{ij_k}$
    $\mathbf{C}_i \leftarrow \frac{1}{K} \sum_{k=1}^{K} \mathbf{C}_{ij_k}$

---

4   Experimental Results

The performance of proposed algorithm is tested on objective functions from the ICSI 2022 benchmark test set. This test set contains 10 black-box test functions, including 3 unimodal functions, 5 multimodal functions and 3 composition functions.

According to the settings of the single-objective optimization competition, each function is tested for 50 repetitions with 10, 20, 50 dimensions. The termination condition is a maximum of 10,000, 30,000 or 70,000 evaluations for 10, 20 or 50 dimensions respectively.

To demonstrate the efficiency of our proposed algorithm, the proposed algorithm is compared with three baselines. LoTFWA [2] is the most efficient one of the main variants of the firework algorithm. CMA-ES [1] is an excellent evolutionary algorithm with outstanding local optimization ability. APGSK-IMODE [6] is a variant of differential evolution algorithm. It has achieved excellent results in the CEC 2021 competition. All these algorithms are tested under the same conditions as the proposed algorithm.

The parameters of all the tested algorithms are set as follows. Its basic settings and parameters are as same as LoTFWA, which includes 5 fireworks and each firework generates 300 sparks in each iteration. In the restart conditions, ϵ_v and ϵ_p are both $1E - 12$, and the maximum number of unimproved iteration $I_{max\_unimprove}$ is 150. The parameters of local adaption is also set to be the same as CMA-ES. As we can see, our algorithm does not choose different parameters according to the specific problem. In order to ensure a fair comparison, the parameters of other algorithms are set according to the settings in their original papers.

The statistical results of the four algorithms are shown in the Table 1, Table 2 and Table 3 for 10 D, 20 D and 50 D respectively. Each table contains the mean errors and the mean standard deviations of four algorithms on ICSI'2022 test set. In addition, these algorithms are ranked according to their mean errors on each function, and the average rankings (ARs) over the 10 functions are presented at the bottom of the table. Their statistical information is shown in Fig. 2.

It can be seen from the experimental results that the performance of the algorithm are excellent, whether it is on unimodal or multimodal functions. The performance of the algorithm on the composition function is slightly worse, because the composition function is more complicated. On high-dimensional problems, it has also achieved very good optimization results compared with other baseline algorithms. This is because the global collaboration strategy of the algorithm can make the algorithm avoid trapping in many local optimal values.

On unimodal functions, both CMA-ES and proposed algorithm performs well because they have strong local exploitation mechanisms. On multimodal functions, our proposed algorithm performs the best. The reason for the excellent performance of our newly proposed algorithm is the algorithm is composed of multiple populations. The collaboration mechanism of them saves search resources thus it can find better global optimum in complex multimodal functions. As for composition functions, due to the complexity of the function, the performance of the newly proposed algorithm is comparable to CMA-ES.

Table 1. Results for 10 D problems

F.	LoTFWA		APGSK-IMODE		CMA-ES		Proposed	
	Mean	Std	Mean	Std	Mean	Std	Mean	Std
1	7.02e+04	7.18e+04	1.24e−01	5.87e−01	1.89e+00	9.99e+00	4.34e−03	7.77e−03
2	9.02e+02	1.50e+03	8.14e−03	2.53e−02	8.87e−14	2.30e−13	3.23e−08	5.94e−08
3	1.45e+00	2.23e+00	6.30e+00	0.00e+00	6.30e+00	1.08e−06	6.30e+00	3.46e−07
4	9.24e+00	2.93e+00	4.38e+00	1.31e+00	8.76e−01	7.87e−01	4.78e−01	5.35e−01
5	3.15e+02	1.20e+02	2.50e+02	1.11e+02	1.96e+01	5.78e+01	1.50e+01	3.57e+01
6	2.23e−01	3.47e−01	2.24e−01	4.49e−01	9.64e−04	2.07e−03	4.58e−09	8.40e−09
7	1.18e−01	4.29e−02	8.40e−02	3.57e−02	1.22e−01	3.91e−02	1.97e−02	5.06e−03
8	2.28e+05	2.04e+05	6.51e+04	1.45e+05	2.99e+05	1.69e+05	2.65e+05	1.90e+05
9	1.59e+03	2.84e+03	2.19e+04	1.67e+04	7.03e−02	1.31e−01	1.01e−03	7.03e−03
10	1.62e+01	9.20e+00	8.38e+00	6.58e+00	2.05e+01	1.23e+01	4.04e−01	2.20e−01
AR.	3.10		2.60		2.80		1.50	

Table 2. Results for 20 D problems

F.	LoTFWA		APGSK-IMODE		CMA-ES		Proposed	
	Mean	Std	Mean	Std	Mean	Std	Mean	Std
1	3.64e+05	1.82e+05	9.82e+04	7.05e+04	6.95e−11	1.80e−10	5.56e−08	1.06e−07
2	1.14e+02	2.34e+02	1.45e+03	6.87e+02	0.00e+00	0.00e+00	9.50e−13	4.11e−13
3	3.62e+00	4.51e+00	9.77e+00	3.69e−06	9.77e+00	1.00e−07	9.77e+00	1.37e−07
4	2.80e+01	6.14e+00	2.31e+01	3.85e+00	2.19e+00	1.50e+00	8.56e−01	7.71e−01
5	1.26e+03	2.17e+02	1.30e+03	2.47e+02	1.78e+02	1.26e+02	9.21e+01	6.62e+01
6	1.80e+00	1.64e+00	2.68e+01	5.70e+00	1.14e−13	0.00e+00	7.22e−11	2.93e−11
7	2.74e−01	5.98e−02	1.83e−01	2.58e−02	1.33e−01	3.72e−02	2.05e−02	5.52e−03
8	4.29e+05	5.93e+03	4.48e+05	5.91e+03	4.03e+05	4.58e+02	4.03e+05	3.67e+02
9	2.21e+04	4.45e+04	1.30e+07	3.91e+06	1.05e+01	6.17e+01	2.43e+06	9.77e+06
10	4.99e+01	2.76e+01	1.11e+01	4.56e+00	1.92e+01	9.84e+00	6.55e−01	3.05e−01
AR.	2.90		3.00		1.80		1.40	

Table 3. Results for 50 D problems

F.	LoTFWA		APGSK-IMODE		CMA-ES		Proposed	
	Mean	Std	Mean	Std	Mean	Std	Mean	Std
1	5.60e+06	1.34e+06	1.62e+07	2.86e+06	0.00e+00	0.00e+00	4.23e−12	9.60e−13
2	8.30e+01	1.31e+02	6.69e+01	5.97e+01	0.00e+00	0.00e+00	1.83e−12	1.59e−13
3	2.71e+01	3.16e+01	2.01e+00	1.02e+01	0.00e+00	0.00e+00	1.93e−12	1.30e−13
4	1.21e+02	1.80e+01	1.84e+02	1.78e+01	6.09e+00	2.74e+00	1.75e+00	9.02e−01
5	4.98e+03	4.55e+02	6.32e+03	2.85e+02	1.15e+03	3.89e+02	3.65e+02	9.69e+01
6	9.61e+00	3.92e+00	3.71e+02	3.94e+01	4.41e−13	2.91e−14	2.11e−10	3.55e−11
7	4.81e−01	7.57e−02	3.39e−01	2.57e−02	1.74e−01	3.48e−02	3.75e−02	7.28e−03
8	4.57e+05	8.85e+03	5.73e+05	9.10e+03	4.07e+05	6.99e+02	4.06e+05	2.62e+02
9	9.62e+05	1.19e+06	2.07e+09	3.77e+08	3.04e−03	5.67e−03	1.73e−09	3.74e−10
10	7.61e+02	2.07e+02	3.62e+02	3.57e+01	3.24e+02	4.97e+01	1.62e+02	8.21e+00
AR.	3.00		3.20		1.60		1.00	

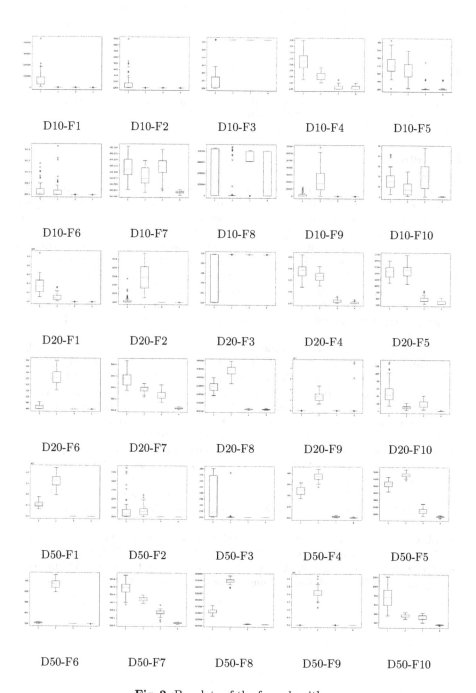

Fig. 2. Boxplots of the four algorithms

In conclusion, the newly proposed algorithm has a significant improvement over LoTFWA. Compared with other classic heuristic algorithms, FWASSP also has an extremely good performance. It is worth further investigation.

Acknowledgment. This work is supported by the National Natural Science Foundation of China (Grant No. 62076010), and partially supported by Science and Technology Innovation 2030 - "New Generation Artificial Intelligence" Major Project (Grant Nos.: 2018AAA0102301 and 2018AAA0100302).

References

1. Hansen, N.: The cma evolution strategy: a comparing review. In: Lozano, J.A., Larrañaga, P., Inza, I., Bengoetxea, E. (eds) Towards a New Evolutionary Computation. Studies in Fuzziness and Soft Computing, vol 192, pp. 75–102 . Springer, Berlin, Heidelberg (2006). https://doi.org/10.1007/3-540-32494-1_4

2. Li, J., Tan, Y.: Loser-out tournament-based fireworks algorithm for multimodal function optimization. IEEE Trans. Evol. Comput. **22**(5), 679–691 (2017)

3. Li, J., Zheng, S., Tan, Y.: Adaptive fireworks algorithm. In: 2014 IEEE Congress on evolutionary computation (CEC), pp. 3214–3221. IEEE (2014)

4. Li, J., Zheng, S., Tan, Y.: The effect of information utilization: introducing a novel guiding spark in the fireworks algorithm. IEEE Trans. Evol. Comput. **21**(1), 153–166 (2016)

5. Li, Y., Tan, Y.: Enhancing fireworks algorithm in local adaptation and global collaboration. In: Tan, Y., Shi, Y. (eds.) ICSI 2021. LNCS, vol. 12689, pp. 451–465. Springer, Cham (2021). https://doi.org/10.1007/978-3-030-78743-1_41

6. Mohamed, A.W., Hadi, A.A., Agrawal, P., Sallam, K.M., Mohamed, A.K.: Gaining-sharing knowledge based algorithm with adaptive parameters hybrid with imode algorithm for solving cec 2021 benchmark problems. In: 2021 IEEE Congress on Evolutionary Computation (CEC), pp. 841–848. IEEE (2021)

7. Tan, Y., Zhu, Y.: Fireworks algorithm for optimization. In: Tan, Y., Shi, Y., Tan, K.C. (eds.) ICSI 2010. LNCS, vol. 6145, pp. 355–364. Springer, Heidelberg (2010). https://doi.org/10.1007/978-3-642-13495-1_44

8. Zheng, S., Janecek, A., Li, J., Tan, Y.: Dynamic search in fireworks algorithm. In: 2014 IEEE Congress on Evolutionary Computation (CEC), pp. 3222–3229. IEEE (2014)

9. Zheng, S., Janecek, A., Tan, Y.: Enhanced fireworks algorithm. In: 2013 IEEE Congress on Evolutionary Computation (CEC), pp. 2069–2077. IEEE (2013)

10. Zheng, S., Li, J., Janecek, A., Tan, Y.: A cooperative framework for fireworks algorithm. IEEE/ACM Trans. Comput. Biol. Bioinform. **14**(1), 27–41 (2015)

Differential Evolution with Biased Parameter Adaptation for ICSI-OC 2022 Competition

Vladimir Stanovov$^{(\boxtimes)}$ and Shakhnaz Akhmedova

Reshetnev Siberian State University of Science and Technology,
Institute of Informatics and Telecommunication, Krasnoyarsk 660037, Russia
vladimirstanovov@yandex.ru, shahnaz@inbox.ru

Abstract. In this paper the differential evolution based algorithm is proposed for solving ICSI-OC'2022 benchmark functions. The presented NL-SHADE-LM algorithm uses several important modifications, such as biased parameter adaptation with generalized Lehmer mean with bias towards larger scaling factors, improved archive set update strategy, crossover rate sorting, as well as rank-based selective pressure. The performed computational experiments show that the proposed algorithm achieves highly competitive results.

Keywords: Differential evolution · Parameter adaptation · Numerical optimization

1 Introduction

In the area of computational intelligence among the evolutionary computation the numerical optimization methods have one of the major roles, as they often serve as parts of other algorithms where optimization problems occur. Popular modern evolutionary optimization methods [1] are particle swarm optimization (PSO), biology inspired and swarm intelligence algorithms (SI) [2], evolutionary strategies (ES), genetic algorithms (GA) and differential evolution (DE) [3]. The differential evolution has recently achieved very promising results in many areas, and is currently one of the most widely used algorithmic schemes due to its simplicity and high efficiency. The success of DE is clearly observed through different optimization competitions, where it takes leading positions last years.

Among the differential evolution algorithms the L-SHADE family of approaches is one of the most often applied due to the efficient parameter adaptation schemes [4]. The L-SHADE and its modifications have been often considered and modified in many studies, and most of the algorithm elements have been altered in the modifications. Still the main problem often considered in DE is the parameter adaptation, as the scaling factor and crossover rate have significant influence on the algorithm performance [5].

© Springer Nature Switzerland AG 2022
Y. Tan et al. (Eds.): ICSI 2022, LNCS 13345, pp. 423–431, 2022.
https://doi.org/10.1007/978-3-031-09726-3_38

In this paper two modifications to the parameter adaptation are considered. First, the modified Lehmer mean is applied for updating memory cells in success-history adaptation, and second the crossover rate sampling is changed by introducing sorting of the generated Cr values. The computational experiments are performed on a set of benchmark functions from the ICSI Optimization Competition 2022 (ICSI-OC'2022) [6], and the results of the modified algorithm are compared to the baseline and other methods.

The rest of the paper is organized as follows: Sect. 2 gives a description of the modern DE methods, Sect. 3 describes the proposed approach, Sect. 4 contains the experimental setup and results, and Sect. 5 concludes the paper.

2   Related Work

2.1   Differential Evolution

The development of evolutionary computation methods has led to emergence of several research directions for different areas. For the numerical optimization problems the real-coded GAs, PSO and DE are often applied, as well as CMA-ES methods. The DE has shown to be capable of achieving promising results, and gained high popularity among researchers. The main idea of differential evolution, originally proposed in [7], is the usage of difference vectors, which are calculated using positions of the points in the search space. The three main parameters of DE are the population size NP, scaling factor F and crossover rate Cr. The population $x_{i,j}$, $i = 1...NP$, $j = 1...D$, where D is the problem dimension, is randomly initialized within boundaries $[xmin_j, xmax_j]$, $j = 1...D$. Next, the main cycle of the algorithm begins with three main operators: mutation, crossover and selection.

The mutation operation in the original DE used three randomly selected vectors, however, more recent studies usually apply *current-to-pbest/1* strategy, originally described in the JADE [8] algorithm:

$$v_{i,j} = x_{i,j} + F(x_{pbest,j} - x_{i,j}) + F(x_{r1,j} - x_{r2,j}). \tag{1}$$

where *pbest* is one of the $pb * 100\%$ top solutions, $r1$ and $r2$ random indexes, are mutually different from each other, *pbest* and current index i. The resulting mutant vector v is then transferred to crossover. The $r2$ index in JADE, SHADE and L-SHADE algorithms can be selected from either current population or archive set. There are two well-known crossover schemes in differential evolution: binomial and exponential, however the binomial crossover is used more often. It uses the mutant vector v to produce the trial vector u using crossover probability parameter Cr as follows:

$$u_{i,j} = \begin{cases} v_{i,j}, & \text{if } rand(0,1) < Cr \text{ or } j = jrand \\ x_{i,j}, & \text{otherwise} \end{cases}. \tag{2}$$

where $jrand$ is a randomly chosen index, which will be taken from the mutant vector. This is required to make sure that at least one component is taken from the mutant vector, otherwise the trial vector could be equal to the target vector.

After trial vector is generated, it should be checked to be within the search space boundaries. This is performed using the bound constraint handling method (BCHM), and one of the most efficient approaches is the midpoint-target, which uses the parent's position to reset the constraint violating components:

$$u_{i,j} = \begin{cases} \frac{xmin_j + x_{i,j}}{2}, & \text{if } u_{i,j} < xmin_j \\ \frac{xmax_j + x_{i,j}}{2}, & \text{if } u_{i,j} > xmax_j \end{cases}. \tag{3}$$

Finally, the selection procedure in DE compares the newly generated trial vector u with the target vector x_i and performs the replacement if it is better:

$$x_{i,j} = \begin{cases} u_{i,j}, & \text{if } f(u_i) \leq f(x_i) \\ x_{i,j}, & \text{if } f(u_i) > f(x_i) \end{cases}. \tag{4}$$

Adjusting the parameters of DE is one of the crucial parts of algorithm implementation, and most commonly used approaches are described in the next subsection.

2.2 Parameter Adaptation in DE

One of the first attempts of applying paramters adaptation in differential evolution was performed in [9]. According to several recent studies [10], the JADE and SHADE based parameter adaptation in DE appears to be one of the most efficient adaptation schemes. The success-history adaptation (SHA) creates a set of H memory cells, each containing a pair of parameters F and Cr. The values in the memory cells $(M_{F,k}, M_{Cr,k})$ are used to sample new parameters for every mutation and crossover operation using Cauchy and normal distribution:

$$F = randc(M_{F,k}, 0.1), randn(M_{Cr,k}, 0.1). \tag{5}$$

where $randc$ is a Cauchy distributed random value, and $randn$ is normally distributed, k is chosen in range $[1, D]$. The sampling of F is restricted in the following way: if $F < 0$, it is generated again, and if $F > 1$ then it is set to $F = 1$. The Cr value is sampled until it falls within the $[0, 1]$ range.

The successful F and Cr values, i.e. the ones that helped improving the solution, are saved into S_F and S_{Cr} arrays, as well as the improvement value $\Delta f = |f(u) - f(x_i)|$. At the end of each generation the weighted Lehmer mean is used to calculate new values:

$$mean_{wL} = \frac{\sum_{j=1}^{|S|} w_j S_j^2}{\sum_{j=1}^{|S|} w_j S_j}. \tag{6}$$

where $w_j = \frac{\Delta f_j}{\sum_{k=1}^{|S|} \Delta f_k}$, and S is either S_F or S_{Cr}. One of the memory cells with index h is then updated using previous values:

$$M_{F,k}^{g+1} = 0.5(M_{F,k}^g + mean_{wL}(S_F)), \quad M_{Cr,k}^{g+1} = 0.5(M_{Cr,k}^g + mean_{wL}(S_{Cr})). \tag{7}$$

where g is the current generation number. The memory cell index h is incremented every generation and reset once it reaches H.

In the L-SHADE algorithm it was proposed to use the linear population size reduction (LPSR), however more recent studies, such as AGSK [11] and NL-SHADE-RSP [12] use non-linear population size reduction:

$$NP_{g+1} = round((NP_{min} - NP_{max})NFE_r^{1-NFE_r} + NP_{max}). \qquad (8)$$

where $NFE_r = NFE/NFE_{max}$ is the ratio of current number of fitness evaluations. At the end of every generation NP_{g+1} is calculated, and if it is less than NP, then one or several worst solutions are removed from the population.

The NL-SHADE-RSP algorithm also introduces the adaptive archive usage, where the probability of an individual to be chosen from the archive for $r2$ index depends on the success of applying the archive. The archive set in L-SHADE is updated every time a target vector is replaced by trial vector. In this case the target vector (parent) is copied to the archive, and if the archive set reaches its maximum size, then a randomly selected solution from the archive is replaced. The probability p_A of using a solution from the archive is initially set to 0.5 and updated every generation based on the number of archive usages n_A, achieved improvements with archive Δf_A and without it Δf_P. The probability is updated as follows:

$$p_A = \frac{\Delta f_A/n_A}{\Delta f_A/n_A + \Delta f_p/(1 - n_A)}. \qquad (9)$$

The p_A value is additionally checked to be within $[0.1, 0.9]$ range. Another feature of NL-SHADE-RSP is the rank-based selective pressure, applied for the $r2$ index. The individuals are ranked according to fitness, and probabilities to be chosen are calculated proportional to rank values, i.e. $pr_i = \frac{R_i}{\sum_{j=1}^{NP} R_j}$, where ranks $R_i = e^{-i/NP}$ were set as indexes of individuals in a sorted array.

3   Proposed Approach

The NL-SHADE-LM algorithm presented in this study is based on the NL-SHADE-RSP algorithm and uses biased parameter adaptation, presented in [13]. The NL-SHADE-LM uses generalized Lehmer mean, presented in [12] for updating the memory cells M_F, and for M_{Cr} p was set to 0.5. The second parameter was set to $m = 1.5$ for both F and Cr:

$$mean_{p,wL}(S) = \frac{\sum_{j=1}^{|S|} w_j S_j^p}{\sum_{j=1}^{|S|} w_j S_j^{p-m}}. \qquad (10)$$

where the p parameter is responsible for the bias, changing this parameter may result in harmonic ($p = 0$), geometric ($p = 0.5$), arithmetic ($p = 1$) and contra-harmonic mean ($p = 2$). NL-SHADE-LM uses $p = 4$, i.e. the calculation of mean is biased towards larger F values.

Algorithm 1. NL-SHADE-LM

1: Set $NP_{max} = 10D$, $NP = NP_{max}$ D, NFE_{max},
2: $H = 15$, $A = \emptyset$, $M_{F,r} = 0.5$, $M_{Cr,r} = 0.9$, $k = 1$
3: $NA = NP$, $p_A = 0.5$, $g = 0$
4: Initialize population $P_0 = (x_{1,j}, ..., x_{NP,j})$ randomly
5: **while** $NFE < NFE_{max}$ **do**
6:    $S_F = \emptyset$, $S_{Cr} = \emptyset$, $n_A = 0$
7:    Rank population according to fitness $f(x_i)$
8:    **for** $i = 1$ to NP **do**
9:       Current memory index $r = randInt[1, H + 1]$
10:       Crossover rates $Cr_i = randn(M_{Cr,r}, 0.1)$
11:       $Cr_i = min(1, max(0, Cr))$
12:       **repeat**
13:          $F_i = randc(M_{F,r}, 0.1)$
14:       **until** $F_i \geq 0$
15:       $F_i = min(1, F_i)$
16:    **end for**
17:    Sort Cr_i according to fitness $f(x_i)$
18:    **for** $i = 1$ to NP **do**
19:       **repeat**
20:          $pbest = randInt(1, NP * p)$
21:          $r1 = randInt(1, NP)$
22:          **if** $rand[0, 1] < p_A$ **then**
23:             $r2 = randInt(1, NP)$
24:          **else**
25:             $r2 = randInt(1, NA)$
26:          **end if**
27:       **until** $i \neq pbest \neq r1 \neq r2$
28:       **for** j=1 to D **do**
29:          $v_{i,j} = x_{i,j} + F(x_{pbest,j} - x_{i,j}) + F(x_{r1,j} - x_{r2,j})$
30:       **end for**
31:       Update Cr_b
32:       Binomial crossover with Cr_b
33:       Calculate $f(u_i)$
34:       **if** $f(u_i) < f(x_i)$ **then**
35:          $x_i \rightarrow A_{randInt[1,NA]}$, $x_i = u_i$
36:          $F \rightarrow S_F$, $Cr \rightarrow S_{Cr}$, $\Delta f_i = f(x_i) - f(u_i)$
37:       **end if**
38:    **end for**
39:    Get NP_{g+1} and NA_{g+1} with NLPSR
40:    **if** $|A| > NA_{g+1}$ **then**
41:       Remove random individuals from the archive
42:    **end if**
43:    **if** $NP_g > NP_{g+1}$ **then**
44:       Remove worst individuals from the population
45:    **end if**
46:    Update $M_{F,k}$, $M_{Cr,k}$, p_A
47:    $k = mod(k, H) + 1$, $g = g + 1$
48: **end while**
49: Return best solution x_{best}

The NL-SHADE-LM also uses crossover rates sorting. For this purpose the Cr values are sampled for all individuals at the beginning of the generation, and sorted by fitness values, so that better individuals receive smaller crossover rates. NL-SHADE-LM uses only binomial crossover. The pseudocode of NL-SHADE-LM is shown in Algorithm 1.

The pb value controlling the greediness of the *current-to-pbest* strategy is linearly increased, $pb = 0.2 + 0.1 \cdot NFE/NFE_{max}$. The archive handling in NL-SHADE-LM is changed according to [14]. If the archive is full, to add a new solution first two random solutions are chosen, and the one having worst fitness is replaced by the new solution. This scheme is similar to tournament selection.

4   Experimental Setup and Results

The experiments in this study were performed on a set of test problems from the ICSI-OC'2022 competition on numerical optimization, which consists of 10 test functions defined for $D = 10$, 20 and 50. The computational resource is set to 10000, 30000 and 70000 respectively. The experiments were performed for three cases: without modified Lehmer mean and without crossover rate sorting (NL-SHADE-LM$_{nlm,ncr}$), with modified Lehmer mean and without Cr sorting (NL-SHADE-LM$_{ncr}$), and with both modifications. Table 1 shows the Mann-Whitney statistical tests comparing these variants and an one of the best algorithm from CEC 2021 competition [13]. The values in the table are the number of wins (+), ties (=) and losses (−). The numbers indicate the number of functions, on which improvements, ties, or losses were observed.

Table 1. NL-SHADE-LM vs other modifications

Algorithms	$10D$	$20D$	$50D$
NL-SHADE-LM vs NL-SHADE-LM$_{nlm,ncr}$	2+/6=/2−	3+/5=/2−	4+/3=/3−
NL-SHADE-LM vs NL-SHADE-LM$_{nlm}$	1+/7=/2−	3+/5=/2−	4+/3=/3−
NL-SHADE-LM vs NL-SHADE-LM$_{ncr}$	4+/6=/0−	4+/5=/1−	4+/5=/1−
NL-SHADE-LM vs NL-SHADE-RSP [13]	7+/2=/1−	8+/2=/0−	9+/1=/0−

Table 1 shows that adding generalized Lehmer mean and sorting crossover rate significantly improves algorithm performance, especially in high-dimensional functions. Tables 2, 3 and 4 contain the results of NL-SHADE-LM for $10D$, $20D$ and $50D$, these values are provided for comparison with alternative approaches.

Table 2. NL-SHADE-LM results, $10D$

No.	Best	Worst	Median	Mean	Std
F1	6.446513e−03	1.084766e+04	1.282133e+02	1.067174e+03	2.684172e+03
F2	3.374623e−04	1.354462e+04	1.921406e+03	2.668539e+03	2.829530e+03
F3	6.300706e+00	6.300706e+00	6.300706e+00	6.300706e+00	6.322624e−11
F4	2.513869e+00	2.094280e+01	7.684952e+00	8.313127e+00	3.754021e+00
F5	2.370496e+02	1.416116e+03	8.617830e+02	8.351595e+02	2.534449e+02
F6	6.310307e−05	1.606420e+00	4.525171e−03	1.084759e−01	3.492718e−01
F7	8.001120e−02	2.448226e−01	1.637408e−01	1.657474e−01	3.537959e−02
F8	6.353788e+04	4.617354e+05	4.502627e+05	4.350200e+05	7.426634e+04
F9	2.914080e+05	3.312098e+06	1.387940e+06	1.358431e+06	6.108474e+05
F10	8.761310e+00	7.691985e+01	2.896645e+01	3.158693e+01	1.602981e+01

Table 3. NL-SHADE-LM results, $20D$

No.	Best	Worst	Median	Mean	Std
F1	9.105971e+01	1.445305e+05	2.733760e+04	3.705226e+04	3.336954e+04
F2	2.607176e+00	6.645190e+03	1.481382e+03	2.070434e+03	1.805933e+03
F3	9.774071e+00	9.780471e+00	9.774072e+00	9.774209e+00	9.048427e−04
F4	8.043015e+00	2.232736e+01	1.451965e+01	1.475832e+01	3.774855e+00
F5	1.392954e+03	3.527395e+03	2.091063e+03	2.151690e+03	4.934307e+02
F6	2.369144e−05	6.880743e−02	2.500354e−03	4.980412e−03	1.029763e−02
F7	1.332610e−01	2.946701e−01	2.062218e−01	2.047659e−01	4.330983e−02
F8	4.665928e+05	4.930627e+05	4.806035e+05	4.805060e+05	5.923694e+03
F9	1.714862e+07	9.137096e+07	5.371399e+07	5.345002e+07	1.522453e+07
F10	2.718386e+01	1.454984e+02	8.601746e+01	8.625255e+01	2.704173e+01

Table 4. NL-SHADE-LM results, $50D$

No.	Best	Worst	Median	Mean	Std
F1	3.960196e+05	2.035145e+06	1.008827e+06	1.080563e+06	3.724943e+05
F2	2.515004e−02	2.000336e+03	1.912894e+02	4.444295e+02	5.249607e+02
F3	4.889025e−01	1.448713e+02	1.167319e+02	1.071055e+02	3.439101e+01
F4	3.255878e+01	1.059859e+02	7.320905e+01	7.240839e+01	1.450227e+01
F5	6.865254e+03	9.174554e+03	7.869923e+03	7.881189e+03	5.262223e+02
F6	1.743882e−01	2.958089e+01	1.530902e+00	4.221346e+00	6.351358e+00
F7	2.413929e−01	4.305074e−01	3.345613e−01	3.363674e−01	4.766772e−02
F8	5.268653e+05	5.676130e+05	5.476659e+05	5.476841e+05	8.483495e+03
F9	5.434750e+08	1.885761e+09	8.411820e+08	9.487407e+08	3.557633e+08
F10	3.610184e+02	6.618797e+02	4.954846e+02	5.071518e+02	7.545588e+01

5 Conclusion

In this study the NL-SHADE-LM algorithm with biased parameter adaptation, crossover rate sorting and improved archive update strategy was proposed. The experiments performed on the ICSI-OC'2022 competition functions have shown that the proposed modifications are capable improving the algorithm performance, especially for high-dimensional cases. The experiments demonstrated that the crossover rate sorting has more effect then modified Lehmer mean.

Acknowledgement. The work was carried out within the framework of the state support program for leading scientific schools (grant of the President of the Russian Federation NSh-421.2022.4).

References

1. Sloss, A.N., Gustafson, S.M.: 2019 Evolutionary algorithms review. In: Banzhaf, W., Goodman, E., Sheneman, L., Trujillo, L., Worzel, B. (eds) Genetic Programming Theory and Practice XVII. Genetic and Evolutionary Computation, pp. 307–340. Springer, Cham (2019) https://doi.org/10.1007/978-3-030-39958-0_16

2. Kar, A.K.: Bio inspired computing-a review of algorithms and scope of applications. Expert Syst. Appl. **59**, 20–32 (2016). https://doi.org/10.1016/j.eswa.2016.04.018

3. Das, S., Mullick, S.S., Suganthan, P.N.: Recent advances in differential evolution - an updated survey. Swarm Evol. Comput. **27**, 1–30 (2016)

4. Tanabe, R., Fukunaga, A.S.: Improving the search performance of SHADE using linear population size reduction. In: Proceedings of the IEEE Congress on Evolutionary Computation, pp. 1658–1665 (2014)

5. Al-Dabbagh, R.D., Neri, F., Idris, N., Baba, M.S.: Algorithmic design issues in adaptive differential evolution schemes: review and taxonomy. Swarm Evol. Comput. **43**, 284–311 (2018)

6. Yifan, L.: Definitions for the ICSI Optimization Competition'2022 on Single Objective Bounded Optimization Problems, Technical report, Peking University (2022)

7. Storn, R., Price, K.: Differential evolution-a simple and efficient heuristic for global optimization over continuous spaces. J. Global Optim. **11**(4), 341–359 (1997). https://doi.org/10.1023/A:1008202821328

8. Zhang, J., Sanderson, A.C.: JADE: adaptive differential evolution with optional external archive. IEEE Trans. Evol. Comput. **13**(5), 945–958 (2009)

9. Qin, A., Suganthan, P.N.: Self-adaptive differential evolution algorithm for numerical optimization. IEEE Congr. Evol. Comput. **2**, 1785–1791 (2005)

10. Piotrowski, A.P., Napiorkowski, J.J.: Step-by-step improvement of JADE and SHADE-based algorithms: success or failure? Swarm Evol. Comput. **43**, 88–108 (2018). https://doi.org/10.1016/j.swevo.2018.03.007

11. Mohamed, A. W., Hadi, A. A., Mohamed, A., Awad, N. H.: Evaluating the performance of adaptive gainingsharing knowledge based algorithm on CEC 2020 benchmark problems. In: 2020 IEEE Congress on Evolutionary Computation, pp. 1–8 (2020)

12. Stanovov, V., Akhmedova, S. Semenkin, E.: NL-SHADE-RSP algorithm with adaptive archive and selective pressure for CEC 2021 numerical optimization. In: 2021 IEEE Congress on Evolutionary Computation, pp. 809–816 (2021). https://doi.org/10.1109/CEC45853.2021.9504959

13. Stanovov, V., Akhmedova, S., Semenkin, E.: Biased parameter adaptation in differential evolution. Inf. Sci. **566**, 215–238 (2021)
14. Stanovov, V., Akhmedova, S., Semenkin, E.: Archive update strategy influences differential evolution performance. Adv. Swarm Intell. **12145**, 397–404 (2020). https://doi.org/10.1007/978-3-030-53956-6_35

Composite Evolutionary Strategy and Differential Evolution Method for the ICSI'2022 Competition

Jakub Kudela[✉] [iD], Tomas Nevoral, and Tomas Holoubek

Institute of Automation and Computer Science, Brno University of Technology,
Brno, Czech Republic
Jakub.Kudela@vutbr.cz

Abstract. In this paper, a composite method for bound constrained optimization called Composite Evolutionary Strategy and Differential Evolution (CESDE) is described. This method combines two well-performing methods from the Congress on Evolutionary Computation Competitions. Through numerical investigation on the ICSI'2022 benchmark set, the favourite scheme for combining the two methods was determined, and it was found that CESDE outperforms both of its "parental" methods on all studied instances.

Keywords: ICSI'2022 competition · Evolutionary strategy · Differential evolution · Numerical optimization · Composite method

1 Introduction

Evolutionary algorithms (EAs) are a highly performing class of metaheuristics used for global optimization. EAs are inspired by various processes of biological evolution: reproduction, mutation, recombination, and natural selection. Among the most widely utilized variants of EAs are genetic algorithms, evolutionary strategy (ES), and differential evolution (DE) [2]. These methods are extensively used in the optimization of complicated problems such as the design of quantum operators [17], stabilization of chaos [9], difficult assignment problems [10], or the hyperparameter optimization in deep learning [15].

In this paper, a composite method utilizing high-performing variants of ES and DE is proposed and is evaluated on the ICSI'2022 benchmark set. Composite methods are popular methods in structural dynamics (and similar areas) which combine different methods at different time steps for the purpose of getting the respective benefits of the different methods [3,4,6,12]. The proposed method is called Composite Evolutionary Strategy and Differential Evolution (CESDE) and its two components are the Hybrid Sampling Evolution Strategy (HSES) [16] and the Linear Population Size Reduction SHADE (LSHADE) [13], two very well-performing methods from the Congress on Evolutionary Computation (CEC) Competitions on bound constrained single objective optimization [11]. It takes advantage of the properties found by investigating the convergence profiles of the two methods [7].

© Springer Nature Switzerland AG 2022
Y. Tan et al. (Eds.): ICSI 2022, LNCS 13345, pp. 432–439, 2022.
https://doi.org/10.1007/978-3-031-09726-3_39

2   Benchmark Problems

A brief summary of the functions used for the ICSI'2022 competitions is shown in Table 1. A more detailed analysis is provided in [14]. Each of the functions is used in a minimization problem on a range between $[-100, 100]^D$, with dimensions $D = [10, 20, 50]$. The maximum allowed number of function evaluations ($MaxFET$) was set to $MaxFET = 10000$ for $D = 10$, $MaxFET = 30000$ for $D = 20$, and $MaxFET = 70000$ for $D = 50$.

Table 1. Summary of the functions used in the ICSI'2022 competition.

No.	Function	F^*
F01	Rotated Shifted High Conditioned Elliptic Function	1000
F02	Rotated and Shifted Bent Cigar Function	1000
F03	Rotated and Shifted Rosenbrock's Function	200
F04	Rotated and Shifted Rastrigin's Function	200
F05	Rotated and Shifted Modified Schwefel's Function	300
F06	Rotated and Shifted Alpine Function	300
F07	Shifted HappyCat Function	500
F08	Composition Function 1	0
F09	Composition Function 2	0
F10	Composition Function 3	0

3   Method Description

HSES. The HSES is a modified evolution strategy that utilizes a well known covariance matrix adaptation strategy (CMA-ES) and a method based on multivariate sampling, which was previously successfully used for separable problems. It takes advantage of both of these methods to improve upon its performance. In the proposed composite method, the version of HSES used was the one developed by [16] for the purpose of the CEC'18 Competition on single objective optimization.

LSHADE. The LSHADE is one of the best-performing variants of DE which was as the basis of several of the best performing algorithm in the recent CEC Competitions on single objective optimization. It is based upon SHADE, a DE method that incorporates success-history based parameter adaptation, and extends it by adding a linear population size reduction, that progressively decreases the size of the population. In the proposed composite method, the version of LSHADE used was the one developed by [1] for the purpose of the CEC'21 Competition on single objective optimization.

Proposed Composite Method. The proposed composite method is based on the evaluation of the behaviour of HSES and LSHADE on the ambiguous benchmark set [7]. It was found that both of these methods performed very well, but had a quite different convergence profile. While HSES was able to "good" solutions very quickly, in some cases it failed to improve upon these solutions and was overtaken by other methods as the number of function evaluations progressed. On the other hand, LSHADE was found to behave in the opposite manner - it showed relatively slower convergence at the start, but once it found a region with a good local optimum it converged quickly. The proposed composite method switches between the evaluation of HSES and LSHADE, based on a predetermined scheme. Various possibilities for the scheme are analyzed in the upcomming section.

Parameter Settings. The parameter settings used by the composite method are the same as the ones that were used in the respective papers [16] and [1]. As the composite scheme is concerned, eight possibilities have been tested to find the one most suitable for the benchmark functions in the competition. The studied schemes are summarized in Table 2 - for instance the C1 scheme uses the fist half of the available iterations to run HSES and the other half to run the LSHADE.

Table 2. Studied composite schemes.

Composite scheme	Progression of iterations
C1	1/2 HSES, 1/2 LSHADE
C2	1/4 HSES, 3/4 LSHADE
C3	1/8 HSES, 7/8 LSHADE
C4	1/4 HSES, 1/4 LSHADE, 1/4 HSES, 1/4 LSHADE
C5	3/4 HSES, 1/4 LSHADE
C6	7/8 HSES, 1/8 LSHADE
C7	3/8 HSES, 1/8 LSHADE, 3/8 HSES, 1/8 LSHADE
C8	7/16 HSES, 1/16 LSHADE, 7/16 HSES, 1/16 LSHADE

4 Results and Discussion

First, the performance of the eight studied composite schemes is compared for the selection of the most promising one. All of the schemes were evaluated for the dimension $D = 10$. The results of this comparison are summarized in Table 3 (showing only the "mean") values, which is color coded for easier comparison - green cells signify better performance, while red values mean worse performance. What is also apparent is that for the particular number of possible function evaluations, having more HSES iterations was preferable. The other thing to notice is that switching between HSES and LSHADE multiple times (as in C4, C7, and C8) did not bring any improvements. For the sake of saving computational time, only the most promising schemes C1, C5, and C6 were subsequently evaluated on problems with $D = 20$. The results of this comparison are shown in Table 4. As a result, only the best performing scheme C6 was selected as the representative of the CESDE method for the comparison with other methods.

Table 3. Comparing mean results of different composite schemes for D = 10.

No.	C1	C2	C3	C4	C5	C6	C7	C8
F01	4.46E+06	7.38E+06	1.72E+07	7.38E+06	3.46E+06	4.22E+06	6.76E+06	3.95E+06
F02	8.21E+08	1.12E+09	1.39E+09	1.12E+09	8.87E+08	8.13E+08	8.55E+08	9.30E+08
F03	1.50E+02	1.98E+02	2.13E+02	1.98E+02	1.15E+02	1.04E+02	1.55E+02	1.56E+02
F04	6.20E+01	6.43E+01	6.51E+01	6.43E+01	6.28E+01	6.16E+01	6.21E+01	6.39E+01
F05	1.27E+03	1.35E+03	1.47E+03	1.35E+03	1.35E+03	1.30E+03	1.30E+03	1.28E+03
F06	5.88E-01	1.28E+01	5.54E+01	1.28E+01	1.04E-02	1.23E-03	3.27E+00	1.15E+00
F07	1.02E+00	1.29E+00	1.58E+00	1.29E+00	1.02E+00	9.24E-01	1.18E+00	1.19E+00
F08	2.78E+05	2.78E+05	2.78E+05	2.78E+05	2.78E+05	2.78E+05	2.78E+05	2.78E+05
F09	8.14E+03	1.75E+06	2.93E+06	1.75E+06	1.46E+04	1.16E+01	5.15E+05	2.18E+05
F10	2.04E+01	9.97E+01	3.56E+02	9.97E+01	7.50E+00	7.28E+00	3.87E+01	3.03E+01

Table 4. Comparing mean results of selected composite schemes for D = 20.

No.	C1	C5	C6
F01	5.05E+07	5.88E+07	4.39E+07
F02	7.31E+09	7.20E+09	6.91E+09
F03	5.08E+02	4.33E+02	3.97E+02
F04	1.53E+02	1.49E+02	1.45E+02
F05	2.32E+03	2.56E+03	2.57E+03
F06	8.82E+00	3.93E-01	6.55E+00
F07	1.97E+00	1.88E+00	1.66E+00
F08	4.00E+05	4.00E+05	4.00E+05
F09	1.96E+03	2.00E+02	2.13E+02
F10	5.93E+03	4.85E+03	5.39E+03

In Table 5 is reported the comparison between the CESDE method (using the composite scheme C6) and its two "parent" methods HSES and LSHADE. For the LSHADE and HSES methods the same parameter setting used was the one reported in the corresponding publication [5] and all algorithms were started from the same random seed [8]. For the evaluation, each of the algorithms was run independently 50 times to obtain statistically relevant results. The chosen testing platform was MATLAB R2021b, and the computations were conducted on a computer with 3.2 GHz i5-4460 CPU and 16 GB RAM, using WIN 10 OS. In Table 5, the Mean distance from the global optimum (which can be found in

Table 1) is reported. This distance is computed as $F_{res} - F^*$, where F_{res} is the objective value of the best individual and F^* is the optimal value. It can be seen that CESDE outperformed both of its "parent" methods on all studied instances. While on some instance it made only slight improvement (when compared to its parent methods) it performed significantly better on F06, F07, and F09 (by more than two orders of magnitude).

Table 5. Comparing CESDE with HSES and LSHADE, mean values of $(F_{res} - F^*)$.

No	D = 10			D = 20			D = 50		
	HSES	LSHADE	CESDE	HSES	LSHADE	CESDE	HSES	LSHADE	CESDE
F01	2.32E+08	2.93E+07	4.22E+06	1.13E+09	1.55E+08	4.39E+07	1.10E+09	1.49E+09	9.75E+08
F02	8.64E+09	3.21E+09	8.13E+08	3.31E+10	1.81E+10	6.91E+09	9.20E+08	9.56E+08	4.25E+08
F03	2.41E+03	3.59E+02	1.04E+02	5.78E+03	1.84E+03	3.97E+02	7.42E+03	1.54E+04	6.51E+03
F04	1.12E+02	7.28E+01	6.16E+01	2.48E+02	2.04E+02	1.45E+02	7.30E+02	6.81E+02	3.54E+02
F05	2.15E+03	1.71E+03	1.30E+03	5.97E+03	4.50E+03	2.57E+03	1.91E+04	1.40E+04	4.81E+03
F06	1.08E+02	7.63E+01	1.23E−03	3.31E+02	2.42E+02	6.55E+00	1.63E+03	9.25E+02	1.44E+01
F07	4.19E+00	2.24E+00	9.24E−01	5.65E+00	3.89E+00	1.66E+00	9.57E+00	6.36E+00	4.25E+00
F08	3.96E+05	4.66E+05	2.78E+05	4.50E+05	4.35E+05	4.00E+05	4.49E+05	4.50E+05	4.00E+05
F09	1.97E+07	2.50E+07	1.16E+01	4.89E+07	4.73E+07	2.13E+02	8.11E+08	8.40E+08	1.64E+05
F10	2.51E+03	2.73E+03	7.28E+00	1.41E+04	1.26E+04	5.39E+03	3.27E+04	4.10E+04	1.91E+04

For the purpose of the ICSI'2022 competition, the detailed statistics from the 50 runs (reporting the best value, the worst value, the median value, the mean value, and the standard deviation) can be found in Tables 6, 7, and 8, for the different dimensions.

Table 6. Results $(F_{res} - F^*)$ for $D = 10$ problems.

No.	Best	Worst	Median	Mean	Std
F01	8.41E−01	1.45E+07	2.71E+01	4.22E+06	5.78E+06
F02	2.69E+01	1.73E+09	8.53E+08	8.13E+08	7.36E+08
F03	6.30E+00	1.53E+02	1.08E+02	1.04E+02	2.92E+01
F04	2.98E+01	8.62E+01	6.11E+01	6.16E+01	1.46E+01
F05	8.01E+02	1.58E+03	1.37E+03	1.30E+03	2.18E+02
F06	2.50E−04	5.98E−03	8.81E−04	1.23E−03	1.28E−03
F07	7.93E−02	1.30E+00	9.58E−01	9.24E−01	3.21E−01
F08	1.74E+05	3.44E+05	2.92E+05	2.78E+05	4.20E+04
F09	2.25E−03	2.17E+02	2.03E−02	1.16E+01	4.84E+01
F10	2.08E+00	1.98E+01	5.77E+00	7.28E+00	4.50E+00

Table 7. Results $(F_{res} - F^*)$ for $D = 20$ problems.

No.	Best	Worst	Median	Mean	Std
F01	9.04E−03	1.28E+08	2.76E+07	4.39E+07	4.75E+07
F02	4.95E+09	1.00E+10	6.67E+09	6.91E+09	1.37E+09
F03	2.73E+02	5.80E+02	3.97E+02	3.97E+02	7.90E+01
F04	1.06E+02	1.96E+02	1.46E+02	1.45E+02	2.02E+01
F05	8.74E+02	4.29E+03	2.54E+03	2.57E+03	8.52E+02
F06	1.02E−07	1.30E+02	4.22E−03	6.55E+00	2.91E+01
F07	7.89E−01	2.08E+00	1.71E+00	1.66E+00	2.97E−01
F08	3.97E+05	4.00E+05	4.00E+05	4.00E+05	6.16E+02
F09	2.51E−08	4.25E+03	2.69E−03	2.13E+02	9.50E+02
F10	2.20E+03	7.19E+03	5.88E+03	5.39E+03	1.38E+03

Table 8. Results $(F_{res} - F^*)$ for $D = 50$ problems.

No.	Best	Worst	Median	Mean	Std
F01	9.25E+05	1.40E+09	1.02E+09	9.75E+08	3.78E+08
F02	2.96E+08	4.89E+08	4.36E+08	4.25E+08	4.28E+07
F03	4.51E+03	9.48E+03	6.44E+03	6.51E+03	1.33E+03
F04	2.59E+01	5.13E+02	3.97E+02	3.54E+02	1.28E+02
F05	2.79E+03	8.07E+03	4.63E+03	4.81E+03	1.54E+03
F06	1.98E+00	3.79E+01	1.23E+01	1.44E+01	9.50E+00
F07	3.99E+00	4.63E+00	4.21E+00	4.25E+00	1.91E−01
F08	4.00E+05	4.00E+05	4.00E+05	4.00E+05	4.45E+01
F09	1.18E+04	9.73E+05	7.98E+04	1.64E+05	2.23E+05
F10	1.51E+04	2.26E+04	1.88E+04	1.91E+04	1.89E+03

5 Conclusions

In this paper, a composite method that combined HSES and LSHADE was proposed. The numerical investigation on the ICSI'2022 benchmark set confirmed that combining the two above mentioned method in a particular way brought advantages of both, and the composite method was able to outperform both of its "parental" methods.

As only two methods and a single benchmark dataset was investigates, there is plenty of room for additional research – different version of EA methods can be combined on various benchmark sets to see the impact of their composition, as well as the impact hyperparameter optimization can have on the performance of these methods.

Acknowledgments. This work was supported by internal grant agency of BUT: FME-S-20-6538 "Industry 4.0 and AI methods", FIT/FSI-J-22-7980, and FEKT/FSI-J-22-7968.

References

1. Biswas, S., Saha, D., De, S., Cobb, A.D., Das, S., Jalaian, B.A.: Improving differential evolution through Bayesian hyperparameter optimization. In: 2021 IEEE Congress on Evolutionary Computation (CEC) (2021)
2. Das, S., Suganthan, P.N.: Differential evolution: a survey of the state-of-the-art. IEEE Trans. Evol. Comput. **15**(1), 4–31 (2010)
3. Hadi, A.A., Mohamed, A.W., Jambi, K.M.: Single-objective real-parameter optimization: enhanced LSHADE-SPACMA algorithm. In: Yalaoui, F., Amodeo, L., Talbi, E.-G. (eds.) Heuristics for Optimization and Learning. SCI, vol. 906, pp. 103–121. Springer, Cham (2021). https://doi.org/10.1007/978-3-030-58930-1_7
4. Ji, Y., Xing, Y.: An optimized three-sub-step composite time integration method with controllable numerical dissipation. Comput. Struct. **231**, 106210 (2020)
5. Kazikova, A., Pluhacek, M., Senkerik, R.: Why tuning the control parameters of metaheuristic algorithms is so important for fair comparison? MENDEL J. **26**(2), 9–16 (2020)
6. Kudela, J., Popela, P.: Two-stage stochastic facility location problem: GA with benders decomposition. In: Mendel, vol. 21, pp. 53–58 (2015)
7. Kudela, J., Matousek, R.: New benchmark functions for single-objective optimization based on a zigzag pattern. IEEE Access **10**, 8262–8278 (2022)
8. Matousek, R., Dobrovsky, L., Kudela, J.: How to start a heuristic? Utilizing lower bounds for solving the quadratic assignment problem. Int. J. Ind. Eng. Comput. **13**(2), 151–164 (2022)
9. Matousek, R., Hulka, T.: Stabilization of higher periodic orbits of the chaotic logistic and Hénon maps using meta-evolutionary approaches. In: 2019 IEEE Congress on Evolutionary Computation (CEC), pp. 1758–1765. IEEE (2019)
10. Matousek, R., Popela, P., Kudela, J.: Heuristic approaches to stochastic quadratic assignment problem: VaR and CVaR cases. In: Mendel, vol. 23, pp. 73–78 (2017)
11. Sallam, K.M., Elsayed, S.M., Chakrabortty, R.K., Ryan, M.J.: Improved multi-operator differential evolution algorithm for solving unconstrained problems. In: 2020 IEEE Congress on Evolutionary Computation (CEC), pp. 1–8 (2020)
12. Schröder, D., Vermetten, D., Wang, H., Doerr, C., Bäck, T.: Chaining of numerical black-box algorithms: warm-starting and switching points. arXiv preprint arXiv:2204.06539 (2022)
13. Tanabe, R., Fukunaga, A.S.: Improving the search performance of shade using linear population size reduction. In: 2014 IEEE Congress on Evolutionary Computation (CEC), pp. 1658–1665 (2014)
14. Yifan, L.: Definitions for the ICSI optimization competition 2022 on single objective bounded optimization problems. Key Laboratory of Machine Perception, Peking University, China, Technical report (2022)
15. Young, S.R., Rose, D.C., Karnowski, T.P., Lim, S.H., Patton, R.M.: Optimizing deep learning hyper-parameters through an evolutionary algorithm. In: Proceedings of the Workshop on Machine Learning in High-Performance Computing Environments, pp. 1–5 (2015)

16. Zhang, G., Shi, Y.: Hybrid sampling evolution strategy for solving single objective bound constrained problems. In: 2018 IEEE Congress on Evolutionary Computation (CEC), pp. 1–7 (2018)
17. Žufan, P., Bidlo, M.: Advances in evolutionary optimization of quantum operators. MENDEL J. **27**(2), 12–22 (2021)

Surrogate-Assisted Differential Evolution-Based Method for the ICSI'2022 Competition

Jakub Kudela[✉][iD], Tomas Holoubek, and Tomas Nevoral

Institute of Automation and Computer Science, Brno University of Technology,
Brno, Czech Republic
Jakub.Kudela@vutbr.cz

Abstract. In this paper, a method called Lipschitz-surrogate Assisted Differential Evolution (LSADE) is described. The method uses two different surrogates: a standard radial basis function one and a specialized one based on a Lipschitz condition. It also uses two optimization methods: differential evolution and sequential quadratic programming. The LSADE method is investigated on the benchmark set of the ICSI'2022 competition and is compared with two other high-performing methods for bound constrained optimization, outperforming them both on the studied benchmark set.

Keywords: ICSI'2022 competition · Surrogate-assisted optimization · Differential evolution · Numerical optimization

1 Introduction

Evolutionary algorithms (EAs) are one of the most efficient metaheuristics used for global optimization. EAs are inspired by the processes of biological evolution, such as reproduction, recombination, mutation, and natural selection. The most widely used variants of these techniques are genetic algorithms, evolutionary strategy, and differential evolution (DE) [2]. These methods are routinely utilized in the optimization of complicated problems such as the design of quantum operators [22], difficult assignment problems [13], stabilization of chaos [12], or the hyperparameter optimization in deep learning [20].

Real-world optimization problems frequently require expensive computations or executions of physical experiments. In such situations the evaluation of objective functions can take a prohibitively long time for conventional optimization methods [4]. To alleviate the computational costs, surrogate models have been widely used in combination with EAs under the name of surrogate-assisted EAs (SAEAs) [5].

SAEAs compute only a reduced number of real objective function evaluations and utilize them for training of surrogate models as approximations for the real functions. The other use of SAEAs is in situations, where the maximum number

© Springer Nature Switzerland AG 2022
Y. Tan et al. (Eds.): ICSI 2022, LNCS 13345, pp. 440–449, 2022.
https://doi.org/10.1007/978-3-031-09726-3_40

of function evaluations is somehow limited. As the problem set in the ICSI'2022 competition heavily restricts the number of possible function evaluations, the utilization of an SAEA for this purpose is well justified.

In this paper, the lipschitz-surrogate assisted differential evolution (LSADE) method is evaluated on the ICSI'2022 test set. The LSADE method uses two different surrogates, the DE algorithm, and a local search method. Its performance on the test set is compared with two other high-performing evolutionary algorithms.

The rest of the paper is structured as follows: Sect. 2 describes the benchmark test set and the competition conditions, in Sect. 3 details of the proposed method are described, in Sect. 4 the results of the computations are reported and discussed, and conclusions are drawn in Sect. 5.

2   Benchmark Problems

The summary of the benchmark functions used for the ICSI'2022 competitions is presented in Table 1 and more details about the functions can be found in [19]. Every one of the 10 functions presents a minimization problem on the range $[-100, 100]^D$, where the considered dimensions are $D = [10, 20, 50]$. The maximum number of function evaluations ($MaxFET$) was capped at $MaxFET = 10000$ for $D = 10$, $MaxFET = 30000$ for $D = 20$, and $MaxFET = 70000$ for $D = 50$.

Table 1. Summary of the functions used in the ICSI'2022 competition.

No.	Function	F^*
F01	Rotated Shifted High Conditioned Elliptic Function	1000
F02	Rotated and Shifted Bent Cigar Function	1000
F03	Rotated and Shifted Rosenbrock's Function	200
F04	Rotated and Shifted Rastrigin's Function	200
F05	Rotated and Shifted Modified Schwefel's Function	300
F06	Rotated and Shifted Alpine Function	300
F07	Shifted HappyCat Function	500
F08	Composition Function 1	0
F09	Composition Function 2	0
F10	Composition Function 3	0

3   Method Description

3.1   Surrogate Models

Among the different surrogate models, radial basis function (RBF) based ones are on of the most widely applied methods [3]. Representative RBFs include

Gaussian function, linear splines, thin-plate splines, cubic splines, and multi-quadrics splines. Given n different sample points $x^{(1)}, \ldots, x^{(n)}$, the RBF surrogate can be written as

$$\hat{f}(x) = \sum_{i=1}^{n} w_i \psi(||x - x^{(i)}||_2), \tag{1}$$

where w_i denotes the weight which is computed using the method of least-squares, and ψ is the chosen basis function.

The use of a Lipschitz constant in optimization has a long history within global optimization and is an active area of research to this day [9]. It is assumed that there is an unknown (or expensive to compute) function f with a finite Lipschitz constant k, i.e.

$$\exists k \geq 0 \text{ s.t. } |f(x) - f(x')| \leq k||x - x'||_2 \ \forall (x, x') \in \mathcal{X}^2. \tag{2}$$

Based on a sample of n evaluations of the function f at points $x^{(1)}, \ldots, x^{(n)}$, the global underestimator f_L of f is constructed by using the following expression

$$f_L(x) = \max_{i=1,\ldots,t} f(x^{(i)}) - k||x - x^{(i)}||_2. \tag{3}$$

This surrogate displays two important properties – it assigns low values to points that are far from the ones that were previously evaluated, and it combines it with the information (in the form of the objective value and the value of the "global" Lipschitz constant) from the closest evaluated point. This means that it can serve as an "uncertainty measure" of prospective points for evaluation.

However, as the objective function f itself is not known, one cannot expect to know the value of the Lipschitz constant k. This is approached by estimating k from the previously evaluated points. To this end, the approach described in [9] was utilized: a nondecreasing sequence of Lipschitz constants $k_{i \in \mathbf{Z}}$ defines a meshgrid on $\mathbf{R}^+$, and the estimate $\hat{k}_t$ of the Lipschitz constant was computed by

$$\hat{k}_t = \inf \left\{ k_{i \in \mathbf{Z}} : \max_{l \neq j} \frac{|f(x^{(j)}) - f(x^{(l)})|}{||x^{(j)} - x^{(l)}||_2} \leq k_i \right\}. \tag{4}$$

Several possible sequences of different shapes could be used. The one employed was a sequence $k_i = (1 + \alpha)^i$ that uses a parameter $\alpha > 0$. For this sequence, the computation (4) of the estimate is simplifies into $\hat{k}_t = (1 + \alpha)^{i_t}$, where

$$i_t = \left\lceil \ln(\max_{l \neq j} \frac{|f(x^{(j)}) - f(x^{(l)})|}{||x^{(j)} - x^{(l)}||_2}) / \ln(1 + \alpha) \right\rceil. \tag{5}$$

3.2   Differential Evolution

In this paper, DE is used as the optimization method due to its relatively straightforward structure, and its proven optimization abilities. Several variants of DE were introduced in order to improve its performance [2]. Generally, there are four steps of DE: initialization, mutation, crossover, and selection. In the proposed method, the DE/best/1 strategy was utilized for the mutation process of DE which, can be expressed as

$$v_i = x_b + F \cdot (x_{i_1} - x_{i_2}),\tag{6}$$

where x_b is the current best solution and x_{i_1} and x_{i_2} are different randomly selected individuals from the population, and F is a scalar. The crossover step of DE is performed after mutation and has the following form:

$$u_i^j = \begin{cases} v_i^j, & \text{if } (U_j(0,1) \le C_r \,|\, j = j_{rand}) \\ x_i^j, & \text{otherwise} \end{cases}\tag{7}$$

where u_i^j the jth component of ith offspring, x_j^i and v_j^i are the jth component of ith parent individual and the mutated individual, respectively. The crossover constant C_r is between 0 and 1, $U_j(0,1)$ indicates a uniformly distributed random number, and $j_{rand} \in [1, \ldots, D]$ is a randomly chosen index which guarantees u_i has at least one component of v_i.

3.3 Proposed LSADE Method

The LSADE method [7] has four parts: DE-based generation of prospective points, RBF evaluation of the prospective points, Lipschitz surrogate evaluation of the prospective points, and the local optimization within a close range of the best solution found so far. The execution of last three parts of the algorithm can be controlled based on chosen conditions.

At the start, the Latin hypercube sampling [16] is utilized to generate the initial population of t individuals, with their objective function is computed exactly. The best individual is found, a parent population of size p is randomly selected from the evaluated points, and a new population is created based on the DE steps described by (6) and (7). If the *RBF evaluation condition* is true, the new population is evaluated based on the RBF surrogate and the best individual based on this model has its objective function evaluated, and is added to the whole population. This part can be thought of as a global search strategy.

If the *Lipschitz evaluation condition* is satisfied, the Lipschitz constant k is estimated by (5) and the new population is evaluated on the Lipschitz surrogate (3). The best individual based from the Lipschitz model has its objective function evaluated, and enters the whole population.

If the *Local optimization condition* is satisfied, a local RBF surrogate model is constructed using the best c solutions found thus far, which are denote as $\hat{x}_1, \ldots, \hat{x}_c$. The bounds used for the local optimization procedure within those c points are:

Algorithm 1. Pseudocode of the LSADE.

1: Construct the initial population of t points $x_1, \ldots, x_t$, evaluate their objective function values, find the best solution x_b.

2: Set $iter = 0$, $NFE = t$ (number of true function evaluations).

3: Estimate k by (5) and construct the RBF surrogate model.

4: Sample p points from the population as parents for DE.

5: Generate children based on the DE rules (6) and (7), .

6: Increase $iter$ by 1.

7: **if** *RBF condition* **then**

8:     Evaluate the children on the RBF surrogate.

9:     Find the child with the minimum RFB surrogate value, evaluate its objective function value, and add it to the whole population. Increase NFE by 1.

10: **if** *Lipschitz condition* **then**

11:     Evaluate the children on the Lipschitz surrogate (3).

12:     Find the child with the minimum Lipschitz surrogate value, evaluate its objective function value, and add it to the whole population. Increase NFE by 1.

13: **if** *Local Optimization condition* **then**

14:     Construct the local RBF surrogate model with the best c solutions found.

15:     Find the bounds for the local optimization (8).

16:     Minimize the local RBF surrogate model within the bounds. Denote the minimum as $\hat{x}_m$ and, if it is not already in the population, evaluate its objective function value, add it to the population. Increase NFE by 1.

17: Find the best solution so far and denote it as x_b.

18: **if** $NFE < NFE_{max}$ **then**

19:     **goto** *3*.

20: **else**

21:     **terminate.**

$$lb(i) = \min_{j=1,\ldots,c} \hat{x}_j(i), \quad i = 1, \ldots, D$$
$$ub(i) = \max_{j=1,\ldots,c} \hat{x}_j(i), \quad i = 1, \ldots, D \tag{8}$$

and a local optimization of this RBF model is performed within $[lb, ub]$. For this local optimization a sequential quadratic programming strategy was adapted, as it was found to be an excellent strategy by the winner of the 2020 CEC Single Objective Bound Constrained Competition [15]. After the local optimum is found it is checked, if it is not already in the population, before evaluating it and adding it to the whole population.

The evaluation of points by the Lipschitz-based surrogate can be seen as the exploration step in the method, while the evaluation of points by the local optimization method can be seen as the exploitation step of the method. The pseudocode for the proposed LSADE method is described in Algorithm 1.

3.4   Parameter Settings

For constructing both the local and the global RBF surrogate models, the SURROGATES toolbox [18] was used with the default settings (multiquadric RBF).

The coefficients of the DE were chosen as $F = 0.5$ and $C_r = 0.5$. The number of initial points were set to 100 for $D = [30, 50]$ and 200 for $D = [100, 200]$. The number of children was set to D. The local optimization used the best $c = 3 \cdot D$ points found so far (or less if there were not enough points evaluated so far), and used the sequential quadratic programming algorithm implemented in the FMINCON function [10] with default parameters. The Lipschitz parameter was chosen as $\alpha = 0.01$.

It is customary to want greater exploration capabilities in the early stages of the algorithm and exploitation capabilities in the later iterations [1,14]. The conditions for the construction of the Lipschitz surrogate and for the local optimization procedure reflect this by being based on the iteration number in the following way:

Lipschitz condition for $D = [10, 20, 50]$: $\qquad \mathrm{mod}\left(iter, \left\lceil \dfrac{8 \cdot iter}{1000} \right\rceil \right) = 0,$

Local optimization condition for $D = [10, 20]$: $\quad \mathrm{mod}\left(iter, \left\lceil \dfrac{8000 - 15 \cdot iter}{1000} \right\rceil \right) = 0,$

Local optimization condition for $D = 50$: $\quad \mathrm{mod}\left(iter, \left\lceil \dfrac{8000 - 10 \cdot iter}{1000} \right\rceil \right) = 0,$

while the condition for the RBF evaluation was set as true for every iteration.

As the evaluation of the LSADE method takes a nontrivial amount of time (both the construction of the surrogates and the local optimization are relatively complex procedures), the method was not used for the whole $MaxFET$ available evaluations. Instead, it was used only for $NFE_{max} = 500$ evaluations in $D = 10$, $NFE_{max} = 1000$ evaluations in $D = 20$, and $NFE_{max} = 1500$ evaluation in $D = 50$. The rest of the available evaluations ($MaxFET - NFE_{max}$) was used by running the DE algorithm (with the same parameters as reported above).

4   Results and Discussion

In order to demonstrate the effectivity of the LSADE method, its performance was compared with two of the best-performing algorithms in [8]. These algorithms were the Linear Population Size Reduction SHADE (LSHADE) [17] and Hybrid Sampling Evolution Strategy (HSES) [21]. For the LSHADE and HSES methods the same parameter setting used was the one reported in the corresponding publication [6] and all algorithms were started from the same random seed [11]. For the evaluation, each of the algorithms was run independently 50 times to obtain statistically relevant results. The chosen testing platform was MATLAB R2021b, and the computations were conducted on a computer with 3.2 GHz i5-4460 CPU and 16 GB RAM, using WIN 10 OS. The results of this comparison are shown in Table 2, where the Mean distance from the global optimum (which can be found in Table 1) of the 50 runs of the methods is reported.

This distance is computed as $F_{res} - F^*$, where F_{res} is the objective value of the best individual and F^* is the optimal value. It can be seen that LSADE is highly competitive as it outperformed both HSES and LSHADE on all studied instances.

Table 2. Comparison between HSES, LSHADE, and LSADE, mean values of $(F_{res} - F^*)$.

No	D = 10			D = 20			D = 50		
	HSES	LSHADE	LSADE	HSES	LSHADE	LSADE	HSES	LSHADE	LSADE
F01	2.32E+08	2.93E+07	8.94E+06	1.13E+09	1.55E+08	2.07E+07	1.10E+09	1.49E+09	8.75E+07
F02	8.64E+09	3.21E+09	9.95E+03	3.31E+10	1.81E+10	7.01E+06	9.20E+08	9.56E+08	2.74E+06
F03	2.41E+03	3.59E+02	1.85E+01	5.78E+03	1.84E+03	5.03E+01	7.42E+03	1.54E+04	2.38E+02
F04	1.12E+02	7.28E+01	3.04E+01	2.48E+02	2.04E+02	7.10E+01	7.30E+02	6.81E+02	1.39E+02
F05	2.15E+03	1.71E+03	1.18E+03	5.97E+03	4.50E+03	2.77E+03	1.91E+04	1.40E+04	8.31E+03
F06	1.08E+02	7.63E+01	2.91E+01	3.31E+02	2.42E+02	6.10E+01	1.63E+03	9.25E+02	1.55E+02
F07	4.19E+00	2.24E+00	3.71E−01	5.65E+00	3.89E+00	5.61E−01	9.57E+00	6.36E+00	5.01E−01
F08	3.96E+05	4.66E+05	3.95E+05	4.50E+05	4.35E+05	4.02E+05	4.49E+05	4.50E+05	4.06E+05
F09	1.97E+07	2.50E+07	6.02E+06	4.89E+07	4.73E+07	2.49E+06	8.11E+08	8.40E+08	2.67E+07
F10	2.51E+03	2.73E+03	3.45E+02	1.41E+04	1.26E+04	1.88E+02	3.27E+04	4.10E+04	2.62E+02

For the purpose of the competition, the detailed statistics from the 50 runs (best value, worst value, median value, mean value, and standard deviation) are shown in Tables 3, 4, and 5, for the different dimensions.

Table 3. Results $(F_{res} - F^*)$ for $D = 10$ problems.

No.	Best	Worst	Median	Mean	Std
F01	8.92E+05	3.70E+07	5.88E+06	8.94E+06	7.09E+06
F02	1.23E+02	2.62E+04	9.47E+03	9.95E+03	6.03E+03
F03	7.62E+00	1.05E+02	1.25E+01	1.85E+01	1.96E+01
F04	4.98E+00	8.20E+01	2.44E+01	3.04E+01	1.71E+01
F05	5.34E+02	1.86E+03	1.19E+03	1.18E+03	3.10E+02
F06	4.64E+00	7.81E+01	2.24E+01	2.91E+01	1.79E+01
F07	1.35E−01	8.25E−01	3.48E−01	3.71E−01	1.36E−01
F08	2.60E+05	4.66E+05	4.10E+05	3.95E+05	5.01E+04
F09	2.44E+05	2.35E+07	5.26E+06	6.02E+06	4.80E+06
F10	1.65E+01	1.49E+03	1.68E+02	3.45E+02	3.61E+02

Table 4. Results ($F_{res} - F^*$) for $D = 20$ problems.

No.	Best	Worst	Median	Mean	Std
F01	3.46E+06	6.42E+07	1.79E+07	2.07E+07	1.30E+07
F02	1.17E+01	2.72E+08	6.10E+03	7.01E+06	3.89E+07
F03	1.28E+01	8.56E+01	5.49E+01	5.03E+01	2.12E+01
F04	3.28E+01	1.53E+02	6.54E+01	7.10E+01	2.82E+01
F05	1.45E+03	3.68E+03	2.80E+03	2.77E+03	5.28E+02
F06	1.25E+01	1.74E+02	4.71E+01	6.10E+01	3.72E+01
F07	2.19E−01	1.37E+00	5.37E−01	5.61E−01	2.18E−01
F08	4.00E+05	4.08E+05	4.00E+05	4.02E+05	3.13E+03
F09	3.98E+04	1.22E+07	1.89E+06	2.49E+06	2.44E+06
F10	3.46E+01	5.74E+02	1.47E+02	1.88E+02	1.23E+02

Table 5. Results ($F_{res} - F^*$) for $D = 50$ problems.

No.	Best	Worst	Median	Mean	Std
F01	3.53E+07	1.89E+08	7.11E+07	8.75E+07	3.84E+07
F02	8.36E−01	1.75E+07	1.57E+03	2.74E+06	5.25E+06
F03	1.66E+02	3.31E+02	2.16E+02	2.38E+02	4.99E+01
F04	7.96E+01	2.44E+02	1.28E+02	1.39E+02	4.14E+01
F05	6.41E+03	1.05E+04	8.31E+03	8.31E+03	1.12E+03
F06	7.51E+01	2.79E+02	1.41E+02	1.55E+02	5.31E+01
F07	3.08E−01	6.69E−01	4.88E−01	5.01E−01	8.31E−02
F08	4.00E+05	4.50E+05	4.01E+05	4.06E+05	1.17E+04
F09	6.56E+06	1.63E+08	1.66E+07	2.67E+07	3.14E+07
F10	1.33E+02	4.13E+02	2.65E+02	2.62E+02	8.85E+01

5   Conclusions

In this paper the LSADE method was described and investigated on the benchmark set for the ISCI'2022 competition. Through numerical investigation it was found that the LSADE method outperformed two other well-known methods, namely HSES and LSHADE, on the studied benchmark set. Further investigations of the LSADE method will encompass the evaluation of different schemes for the Lipschitz, RBF, and Local optimization conditions. Also interesting would be the possibility of choosing a different optimizer instead of the currently used DE.

Acknowledgments. This work was supported by internal grant agency of BUT: FME-S-20-6538 "Industry 4.0 and AI methods", FIT/FSI-J-22-7980, and FEKT/FSI-J-22-7968.

References

1. Bujok, P.: Three steps to improve jellyfish search optimiser. MENDEL J. **27**(1), 29–40 (2021)
2. Das, S., Suganthan, P.N.: Differential evolution: a survey of the state-of-the-art. IEEE Trans. Evol. Comput. **15**(1), 4–31 (2010)
3. Forrester, A.I., Keane, A.J.: Recent advances in surrogate-based optimization. Prog. Aerosp. Sci. **45**(1–3), 50–79 (2009)
4. Jin, Y., Sendhoff, B.: A systems approach to evolutionary multiobjective structural optimization and beyond. IEEE Comput. Intell. Mag. **4**(3), 62–76 (2009)
5. Jin, Y., Wang, H., Chugh, T., Guo, D., Miettinen, K.: Data-driven evolutionary optimization: an overview and case studies. IEEE Trans. Evol. Comput. **23**(3), 442–458 (2018)
6. Kazikova, A., Pluhacek, M., Senkerik, R.: Why tuning the control parameters of metaheuristic algorithms is so important for fair comparison? MENDEL J. **26**(2), 9–16 (2020)
7. Kudela, J., Matousek, R.: Lipschitz-based surrogate model for high-dimensional computationally expensive problems. arXiv preprint (2022)
8. Kudela, J., Matousek, R.: New benchmark functions for single-objective optimization based on a zigzag pattern. IEEE Access **10**, 8262–8278 (2022)
9. Malherbe, C., Vayatis, N.: Global optimization of Lipschitz functions. In: International Conference on Machine Learning, pp. 2314–2323. PMLR (2017)
10. MathWorks: MATLAB Optimization Toolbox, 2020b edn. Natick, MA, USA (2020)
11. Matousek, R., Dobrovsky, L., Kudela, J.: How to start a heuristic? Utilizing lower bounds for solving the quadratic assignment problem. Int. J. Ind. Eng. Comput. **13**(2), 151–164 (2022)
12. Matousek, R., Hulka, T.: Stabilization of higher periodic orbits of the chaotic logistic and Hénon maps using meta-evolutionary approaches. In: 2019 IEEE Congress on Evolutionary Computation (CEC), pp. 1758–1765. IEEE (2019)
13. Matousek, R., Popela, P., Kudela, J.: Heuristic approaches to stochastic quadratic assignment problem: VaR and CVaR cases. In: Mendel, vol. 23, pp. 73–78 (2017)
14. Pluháček, M., Kazikova, A., Kadavy, T., Viktorin, A., Senkerik, R.: Relation of neighborhood size and diversity loss rate in particle swarm optimization with ring topology. MENDEL J. **27**(2), 74–79 (2021)
15. Sallam, K.M., Elsayed, S.M., Chakrabortty, R.K., Ryan, M.J.: Improved multi-operator differential evolution algorithm for solving unconstrained problems. In: 2020 IEEE Congress on Evolutionary Computation (CEC), pp. 1–8 (2020)
16. Stein, M.: Large sample properties of simulations using Latin hypercube sampling. Technometrics **29**(2), 143–151 (1987)
17. Tanabe, R., Fukunaga, A.S.: Improving the search performance of shade using linear population size reduction. In: 2014 IEEE Congress on Evolutionary Computation (CEC), pp. 1658–1665 (2014)
18. Viana, F.: SURROGATES Toolbox User's Guide. Gainesville, FL, USA, version 2.1 edn. (2010). http://sites.google.com/site/felipeacviana/surrogatestoolbox
19. Yifan, L.: Definitions for the ICSI optimization competition'2022 on single objective bounded optimization problems. Technical report, Key laboratory of Machine Perception, Peking University, China (2022)
20. Young, S.R., Rose, D.C., Karnowski, T.P., Lim, S.H., Patton, R.M.: Optimizing deep learning hyper-parameters through an evolutionary algorithm. In: Proceedings of the Workshop on Machine Learning in High-Performance Computing Environments, pp. 1–5 (2015)

21. Zhang, G., Shi, Y.: Hybrid sampling evolution strategy for solving single objective bound constrained problems. In: 2018 IEEE Congress on Evolutionary Computation (CEC), pp. 1–7 (2018)
22. Žufan, P., Bidlo, M.: Advances in evolutionary optimization of quantum operators. MENDEL J. **27**(2), 12–22 (2021)

Author Index

Printed in the United States
by Baker & Taylor Publisher Services